NANOCELLBIOLOGY

edited by
Bhanu P. Jena and Douglas J. Taatjes

NANOCELLBIOLOGY
Multimodal Imaging in Biology and Medicine

PAN STANFORD PUBLISHING

Published by

Pan Stanford Publishing Pte. Ltd.
Penthouse Level, Suntec Tower 3
8 Temasek Boulevard
Singapore 038988

Email: editorial@panstanford.com
Web: www.panstanford.com

British Library Cataloguing-in-Publication Data
A catalogue record for this book is available from the British Library.

NanoCellBiology: Multimodal Imaging in Biology and Medicine

ISBN 978-981-4411-79-0 (Hardcover)
ISBN 978-981-4411-80-6 (eBook)

Printed in the USA

Dedicated to my students who have made my journey through science greatly rewarding and filled with excitement.

—Bhanu P. Jena

Contents

Preface

Just as the ultimate goal in biology is to unravel the structure and dynamics of a living cell at the atomic level, so is the major challenge in medicine to treat and ameliorate diseases noninvasively. This requires the targeting, imaging, and destruction of pathogen and diseased cells without harm to healthy cells and tissues. On close examination, however, the above two objectives seem inseparable, and an understanding of cellular structure-function is invaluable to the success in drug design, development, and therapy. Similarly, our understanding of the fundamental life processes or the treatment of diseases greatly relies on key technological advancements, an extraordinary example being the invention of the atomic force microscope (AFM), which gave birth to nanotechnology. In this book, applications of the AFM in the discovery of a new cellular structure, the *"porosome,"* in our understanding of cellular and molecular processes, and in the design and development of novel modalities in disease detection and treatment are presented. Similarly, novel approaches to understand molecular evolution, and the surprising involvement of mRNA nanomachines in disease processes, are also discussed.

Secretion is a fundamental cellular process involved in neurotransmission, and the release of hormones and digestive enzymes. Impaired secretion results in diseases such as diabetes and neuronal or digestive disorders. In Chapter 1, a brief commentary is provided on a new cellular structure, the "porosome," a secretory nanomachine discovered nearly 16 years ago and demonstrated to be the universal secretory portal in cells. In Chapters 2 and 3, porosomes in hair cells and in brain neurons of different species of mammals are presented. The porosome discovery has clarified our understanding of the generation of partially empty secretory vesicles in cells following secretion, providing a molecular understanding of the process, and resulting in a paradigm-shift in our understanding of cell secretion. In Chapter 4, the biogenesis of secretory vesicles and their distribution and dynamics, so critical to cell secretion, is elaborated. In Chapter 5, the authors introduce the reader to the application of the AFM in investigating protein assembly, biomineralization, and biomolecular interactions. Similarly, in Chapter 6, using a combination of AFM, fluorescence microscopy, and circular dichroism spectroscopy, the authors provide an insight into amylin aggregation, trafficking, and toxicity. These studies will provide the basis for the treatment of amyloid pathology such as amyloid deposits in type 2 diabetes. As in the case of amyloid pathophysiology, the utilization of the AFM to investigate novel treatment strategies is also in the advance. In Chapter 7, the use of AFM to investigate an autoimmune thrombotic conditions known as the antiphospholipid syndrome is discussed. Biological function of proteins resides in their three-dimensional shape, and an understanding of this three-dimensional protein structure is also essential for the design and development of drugs to regulate the protein. In Chapter 8, a novel approach to determine the molecular evolution of proteins using antibodies as nanoprobes, is presented. Similarly, the interaction of molecules and their assembly and diassembly within cells dictate cellular responses. In cardiac arrest and stroke, brain ischemia resulting from reduced blood flow to the brain leads to brain injury and even death. In Chapter 9, the authors describe a unique buildup of nucleoprotein granules of unknown composition in ischemic brain neurons. An understanding of the structure and composition of these granules promises a new approach for the treatment of brain ischemia. Magnetic nanoparticles have begun to show great promise in targeting, imaging, and destruction of pathogen and diseased cells without harm to healthy tissues. In Chapter 10, the emergence of magnetic nanoparticles for transformative application in medicine and therapy is discussed. Nano gene therapies using nanopolymers and virus-based therapy are rapidly being developed. In Chapter 11, AFM has been used to study DNA release dynamics from biopolymer-based nanosystems. New and novel methods and approaches to evaluate and image both biological and non-biological material are constantly in progress. In Chapter 12, the final chapter, the rapidly developing application of electrical impedance spectroscopy in biology is discussed. It is clear from the studies and findings discussed in this book that the development of nanotechnology and its use in unraveling fundamental biological principles are key to medical breakthroughs and in the effective diagnosis and treatment of diseases.

Bhanu P. Jena
Douglas J. Taatjes

Chapter 1

Porosomes—The Universal Secretory Portals in Cells: A Brief Essay

Bhanu P. Jena

Department of Physiology, Wayne State University School of Medicine, Detroit, MI 48201, USA

bjena@med.wayne.edu

In the mid-1990s, studies using the atomic force microscope (AFM) on live pancreatic acinar cells demonstrated for the first time the presence of 100–180 nm in size cup-shaped lipoprotein secretory portals called "porosomes" at the cell plasma membrane. Subsequent studies using both AFM and electron microscopy (EM) confirmed the presence of porosomes in all secretory cells, including beta cells of the endocrine pancreas, growth hormone–secreting cells of the pituitary gland, chromaffin cells, hair cells of the inner ear, astrocytes, and neurons. While porosomes measure 100–180 nm in the exocrine pancreas and in endocrine and neuroendocrine cells, porosomes in neurons and astrocytes are much smaller, measuring just 10–17 nm. AFM and EM studies further demonstrate the docking and transient fusion (kiss-and-run) of secretory vesicles at the porosome base via SNAREs and calcium. v-SNARE present in secretory vesicle membrane, and t-SNAREs at the porosome base interact in a circular array to establish continuity or fusion pore

NanoCellBiology: Multimodal Imaging in Biology and Medicine
Edited by Bhanu P. Jena and Douglas J. Taatjes
Copyright © 2014 Pan Stanford Publishing Pte. Ltd.
ISBN 978-981-4411-79-0 (Hardcover), 978-981-4411-80-6 (eBook)
www.panstanford.com

formation between the opposing bilayers. The size of the fusion pore formed by the t-/v-SNARE ring complex is dependent on the curvature of the secretory vesicle and the porosome base. To enable the expulsion of intravesicular contents through the porosome, secretory vesicles docked at the porosome base undergo volume increase primarily via water channels or aquaporins (AQPs), resulting in an increae in intravesicular pressure. The greater the intravesicular pressure, the greater is the amount of secretion. These findings established in the past 16 years clarify our understanding of the generation of partially empty secretory vesicles in cells following secretion, as opposed to the complete merger of the secretory vesicle membrane at the cell plasma membrane. The porosome discovery has resulted in a paradigm shift in our understanding of the secretory process in cells. It now makes sense why a fast secretory cell like the neuron would require neurotransmitter transporters at the synaptic vesicles membrane, to be able to rapidly replenish spent synaptic vesicles following neurotransmitter release. It seemed illogical to believe that in mammalian cells, secretory vesicles completely merge at the cell plasma membrane, while single cell organisms and bacteria all have well developed secretory machinery for the precise delivery of secretory products. In this chapter, the structure, dynamics, isolation, reconstitution, and composition of the *porosome—the universal secretory portal* at the cell plasma membrane—are briefly summarized.

The porosome was first discovered in acinar cells of the exocrine pancreas.[1] Pancreatic acinar cells are polarized secretory cells that synthesize digestive enzymes, which are stored within 0.2–1.2 µm in diameter apically located membranous sacs or secretory vesicles, called zymogen granules (ZGs). During secretion, ZGs dock and fuse with the apical plasma membrane to release their contents to the outside. Isolated live pancreatic acinar cells in physiological buffer, when imaged using the AFM, reveal defined structures at the apical pole of the cell where secretion occurs. A group of circular "pits" measuring 0.4–1.2 µm in diameter, contain smaller 100–180 nm in diameter "depressions" were first identified at the apical plasma membrane of pancreatic acinar cells more than 16 years ago (Figs. 1.1–1.3). These "depression structures" were subsequently named "porosomes" or secretory portals, following discovery of their involvement in secretory vesicle docking, fusion, and content release from cells. Typically 3–4 porosomes are located within each pit structure at the apical plasma membrane of pancreatic acinar cells. In contrast, the basolateral membrane of acinar cells is devoid of pits and porosomes.[1] High-resolution AFM micrographs of porosomes in live acinar cells further reveal their 100–180 nm in diameter cup-shaped basket-like morphology. Exposure of acinar cells to a secretagogue results in a time-dependent 20–45% increase in both the diameter and relative depth of porosomes, and a return to resting size on completion of secretion[1,2] (Fig. 1.4). Exposure of cells to cytochalasin B, a fungal toxin that inhibits actin polymerization and secretion, results in a 15–20% decrease in porosome size and a consequent 50–60% loss in cell secretion.[1] These reports were the first to suggest porosomes to be secretory portals in pancreatic acinar cells, and to be functionally regulated by actin. Recent studies using high-resolution electron microscopy provide in greater detail the morphology of the native porosome complex in pancreatic acinar cells.[3] Similar to pancreatic acinar cells, porosomes in growth hormone (GH) cells measuring approximately 150 nm in diameter have also been discovered.[3] Following exposure to a secretory

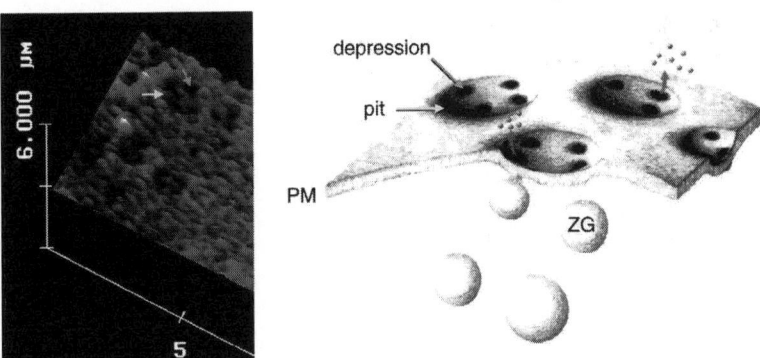

Figure 1.1   Porosomes were previously referred to as "depressions" at the plasma membrane in pancreatic acinar cells (*Proc. Natl. Acad. Sci.,* 1997, **94**: 316–321). AFM micrograph depicting "pits" and "depressions," or "porosomes" within, at the apical plasma membrane in a live pancreatic acinar cell. To the right is a schematic drawing depicting depressions or porosomes at the cell plasma membrane (PM), where membrane-bound secretory vesicles called zymogen granules (ZG) dock and fuse to release intravesicular contents to the outside.

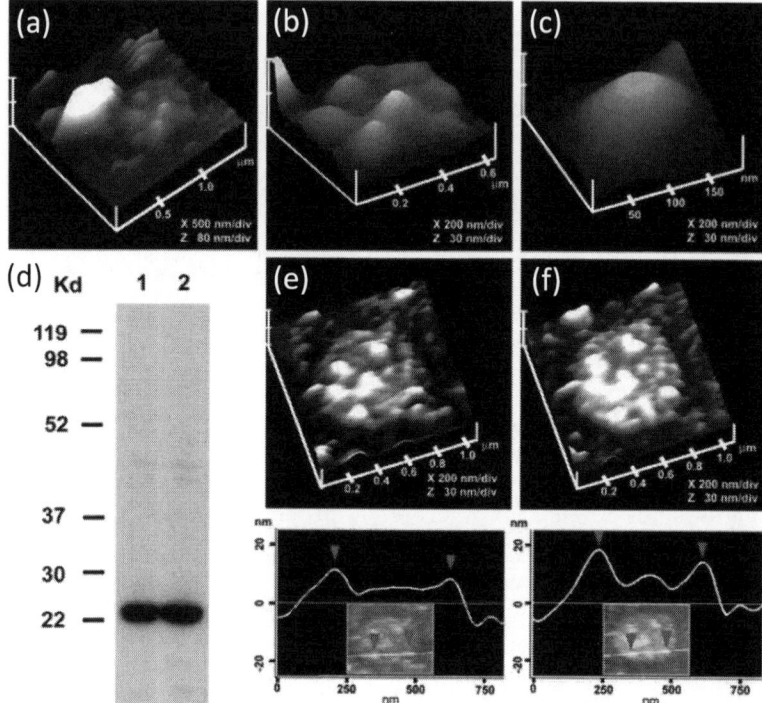

Figure 1.2 Morphology of the cytosolic compartment of the porosome complex revealed in AFM studies on isolated pancreatic plasma membrane preparations. (a) AFM micrograph of isolated plasma membrane preparation reveals the cytosolic compartment of a pit with inverted cup-shaped porosomes. Note the 600 nm in diameter ZG to the left hand corner of the pit. (b) Higher magnification of the same pit demonstrates the presence of 4–5 porosomes within (c). The cytosolic side of a single porosome is depicted in the AFM micrograph. (d) Immunoblot analysis of 10 μg and 20 μg respectively of pancreatic plasma membrane proteins using SNAP-23 antibody, demonstrates specificity of the antibody as reflected by a single 23 kDa immunoreactive band in the Western. (e and f) The cytosolic compartment of the plasma membrane demonstrates the presence of a pit with a number of porosomes within, shown before (e) and after (f) addition of the SNAP-23 antibody. Note the increase in height of the porosome base revealed by section analysis (bottom panel), demonstrating localization of SNAP-23 antibody to the base of the porosome, and establishing them to be ZG docking and fusion sites (*Biophys. J.*, 2003, **84**: 1–7).

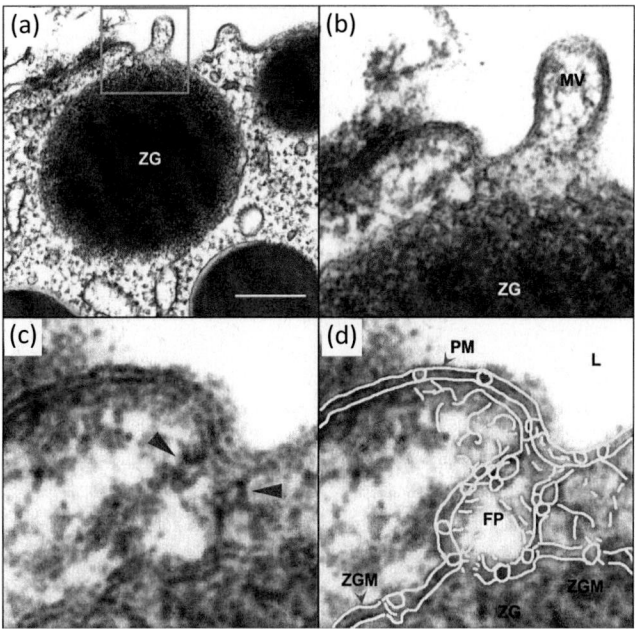

Figure 1.3   Transmission electron micrograph of a porosome associated with a docked secretory vesicle at the apical end of a pancreatic acinar cell. (a) Part of the apical end of a pancreatic acinar cell demonstrating within the square the presence of a fusion pore or porosome and an associated zymogen granule (ZG), the electron dense secretory vesicle of the exocrine pancreas. (Bar = 400 nm). (b) The area within the green square in (a) has been enlarged to show the apical microvilli (MV) and a section through the porosome and the ZG. Note the ZG membrane (ZGM) bilayer is attached directly to the base of the porosome cup. A higher magnification of the porosome in (c) depicts in greater detail the porosome bilayer and cross section through the three protein rings, with the thicker ring (arrowhead) present close to the opening of the porosome to the outside. This ring is likely composed of actin to regulate the opening and closing of the porosome opening to the outside. The third and lowest ring away from the porosome opening is in contact with the ZGM. (d) Outline of the porosome (FP) membrane shows the continuity with the plasma membrane (PM) at the apical end of the pancreatic acinar cell facing the lumen (L), and also defines the exact points of contact of the ZGM with the porosome membrane (*Biophys. J.,* 2003, **85**: 2035–2043).

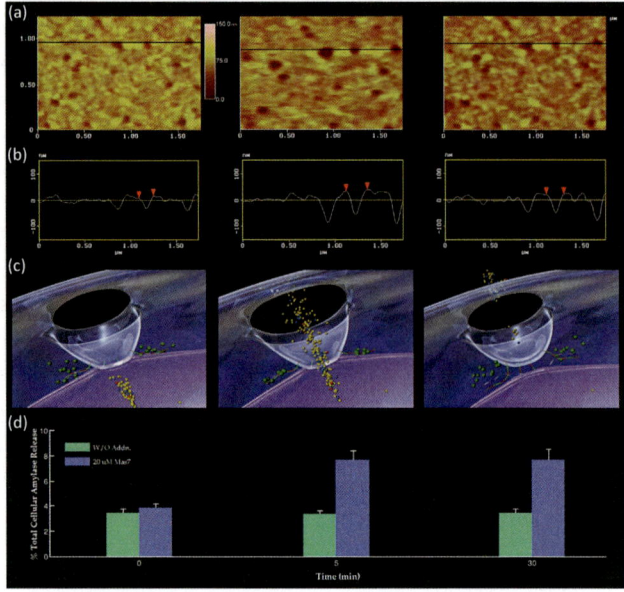

Figure 1.4   Porosome dynamics in pancreatic acinar cells following stimulation of cell secretion. (a) Several porosomes within a pit are shown at zero time, 5 min and 30 min following stimulation of cell secretion. (b) Section analysis across three porosomes in the top panel is represented graphically in the second panel and defines the diameter and relative depth of each of the three porosomes during secretion. The porosome at the center is shown by red arrowheads. (c) The third panel is a 3D rendition of the porosome complex at different times following stimulation of secretion. Note the porosome as a blue cup-shaped structure with black opening to the outside, and part of a secretory vesicle (violet) is shown docked at its base via t-/v-SNAREs. (d) The bottom panel represents % total cellular amylase release in the presence and absence of the secretagogue Mas7 (blue bars). Note an increase in porosome diameter and relative depth, correlating with an increase in total cellular amylase release at 5 min following stimulation of secretion. At 30 min following a secretory stimulus, there is a decrease in diameter and relative depth of porosomes and no further increase in amylase release beyond the 5-min time point. No significant changes in amylase secretion (green bars) or porosome diameter were observed in control cells in either the presence or absence of the non-stimulatory mastoparan analogue (Mas17). High-resolution images of porosomes were obtained before and after stimulation with Mas7, for up to 30 min (*Proc. Natl. Acad. Sci.,* 1997, **94**: 316–321).

stimulus, GH cells register nearly a 40% increase in porosome diameter and a concomitant increase in GH secretion.[4] Similar to acinar cells of the exocrine pancreas, the enlargement of porosome diameter during GH secretion and its subsequent decrease, and the loss in secretion following actin depolymerization,[4] also suggest GH porosomes to be the secretory portals at the plasma membrane in that cell type. Immuno-AFM studies first in the exocrine pancreas,[2] followed by GH cells,[4] directly demonstrated porosomes to represent the secretory portals at the cell plasma membrane. Porosomes in insulin secreting β-cells of the endocrine pancreas also demonstrate similar structures at the cell plasma membrane for regulated insulin release. In pancreatic β-cells, porosomes range in size from 100–130 nm, and similar to GH cells and pancreatic acinar cells, exposure of β-cells to a secretagogue, in this case 25 mM glucose, results in dilation of the porosome opening and increased insulin release. Similarly, exposure of β-cells to 20 μM cytochalasin B, results in a decrease in the porosome opening and a concomitant loss in insulin secretion, suggesting the involvement of actin in porosome regulation and β-cell secretion. Cup-shaped porosomes have also been demonstrated in neurons,[5-7] astrocytes,[8] and hair cells.[9] In contrast to the 100–180 nm porosomes present in exocrine pancreas and in endocrine and neuroendocrine cells, porosomes in neurons, astrocytes, and hair cells are an order of magnitude smaller, measuring just 10–17 nm. Porosomes in neurons[5-7] (Fig. 1.5), astrocytes,[8] and hair cells,[9] are all cup-shaped structures sporting a central plug, which has been implicated in the rapid opening and closing of the structure.

In the past decade, porosomes have been immunoisolated and reconstituted into artificial lipid membrane to further determine their structure function and regulation. Porosomes isolated from the exocrine pancreas when reconstituted into artificial lipid membranes, form 150–200 nm cup-shaped basket-like structures as observed in EM micrographs. The morphology of reconstituted porosomes appears similar to their structure in the native state at the plasma membrane in cells. To further determine if isolated porosomes are intact and functional, they have been reconstituted into lipid bilayer membranes for electrophysiological assessments. Using an electrophysiological lipid bilayer setup that enables continuous monitoring of changes in the porosome-reconstituted membrane under various experimental conditions, the function and regulation of isolated porosomes have been examined.[5,10] Addition of isolated ZG's and calcium to the *cis* compartment of the bilayer chamber results in docking and fusion of ZG's at the porosome-reconstituted bilayer, and a consequent increase in capacitance and conductance, and release of intravesicular contents to the *trans* chamber (Fig. 1.6). It has also been demonstrated that chloride channel activity is associated with the porosome complex, and the chloride channel blocker DIDS inhibits current activity and secretion in the porosome-reconstituted bilayer,[10] demonstrating its requirement in the proper functioning of the pancreatic porosome complex. Similar to the reconstitution of isolated porosomes from the exocrine pancreas into artificial lipid membrane,[10] neuronal porosomes have also been structurally and functionally reconstituted[6] in lipid membrane. To test whether complete and functional neuronal porosomes have been immunoisolated, the immunoisolates have been reconstituted into lipid membrane prepared using brain dioleoylphosphatidyl-choline (DOPC) and dioleoylphosphatidylserine (DOPS) in a ratio of 7:3.[5] These studies demonstrated the presence of porosomes in patches similar to those observed at the presynaptic membrane in intact synaptosome preparations. Furthermore, imaging of the reconstituted immunoisolate at ultrahigh resolutions using AFM and EM, demonstrated the presence of 10–15 nm porosomes having a 2 nm in diameter central plug as observed in their native state.[5,11] These studies confirmed the complete isolation and structural reconstitution of the neuronal porosome complexes into artificial lipid bilayers.[5] To further evaluate the functionality of the reconstituted neuronal porosome preparations, electrophysiological fusion assays using isolated synaptic vesicle preparations and bilayer-reconstituted porosomes have been utilized.[5] Addition of synaptic vesicles to neuronal porosome-reconstituted bilayer results in a large number of synaptic vesicle fusion events at the bilayer, reflected in the stepwise increase in membrane capacitance.[5] Exposure of docked synaptic vesicles at porosomes to 50 μM ATP, results in t-/v-SNARE disassembly and release of vesicles, followed by the return of the bilayer membrane capacitance to resting levels. These results demonstrate transient fusion of synaptic vesicles at the bilayer-reconstituted porosome complex.[5] To further assess if such a release mechanism of docked synaptic vesicles following ATP exposure occur in synaptosome, synaptosomal

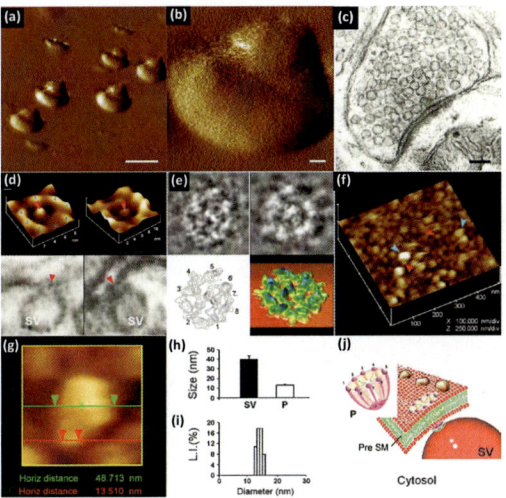

Figure 1.5   Structure and organization of the neuronal porosome complex at the nerve terminal. (a) Low resolution AFM amplitude image (Bar = 1 µm) (a) and high-resolution AFM amplitude image (Bar = 100 nm) (b) of isolated rat brain synaptosomes in buffered solution. (c) Electron micrograph of a synaptosome (Bar = 100 nm). (d) Structure and arrangement of the neuronal porosome complex facing the outside (top left), and the arrangement of the reconstituted complex in PC:PS membrane (top right). Lower panels depict two transmission electron micrographs demonstrating synaptic vesicles (SV) docked at the base of a cup-shaped porosome, having a central plug (red arrowhead). (e) EM, electron density, and 3D contour mapping demonstrate at the nanoscale the structure and assembly of proteins within the complex. (f) AFM micrograph of inside-out membrane preparations of isolated synaptosome. Note the porosomes (red arrowheads) to which synaptic vesicles are found docked (blue arrowhead). (g) High-resolution AFM micrograph of a synaptic vesicle docked to a porosome at the cytoplasmic compartment of the presynaptic membrane. (h) AFM measurements (n = 15) of porosomes (P, 13.05 ± 0.91) and synaptic vesicles (SV, 40.15 ± 3.14) at the cytoplasmic compartment of the presynaptic membrane. (i) Photon correlation spectroscopy (PCS) of immunoisolated neuronal porosome complex, demonstrating a size distribution from 12–16 nm. (j) Schematic illustration of a neuronal porosome at the presynaptic membrane, demonstrating the eight ridges connected to the central plug (*Cell Biol. Int.,* 2004, **28**: 699–708; *Cell Biol. Int.,* 2010, **34**: 1129–1132; *J. Microscopy,* 2008, **232**: 106–111).

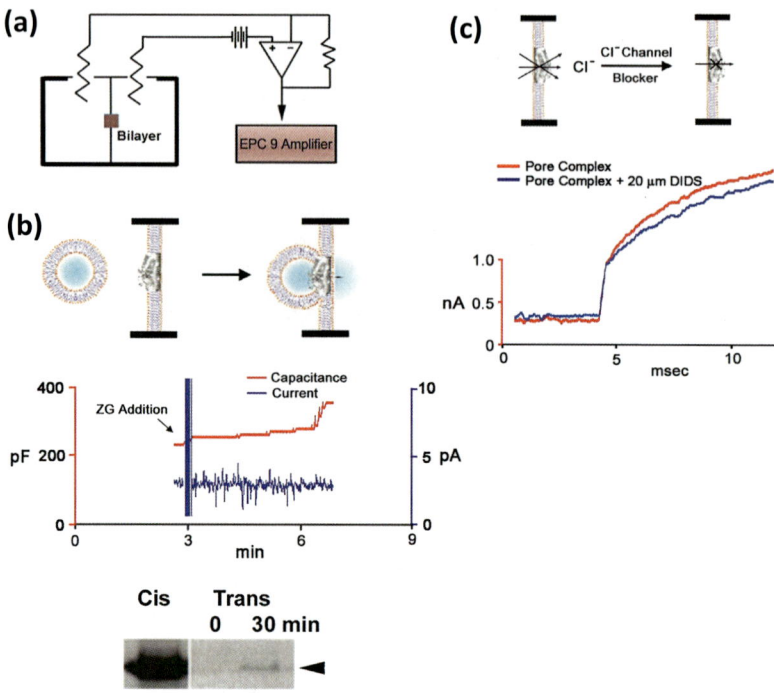

Figure 1.6 Lipid bilayer-reconstituted porosome complex is functional. (a) Schematic drawing of the bilayer setup for electrophysiological measurements. (b) Zymogen granules (ZGs) added to the *cis* compartment of the bilayer fuse with the reconstituted porosomes, as demonstrated by an increase in capacitance and current activities, and a concomitant time dependent release of amylase (a major ZG content protein) to the *trans* compartment of the lipid membrane. The movement of amylase from the *cis* to the *trans* compartment of the chamber was determined by immunoblot analysis of the contents in the *cis* and the *trans* chamber over time. (c) As demonstrated by immunoblot analysis of the immuno-isolated complex, electrical measurements in the presence and absence of the chloride ion channel blocker DIDS, indicate the presence of chloride channels in association with the porosome complex (*Biophys. J.*, 2003, **85**: 2035–2043).

membrane preparations have been exposed to 50 µM ATP, and the supernatant fraction assessed for synaptic vesicle association by monitoring levels of synaptic vesicle proteins SV2 and VAMP-2. Results from these studies demonstrate that both SV2 and VAMP-2 proteins are enriched in supernatant fractions following exposure of isolated synaptosomal membrane to ATP, demonstrating release of synaptic vesicles. This mechanism would allow multiple rounds of transient docking–fusion and release events that synaptic vesicles undergo during neurotransmitter release, without compromising vesicle identity and integrity.[12] Isolated inside-out synaptosomal membrane preparation when placed on mica and imaged by the AFM in near physiological buffer, demonstrates the presence of 40–50 nm synaptic vesicles docked to neuronal porosomes at the presynaptic membrane. Exposure of synaptic vesicles to GTP resulted in an increase in synaptic vesicle swelling, and further exposure to Ca^{2+} results in the transient fusion of synaptic vesicles at the presynaptic membrane, expulsion of intravesicular contents, and the consequent decrease in synaptic vesicle size.[5] These studies demonstrated that synaptic vesicles transiently dock and fuse rather than completely collapse at porosomes in the cell plasma membrane to release neurotransmitters.[5,13]

AFM, EM, and electron density measurements followed by contour mapping, and 3D topography of the neuronal porosome (Fig. 1.5) further provided an understanding of the arrangement of proteins at nanometer resolution within the complex.[11] Results from these studies demonstrated that proteins at the central plug of the porosome interact with proteins at the periphery of the complex, conforming to its eightfold symmetry (Fig. 1.5). At the center of the porosome complex representing the porosome base, where synaptic vesicles dock and transiently fuse, SNARE proteins are assembled in a ring conformation. In neurons, this SNARE ring at the porosome base is composed of just three SNARE pairs[14,15] having a 1–1.5 nm in diameter channel, for the express release of neurotransmitters from fused synaptic vesicles via the porosome to the synaptic cleft. Photon correlation spectroscopy (PCS) of isolated porosome complexes further confirmed that neuronal porosomes measure on average 12–15 nm (Fig. 1.5). In PCS measurements, the size distribution of isolated neuronal porosome complexes is obtained from plots of the relative intensity of light scattered by particles of known sizes and a calculation of their correlation function. Negative staining EM performed using low electron dose, in a Tecnai 20 electron microscope operating at 200 kV further confirmed the porosome size, and demonstrated that proteins at the central plug of the neuronal porosome complex interact with proteins at the periphery of the structure.[11] Similar to AFM micrographs, approximately eight interconnected protein densities are observed at the lip of the porosome complex in EM micrographs. The eight interconnected protein densities are also connected to the central plug, via spoke-like elements.

To understand at the molecular level how the porosome complex mediates secretory content release requires an inventory of the integral and associated molecules composing the structure. Knowledge of how each molecule contributes to the overall architecture of the porosome will further help elucidate how the structure mediates cell secretion. Biochemical analysis of isolated porosomes, demonstrated them to be composed of SNAP, syntaxin, cytoskeletal proteins actin, α-fodrin, and vimentin, calcium channels β3 and α1c, together with the SNARE regulatory protein NSF.[10,16] Chloride ion channels ClC2 and ClC3 have also been identified as components of the porosome complex, critical to its function. Isoforms of the various proteins identified within the porosome complex, have also been determined using 2D-BAC gels electrophoresis. For example, three isoforms each of the calcium ion channel and vimentin have been found in porosomes. Using yeast two-hybrid analysis and immunoisolation studies, the presence and direct interaction between some of these proteins with t-SNAREs within the porosome complex have also been established.[17] Besides proteins, studies report that the neuronal porosome assembly requires membrane cholesterol.[18] Results from studies[18] demonstrate a significant inhibition in interactions between porosome-associated t-SNAREs and calcium channels following depletion of membrane cholesterol. Since calcium is critical to SNARE-induced membrane fusion, the loss of interaction between SNAP-25, Syntaxin-1, and calcium channels at the neuronal porosome complex, would seriously compromise or even abrogate neurotransmitter release at the nerve terminal. Recent studies[19] using immunoisolation and gel filtration chromatography, followed by tandem mass spectrometry, provided further provides information on the composition of the neuronal porosome complex. Results from the study demonstrated that nearly 40 proteins to constitute the neuronal porosome proteome.[19] Furthermore, interaction of proteins within the porosome and their resulting arrangement was predicted in the study.[19] The association and dissociation of proteins at the porosome following stimulation of cell secretion further demonstrates the dynamic nature of the organelle.[19] It is of interest to note that plasma membrane calcium-transporting ATPase 1 and 2 are both found to be present in the neuronal porosome complex, suggesting that they may be involved in ATP-mediated expulsion of the extracellular calcium that enter cells during secretion. Similarly, the presence of sodium/potassium-transporting ATPase subunit alpha-3 with the porosome suggests the role of porosomes in ATP coupled exchange of sodium and potassium across the plasma membrane in opposite directions, creating the required electrochemical gradient required for maintaining the membrane resting potential as well as for the regulation of cell volume during secretion. Similarly, 2′,3′-cyclic-nucleotide 3′-phosphodiesterase, which has been reported to associate with microtubules and has microtubule-associated protein-like activity, is present in the neuronal porosome. The 2′,3′-cyclic-nucleotide 3′-phosphodiesterase in the porosome could also link tubulin to the cell membrane and participate in regulating the distribution of microtubules in the cytoplasm.[20] Hence, the presence of 2′,3′-cyclic-nucleotide 3′-phosphodiesterase at the porosome would be critical to the structural integrity of the complex. Dihydropyrimidinase-related protein present in the porosome proteome is known to play a role in cytoskeletal remodeling and cell polarity,[21] and exactly how this protein would be involved in neurotransmitter release at the porosome, remains to be established. In contrast, the involvement of NSF and dynamin at the porosome complex is predictable, and would likely be required for t-/v-SNARE complex disassembly by NSF, and fission of the established continuity between the lipid vesicle membrane and the porosome base by dynamin, following the transient fusion of synaptic vesicles at the base of the neuronal porosome (Fig. 1.7).

In summary, the discovery of porosomes has resulted in a paradigm shift in our understanding of the secretory process in cells. Studies in the past 16 years have further demonstrated that porosomes are a permanent supramolecular lipoprotein organelle at the cell plasma membrane, where membrane-bound secretory vesicles transiently dock and fuse at its base, to release intravesicular contents to the outside during cell secretion. Since porosomes are present in all secretory cells examined to date, they are the universal secretory portals in cells. It is interesting to note that while the 10–17 nm neuronal porosome complex is composed of nearly 40 different proteins,[19] the nuclear pore complex measuring nearly

120 nm is known to comprise approximately 400 proteins. In yeast, for example, the nuclear pore complex is comprised of some 30 proteins each represented in 16 copies in the structure.[22] Future work will involve understanding the interactions between each of the core neuronal porosome proteins, and mapping of the precise position of each core protein within the neuronal porosome complex.

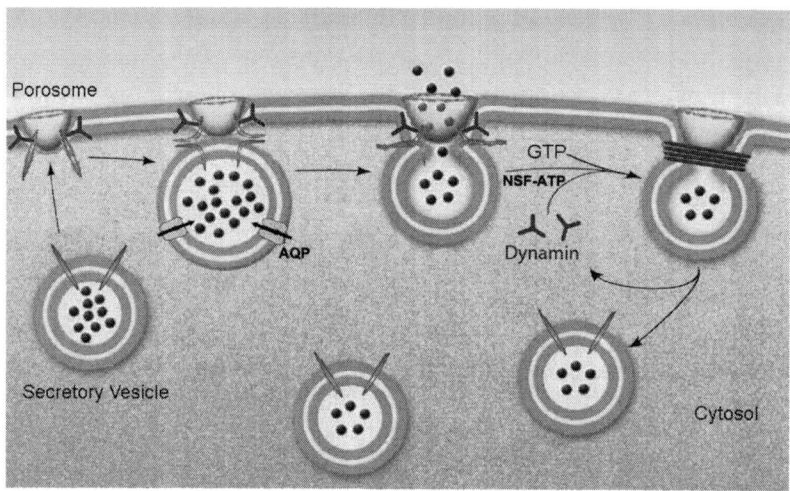

Figure 1.7 Schematic drawing depicting secretory vesicle docking and transient fusion at the porosome base, release of secretory products, and the disengagement of partially empty vesicles from the porosome. Following stimulation of cell secretion, secretory vesicles would dock at the porosome base, develop intravesicular pressure via active transport of water through water channels or aquaporins (AQP) at the vesicle membrane, transiently fuse at the porosome base via SNAREs and calcium, and expel neurotransmitters. Following secretion, NSF (an ATPase), and dynamin (a GTPase), would enable the disassembly t-/v-SNARE complexes and fission the neck of fused vesicles at the porosome base respectively. This process will generate partially empty secretory vesicles, and these vesicles could go through multiple rounds of docking–fusion–expulsion–dissociation cycles. Unlike protein and peptide containing vesicles, synaptic vesicles have neurotransmitter transporters at the vesicle membrane to rapidly replenish spent vesicles following neurotransmitter release (*J. Proteomics,* 2012, **75**: 3952–3962).

The author thanks the many students and collaborators who have participated in the studies discussed in this article. Support from the National Institutes of Health (USA), the National Science Foundation (USA), and Wayne State University, is greatly appreciated.

References

1. Schneider, S. W., Sritharan, K. C., Geibel, J. P., Oberleithner, H., Jena, B. P. (1997). Surface dynamics in living acinar cells imaged by atomic force microscopy: identification of plasma membrane structures involved in exocytosis. *Proc Natl Acad Sci USA,* **94,** 316–321.

2. Cho, S.-J., Quinn, A. S., Stromer, M. H., Dash, S., Cho, J., Taatjes, D. J., Jena, B. P. (2002). Structure and dynamics of the fusion pore in live cells. *Cell Biol Int,* **26,** 35–42.

3. Craciun, C., Barbu-Tudoran, L. (2012). Identification of new structural elements within 'porosomes' of the exocrine pancreas: a detailed study using high-resolution electron microscopy. *Micron* DOI.org/10.1016/j.micron.2012.05.01

4. Cho, S.-J., Jeftinija, K., Glavaski, A., Jeftinija, S., Jena, B. P., Anderson, L. L. (2002). Structure and dynamics of the fusion pores in live GH-secreting cells revealed using atomic force microscopy. *Endocrinology,* **143,** 1144–1148.

5. Cho, W. J., Jeremic, A., Rognlien, K. T., Zhvania, M. G., Lazrishvili, I., Tamar, B., Jena, B. P. (2004). Structure, isolation, composition and reconstitution of the neuronal fusion pore. *Cell Biol Int,* **28,** 699–708.

6. Okuneva, V. G., Japaridze, N. D., Kotaria, N. T., Zhvania, M. G. (2012). Neuronal porosome in the rat and cat brain. *Cell and Tissue Biol,* **6,** 69–72.

7. Japaridze, N. J., Okuneva, V. G., Osovreli, M. G., Surmava, A. G., Lordkipanidze, T. G., Kiladze, M. T., Zhvania, M. G. (2012). Hypokinetic stress and neuronal porosome complex in the rat brain: the electron microscopic study. *Micron,* **43,** 948–953.

8. Lee, J.-S., Cho, W. J., Jeftinija, K., Jeftinija, S., Jena, B. P. (2009). Porosome in astrocytes. *J Cell Mol Med,* **13,** 365–372.

9. Drescher, D. G., Cho, W. J., Drescher, M. J. (2011). Identification of the porosome complex in the hair cell. *Cell Biol Int Rep,* **18,** 31–34.

10. Jeremic, A., Kelly, M., Cho, S.-J., Stromer, M. H., Jena, B. P. (2003). Reconstituted fusion pore. *Biophys J,* **85,** 2035–2043.

11. Cho, W. J., Ren, G., Jena, B. P. (2008). EM 3D contour maps provide protein assembly at the nanoscale within the neuronal porosome complex. *J Microscopy,* **232,** 106–111.

12. Aravanis, A. M., Pyle, J. L., Tsien, R. W. (2003). Single synaptic vesicles fusing transiently and successively without loss of identity. *Nature,* **423,** 643–647.

13. Kelly, M., Cho, W. J., Jeremic, A., Abu-Hamdah, R., Jena, B. P. (2004). Vesicle swelling regulates content expulsion during secretion. *Cell Biol Int,* **28,** 709–716.

14. Cho, W.-J., Shin, L., Ren, G., Jena, B. P. (2009). Structure of membrane-associated neuronal SNARE complex: implication in neurotransmitter release. *J Cell Mol Med,* **13,** 4161–4165.

15. Mohrmann, R., de Wit, H., Verhage, M., Neher, E., Sørensen, J. B. (2010). Fast vesicle fusion in living cells requires at least three SNARE complexes. *Science,* **330,** 502–505.

16. Jena, B. P., Cho, S.-J., Jeremic, A., Stromer, M. H., Abu-Hamdah, R. (2003). Structure and composition of the fusion pore. *Biophys J,* **84,** 1–7.

17. Cho, W. J., Jeremic, A., Jena, B. P. (2005). Direct interaction between SNAP-23 and L-type calcium channel. *J Cell Mol Med,* **9,** 380–386.

18. Cho, W. J., Jeremic, A., Jin, H., Ren, G., Jena, B. P. (2007). Neuronal fusion pore assembly requires membrane cholesterol. *Cell Biol Int,* **31,** 1301–1308.

19. Lee, J.-S., Jeremic, A., Shin, L., Cho, W. J., Chen, X., Jena, B. P. (2012). Neuronal Porosome proteome: molecular dynamics and architecture. *J Proteomics,* **75,** 3952–3962.

20. Bifulco, M., Laezza, C., Stingo, S., Wolff, J. (2002). 2′,3′-Cyclic nucleotide 3′-phosphodiesterase: a membrane-bound, microtubule-associated protein and membrane anchor for tubulin. *Proc Natl Acad Sci USA,* **99,** 1807–1812.

21. Goshima, Y., Nakamura, F., Strittmatter, P., Strittmatter, S. M. (1995). Collapsin-induced growth cone collapse mediated by an intracellular protein related to UNC-33. *Nature,* **376,** 509–514.

22. Rout, M. P., Aitchison, J. D., Suprapto, A., Hjertaas, K., Zhao Y., Chait, B. T. (2000). The yeast nuclear pore complex: composition, architecture, and transport mechanism. *J Cell Biol,* **148,** 635–651.

Chapter 2

The Hair Cell Porosome: Molecular and Synaptic Implications

Dennis G. Drescher

Wayne State University School of Medicine, Detroit, MI 48201, USA

ddresche@med.wayne.edu

Porosomes are inverted cup-shaped, presynaptic structures where secretory vesicles transiently fuse and expel their contents to the extracellular space. It is known that hair cells of hearing and balance secrete transmitter from synaptic vesicles during sensory signal transduction, but it was previously unknown whether these sensitive mechanosensory cells possess porosome structures that could participate in the secretory process. In this chapter, we provide evidence that porosome structures indeed exist in the hair cell, suggesting a mechanism of hair-cell transmitter secretion different from that of the exocytotic process currently proposed. We also discuss molecular mechanisms consistent with the porosome model.

2.1 Introduction

Cell secretion is a fundamental process of living cells, playing a central role in cell division, exocrine and endocrine function, and

NanoCellBiology: Multimodal Imaging in Biology and Medicine
Edited by Bhanu P. Jena and Douglas J. Taatjes
Copyright © 2014 Pan Stanford Publishing Pte. Ltd.
ISBN 978-981-4411-79-0 (Hardcover), 978-981-4411-80-6 (eBook)
www.panstanford.com

neurotransmitter release. The elements of secretion, particularly of neurosecretion, involve processes in which vesicles contact the cell membrane and expel their contents to the outside.

The pioneering electrophysiologists of the 1950s and 1960s[1] laid groundwork for the idea that vesicles secreted their contents into the extracellular space by the process of exocytosis and then were recycled.[2] In exocytosis, the vesicle membrane was thought to fuse completely with the plasma membrane. This increase in total plasma membrane would be reclaimed as vesicles by endocytosis, which could occur at a site distant from where the vesicles were secreted.[3-5] For auditory and vestibular hair cells, results of electrophysiological studies showing an increase in hair-cell membrane capacitance after stimulation[6,7] were interpreted as supporting the classical exocytotic mechanism.

The early investigators noted, however, that the stated mechanism of exocytosis did not account entirely for vesicle neurosecretion. Capacitance changes, used to measure the increase in total cell plasma membrane, did not perfectly match the estimated membrane increase calculated for the vesicles secreted, being less than what would be expected from a pure fusion mechanism.[8] It was further observed, in photomicrographs, that after vigorous stimulation, vesicles still remained.[3,9] Thus the concept of "kiss and run" emerged, with the proposal that vesicles transiently dock to release their contents and then detach to return intracellularly.[10] However, no clear morphological correlates were put forward corresponding to a kiss and run mechanism.

In 1997, Schneider et al.,[11] studying the acinar cells of the exocrine pancreas, reported the presence of a stable morphological complex (now termed the "porosome") in the presynaptic membrane that could account for the observed behavior of vesicles and membrane. The proposed porosome mechanism, whereby the vesicle attached, released, and broke off to return intracellularly for refilling, could occur in place of or in addition to classical exocytosis. This process would offer an energetically efficient model for support of rapid and continuous neurosecretion, which is particularly characteristic of auditory/vestibular hair cells.

In the last two decades, the porosome structure has been described in a number of neuroendocrine tissues and neurons,[12] but until recently has not been identified in any sensory secretory cell. This chapter describes the identification of the porosome structure in the hair cell, and the synaptic mechanisms and molecules that are likely to be associated with the porosome. Areas for further research are also proposed.

2.2 Methods

In studies on the identification of the porosome structure in the hair cell, transmission electron microscopy was employed. Saccular maculae from rainbow trout (*Oncorhynchus mykiss*) were dissected to yield a cell layer in which the hair cell was the only intact cell type[13] and transferred to Trump's fixative consisting of 1% glutaraldehyde, 4% formalin, 0.1 M sodium phosphate, pH 7.2.[14] Tissues were post-fixed in 1% osmium tetroxide for 1 h, dehydrated, and embedded in Embed 812 (Electron Microscopy Sciences, Fort Washington, PA). Pale gold-to-silver sections (65–70 nm thick) were placed on 200-mesh copper grids, post-stained with aqueous uranyl acetate and Reynolds lead citrate,[15] examined with a Zeiss EM10-CA transmission electron microscope, and photographed. Electron micrograph photos representing a magnification of 200,000× actual size were quantitatively analyzed with Bioquant II software (R & M Biometrics, Nashville, TN).

A technique that is likely to become important in future studies of the biochemistry of the hair-cell porosome, and of porosomes from other tissues, is surface plasmon resonance (SPR). Our laboratory has previously employed SPR to define interactions between major proteins of the synaptic complex,[16] including those of the calcium-binding domains of otoferlin, a hair-cell synapse-associated protein.[17] (Otoferlin may take the place of synaptotagmin as the main calcium-sensing protein of the hair-cell synaptic complex.) SPR, in addition to such well-established methods as pull-down and yeast two-hybrid analysis, is useful for determining the interaction between two proteins at nanomolar concentrations, and has the added advantage that it is quantitative. In performing the assay, affinity-purified fusion proteins are first immobilized by an amine-coupling reaction on a CM5 sensor chip (Biacore, Piscataway, NJ). The CM5 chip has a gold surface coated with carboxymethylated dextran,[18] forming a matrix of high binding capacity and low non-specific binding.[19] The chip is inserted into the flow chamber of a Biacore 3000 instrument (Biacore, Uppsala,

Sweden). Immobilization to the chip surface of one protein-binding partner, the ligand, involves activation of carboxymethyl groups on the chip by reaction with *N*-hydroxysuccinimide to form a highly reactive succinimide ester, followed by covalent bonding to the chip through reaction of its amine groups to form amide linkages and blockage of excess activated esterified carboxyls with ethanolamine.[19] Reference surfaces are prepared in the same manner, except that all carboxyls are blocked and no ligand is added. During analysis,[20] each cell on the chip containing an immobilized fusion polypeptide is paired with an adjacent cell, the latter serving as a reference. Addition of a second fusion protein to the chamber, the flow-through analyte, can result in binding to the immobilized ligand, producing a small change in refractive index at the detector's gold surface,[21] which can be quantitated with precision.[22] The chips can be re-used after regeneration with 0.5–1.0 M NaCl at neutral pH, a procedure which retains the activity of the bound ligand for the next analysis.[20]

2.3 SNAREs

Central to cell secretion are proteins termed SNARES (soluble *N*-ethylmaleimide-sensitive fusion protein attachment protein receptors), which play an essential role in vesicle docking, priming, fusion, and synchronization of neurotransmitter release.[23] The hair-cell ribbon synapse, similarly to other secretory sites, is characterized by the presence of SNAREs,[24] and the basic SNARE mechanisms are thought to occur for hair cells as well as for neurons and other secretory cells. The SNARE hypothesis constitutes a mechanistic model of membrane fusion based on the characteristics of plasma membranes and vesicles. Most of the mechanistic steps necessary for transmitter release from neurons and sensory receptor cells occur at the presynaptic region. The SNARE proteins present in the acceptor (plasma membrane) and donor (vesicle) membranes are thought to mediate the spatial specificity of the interaction between the vesicle and presynaptic membrane preceding fusion.[25] Extensive studies have shown that the SNARE complex comprises two main classes of components: (1) the *v-SNAREs*, the SNARE proteins present in the vesicles (predominantly synaptobrevin[26]) and (2) the *t-SNAREs*, the proteins present on the target presynaptic plasma membrane (predominantly syntaxin and synaptosomal-associated proteins such as SNAP-25). Interaction between these two groups of proteins occurs through the highly conserved SNARE motifs present in these molecules that form an extremely stable four-helix bundle and bring together the vesicle and plasma membranes, thereby facilitating their fusion and release of the vesicle contents. Because of the characteristic complex formed by the three core proteins, synaptobrevin, syntaxin, and SNAP-25/23 (SNAP contributes two helices), SNARE proteins are thought to catalyze the steps involved in the release by reducing the energy barrier[27] and increasing the specificity of vesicle fusion, as well as by directly facilitating pore formation by inducing distortion in the membranes.

Four main steps lead to the fusion of vesicles prior to secretion (Fig. 2.1). The hair cell utilizes synaptic ribbons or bodies, apparently to guide its secretory vesicles to presynaptic sites. The first step preceding fusion is *tethering*, where the vesicles are brought to the active zone to be attached to protein complexes at the presynaptic membrane,[28] facilitating contact between v-SNARE and t-SNARE proteins. At the presynaptic membrane of neurons, syntaxin 1A is attached to Munc-18, forming a closed structure.[29] The tethering process is thought to detach this complex and open syntaxin 1A for SNARE interaction. Munc-18 activates exocytosis/neurosecretion[30] with its dissociation from the syntaxin 1A closed form, freeing the SNARE motif for complex formation.[31,32] There is evidence for a continued association of Munc-18 with the amino terminus of syntaxin 1A in the assembled SNARE complex. Munc-18 may also be involved in chaperoning syntaxin 1A to the membrane.[32] The tethering step is followed by the *docking* of vesicles, where SNARE proteins come in contact with each other via the SNARE motifs in a calcium-independent manner. Docking is followed by vesicle *priming*, where the SNARE proteins form a stable complex via their SNARE motifs, rendering the vesicles competent for fusion. Neher and Sakaba[33] have defined two kinds of synaptic-vesicle priming, "molecular priming" and "positional priming." Molecular priming involves the association of the vesicle with the SNARE complex. Positional priming describes the process in which primed vesicles situate themselves optimally near the voltage-gated calcium channels.[34] Both of these modes of priming are calcium-dependent.

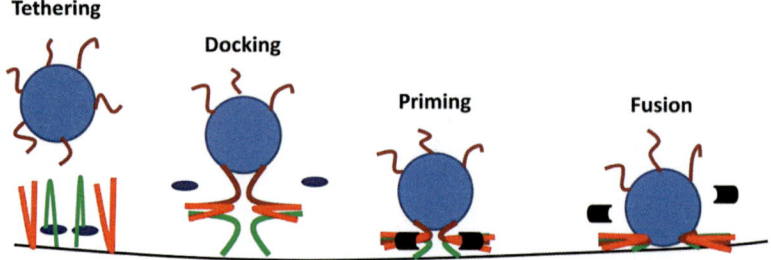

Tethering

Docking

Priming

Fusion

Steps in Vesicle Fusion

— SNAP-25
— SYNTAXIN 1   — SYNAPTOBREVIN
● MUNC 18   ◖ COMPLEXIN

Figure 2.1   Steps in synaptic vesicle fusion include tethering, docking, priming, and fusion. These events are driven by high-affinity interaction between v-SNARE and t-SNARE proteins, regulated by calcium and calcium-binding proteins through their interaction with the SNARE complexes. Major synaptic complex proteins are listed. (This research[23] was originally published in *Molecular and Cellular Neuroscience*: Ramakrishnan, N. A., Drescher, M. J., and Drescher, D. G. (2012) The SNARE complex in neuronal and sensory cells, *Mol. Cell. Neurosci.*, **50**, 58–69. © Elsevier.)

The *fusion* of vesicles is triggered by high calcium surrounding the active zone. Active zones possess clusters of voltage-gated calcium channels, and the t-SNARE proteins are thought to surround and directly interact with these channels.[17] During generation of the action potential or receptor potential, the cells undergo depolarization, prompting the voltage-gated calcium channels to open, resulting in an influx of calcium and formation of a calcium micro-domain surrounding the presynaptic region. Calcium is thought to mediate several molecular interactions between the vesicle proteins and the presynaptic plasma-membrane proteins, resulting in structural changes in the SNARE complex.[35] In neurons, synaptotagmins, the vesicle-bound calcium sensors, also play an important role in vesicle fusion by their direct interaction with the lipid membrane. In hair cells, otoferlin or some other protein may replace synaptotagmin as the calcium sensor, since the hair-cell synapse has been found to lack synaptotagmins 1 and 2.[24] The SNARE structural changes lead to a "puncturing" of the membrane, thus forcing fusion and creating a pore for release of transmitters. Synaptotagmins sense calcium, and complexins act as calcium-dependent switches, facilitating synchronized vesicle fusion via their interaction with the SNAREs.[36-38] After releasing transmitters, the vesicles can be detached by dissociation of the SNARE complex in a process that requires the ATPase in the NSF complex (*N*-ethylmaleimide sensitive fusion protein). It is thought that fusion-competent conformation of SNARE molecules is maintained by molecular chaperone complexes composed of cysteine string protein α (CSPα), Hsc70 (heat shock 70 kDa protein 8) and SGT (small glutamine-rich tetratricopeptide repeat-containing protein). Genetic deletion of CSPα leads to degradation of SNAP-25 and decreased SNARE complex assembly.[39]

The SNARE *core complex* is a three-molecule, extremely stable four-helix complex sometimes termed the *SNAREpin*.[27,40,41] The SNAREpin bridges the vesicle membrane and the plasma membrane. The SNARE core complex is formed by SNARE helical motifs, each of approximately 60 amino acids, with synaptobrevin and syntaxin individually contributing one helix and SNAP-25 two helices from the same molecule.[41] The high-resolution structure of the complex reveals parallel alpha-helices that twist around each other and create a leucine-zipper-like assembly with an embedded ionic layer consisting of repeating modules of an arginine residue and three glutamine residues.[41] Grooves containing distinct hydrophobic, hydrophilic and charged regions are also present in the core that may facilitate interaction with regulatory factors. In the neuronal SNARE core complex, SNARE proteins form a continuous helical bundle that is stabilized by side-chain interactions in the linker regions.[42] The structural extension of the core complex into the lipid bilayer[42] is taken as direct evidence of the involvement of SNARE complexes in vesicle fusion.

"Zippering" of the SNAREpin, in which the SNAREpin coils tighten and come together,[43] forces contact between the vesicle and plasma membranes.[40,44] Conformational changes and the effect of the specific amino acid sequences of the transmembrane domains of syntaxin and synaptobrevin mechanistically drive membrane fusion.[45] It is estimated that three or more SNAREpins should provide sufficient energy for the fusion of vesicle and plasma membrane.[27,46] Consistent with this idea, it has been proposed that 5–7 syntaxin transmembrane domains form the outer rim of the fusion pore.[47] Alternatively, it is hypothesized that a stable hemi-fusion product leads to fusion pore formation.[48] Conformational changes in the SNARE complex and the zipper-tightening activity of the complex are hypothesized to force the membranes on each side to distort and form a pore.[49] Complexin, a calcium-dependent regulator of neuronal exocytosis, has been shown to bind to the groove between the synaptobrevin and syntaxin helices in the SNARE complex in an anti-parallel alpha-helical conformation, thus stabilizing the interphase between the two helices. Because of their apposing positions, the vesicle and presynaptic membranes exert mutually repulsive forces that are counteracted by the v-SNARE and t-SNARE helices bound to complexin.[50] Synaptotagmin (and possibly otoferlin in the hair cell), along with calcium and phospholipids, bind SNARE proteins[36] to facilitate calcium-dependent exocytosis.[51] Complexins are thought to function in association with synaptotagmin as a molecular switch in "clamping" the vesicle fusion complex prior to calcium-dependent activation via synaptotagmin.[52] Evidence in support of the role of SNAREs in fusion-pore formation is also provided by studies on the v-SNARE, synaptobrevin 2. It has been suggested that perturbation of the vesicle membrane caused by the C-terminal transmembrane domain of synaptobrevin 2 during zippering causes the vesicle to open.[53] Engineered addition of two charged amino acids (lysine or glutamate) to the C-terminus of synaptobrevin has been reported to inhibit fusion and exocytosis. The latter finding suggests that the SNARE complex provides both the energy (in the zippering process) as well as physical means to perturb the lipid bilayer during membrane fusion.

2.4   The Porosome

Is there a stable structural component, or module, that organizes multiple SNAREs in the pre-synaptic area to facilitate uninterrupted vesicle fusion and neurosecretion? Many in vitro and in vivo studies show that 3–15 SNAREs are required for exocytosis/neurosecretion.[54] For the hair cell in particular (see below), one or few SNAREs may not efficiently catalyze membrane fusion. It has yet to be seen how many SNARE complexes do participate in fast exocytosis/neurosecretion from sensory cells, such as hair cells or photoreceptors. Because of the need for tight regulation and fast dynamics in sustained and high-fidelity neurosecretion, each hair-cell ribbon synapse may employ at least several SNAREs to meet the turnover of hundreds of vesicle fusions per second. Between priming, secretion, and endocytosis, if one hypothesizes a classic exocytotic mechanism, remaining protein cargo of the synaptic membrane and the SNARE complex must be directed toward re-use. A "site-clearing" step has been postulated to involve patch diffusion or declustering-reclustering of vesicle cargo proteins toward the endocytic retrieval site.[5] These reclaiming steps would seem to require significant energy. Energetic efficiency may be gained by the utilization of a more stable structure for vesicle docking, and may be the most economical way to accomplish fast exocytosis/neurosecretion.

The ultrastructure of the porosome[11,55–58] has been proposed to be a stable 8–12 protein umbrella- or cup-shaped transmembrane complex that has multiple conformation states,[59] depending on whether the porosome complex is resting or in active exocytosis.[11] Atomic force microscopy has revealed surface features of the porosome, and transmission electron microscopy has provided supporting evidence for the porosome at the vesicle and cell membrane interface.[60] In addition to its initial identification in pancreatic acinar cells, the porosome structure has been documented in pituitary growth hormone-secreting cells,[61] adrenal chromaffin cells,[62] beta-cells of the endocrine pancreas,[63] neurons,[60, 64,65] and astrocytes.[66] Recent work, described herein, has shown that porosomes are found in the sensory hair cell.[67] Since porosomes have now been found in many and varied cell types, these structures have been proposed to be the universal secretory machinery of the cell plasma membrane.

2.5 Discovery of the Hair-Cell Porosome

Hair cells are highly sensitive mechanoreceptor cells present in the organs of hearing and balance in vertebrates. The characteristic ribbon synapses of hair cells are likely to play an important role in maintaining this receptor cell's remarkable ability for rapid, sustained, and graded exocytosis/neurosecretion. However, the detailed cellular and molecular mechanisms supporting these extraordinary properties of hair cells have yet to be elucidated. Vesicle recruitment at hair-cell ribbons appears to be fast, and apparently, inexhaustible.[68] A rapid burst of transmitter release occurs at depolarization onset, followed by sustained release with continued depolarization.

Since continuous exocytotic plasma-membrane recycling as a sole mechanism seems inefficient in light of the stringent demands of hair-cell neurosecretion, we analyzed the afferent and efferent synapses of the hair cells of the rainbow-trout sacculus, an organ of hearing and balance in teleosts, to determine if porosome-like structures were present in this sensory receptor cell. When we examined transmission electron micrographs of the synaptic structure of trout hair cells, the possibility of a mechanism for receptoneural secretion apart from classic exocytosis emerged.

Hair cells contain dense synaptic bodies, each surrounded by a halo of clear vesicles.[69] These bodies with vesicles characterize the excitatory afferent ribbon synapse, present in hair cells and other receptor cells, thought to aid in fast, synchronous release[70] by directing vesicles to their site of secretion. Efferent endings on hair cells are filled with clear vesicles and modify the afferent signal of the teleost saccular hair cell in an inhibitory manner. The average diameter of the afferent vesicles encircling synaptic bodies is 46.4 ± 0.2 nm ($n = 640$), whereas the average diameter of the efferent vesicles is 53.1 ± 0.3 nm ($n = 433$).[71]

The structure of synaptic vesicles at the teleost hair-cell afferent synapse is indicated in Fig. 2.2. An overview of the synaptic region is shown in Fig. 2.2a. Two different, representative docked vesicles comprising putative porosome structures are shown in Figs. 2.2b,c. Figure 2.2d (from Fig. 2.2c) outlines the hair-cell vesicle (yellow), including its associated porosome region (orange) and pre- and post-synaptic membranes (yellow). A central plug is indicated by the blue arrowhead. The observed porosome-like structure at the hair-cell synapse in Fig. 2.2 is remarkably similar to that described for neuronal cells.[64]

Magnified presumptive porosome structures present in efferent endings synapsing on hair cells are shown in Fig. 2.3. Similarly to Fig. 2.2, an overview of the synaptic region is shown in Fig. 2.3a, and two representative views of the porosome structures are presented in Figs. 2.3b,c. A colored, outlined version of Fig. 2.3c is shown in Fig. 2.3d. The characteristic porosome configuration is again apparent (outlined in orange), including the presence of a central plug (blue arrowhead).

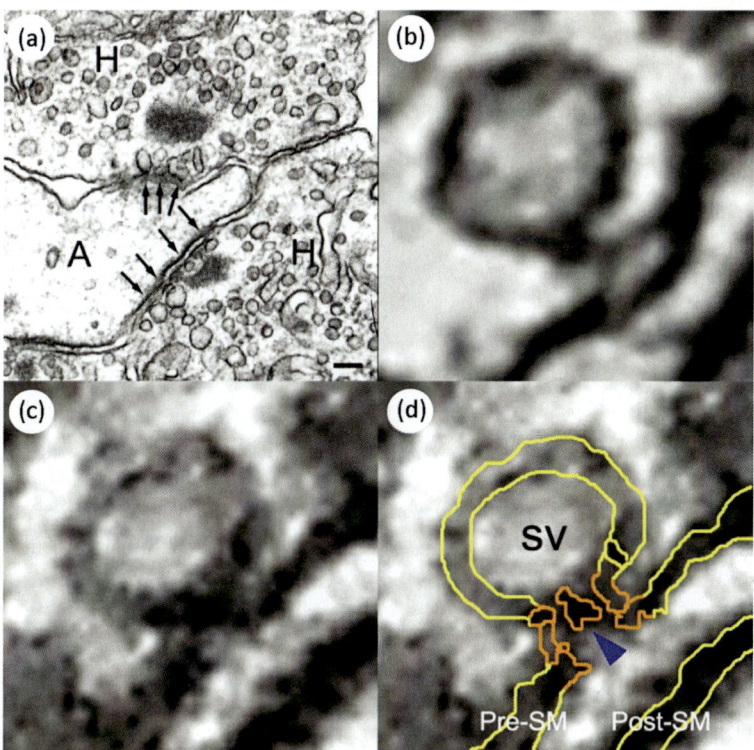

Figure 2.2   Porosomes at the hair-cell plasma membrane of the afferent synapse in the trout saccular macula. (a) Electron micrograph shows porosome locations (arrows) at the afferent nerve ending, with 40–50 nm synaptic vesicles, synaptic body, and intact presynaptic and postsynaptic membranes. Scale bar = 100 nm. (b,c) Two representative porosomes, shown at higher magnification. Vesicle diameter is approximately 46 nm, here and in Fig. 2.3, b and c. (d) Outline of the porosome region from panel c. Orange lines designate the 10–12 nm-diameter porosome structure observed at the presynaptic membrane (Pre-SM), and yellow lines show an approximately 45 nm synaptic vesicle (SV) docked at the porosome base (base = upper portion of porosome in the figure, the bottom of a cup structure; see Cho et al.[59]) and the postsynaptic membrane (Post-SM). Blue arrowhead indicates the porosome's central plug. (This research[67] was originally published in *Cell Biology International Reports*: Drescher, D. G., Cho, W. J., and Drescher, M. J. (2011) Identification of the porosome complex in the hair cell, *Cell Biol. Int. Rep.*, **18**, 31–34. © Portland Press Limited and the International Federation for Cell Biology.)

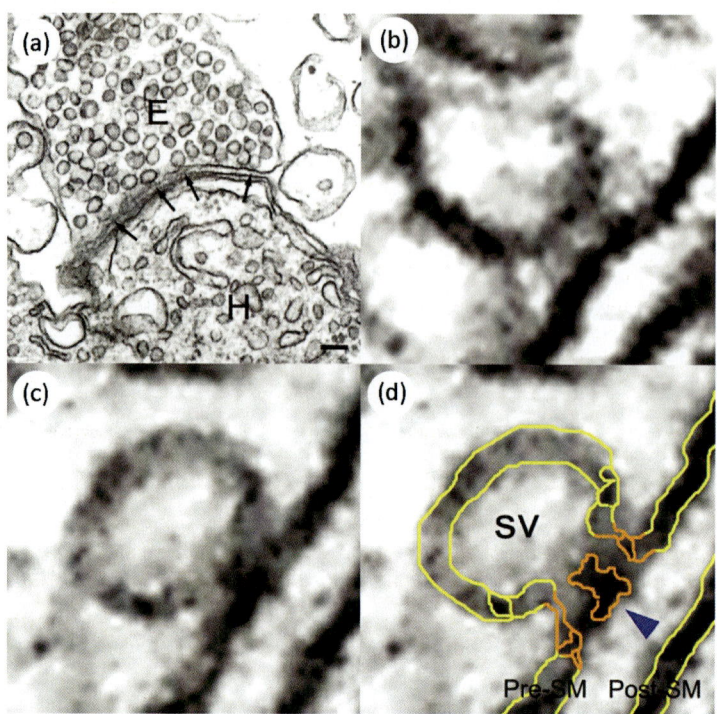

Figure 2.3   Porosomes at the plasma membrane of an efferent (E) nerve ending synapsing on the trout saccular hair cell. (a) Electron micrograph shows porosomes (arrows). (b,c) Two representative efferent porosomes, shown at higher magnification, with intact presynaptic and postsynaptic membranes. (d) Outline of the porosome region from panel c. Format is similar to that of Fig. 2.2. (This research[67] was originally published in *Cell Biology International Reports*: Drescher, D. G., Cho, W. J., and Drescher, M. J. (2011) Identification of the porosome complex in the hair cell, *Cell Biol. Int. Rep.*, **18**, 31–34. © Portland Press Limited and the International Federation for Cell Biology.)

We conclude from our results[67] that there is support for the presence of porosome-like structures in the teleost vestibular hair cell. We found these structures many times, both for hair-cell and efferent synapses (24 times for hair-cell vesicles, 24 times for efferent vesicles in the cited study). This work constitutes the first description of porosomes in the saccular hair cells, and also the first observation of porosomes in a sensory receptor cell. These findings suggest a mechanism of transmitter secretion different from the conventional sole exocytotic process currently proposed for hair cells. The current morphological observation bolsters the hypothesis that porosome structures, and their implied fusion mechanism, may indeed be present universally in secretory cells.

2.6   Conclusions and Future Studies

Our existing knowledge from other tissues concerning functional and structural proteins of porosome complexes should help to design experiments to look for similar elements in the hair-cell porosome. Antibodies are a sensitive means to examine the synaptic complex. It may be necessary to harvest large numbers of hair cells, where large numbers of the cells are available (such as the sacculus of the rainbow trout), and then sub-fractionate to the level of entities such as synaptosomes. Affinity chromatography with bound ligands associated with the porosome, such as the t-SNARE, SNAP-25, can be used to isolate hair-cell porosomes, and mass spectroscopic analysis of digests of porosome proteins can yield information about the protein identities.[72] The hair-cell porosome is likely to contain at least the major proteins found in other porosomes, such as the t-SNARES (syntaxin and SNAP23/25) and the voltage-gated calcium channel. The hair-cell porosome complex probably also encompasses an array of additional structural and synaptic complex proteins. The composition and configuration of fusion pores from other tissue sources, as outlined above, have been addressed in published studies.[56,64] Cytoskeletal proteins such as actin and vimentin, as well as NSF, are considered to be part of the porosome. The SNARE proteins appear to surround the porosome cup, ready for vesicle docking and fusion[12,73] (Fig. 2.4). Another complementary approach for determining porosome composition is to construct cDNA libraries with hair-cell synaptosomes and porosomes, and probe the libraries with sequences representing known or suspected mRNA messages, including those corresponding to porosome proteins identified by mass spectroscopy. Interaction between identified proteins of the hair-cell porosome can be determined quantitatively by SPR analysis (see Section 2.2 and Fig. 2.5). Tight binding between protein candidates for components of the hair-cell porosome complex will provide support for their structural and functional participation in hair-cell porosome mechanisms.

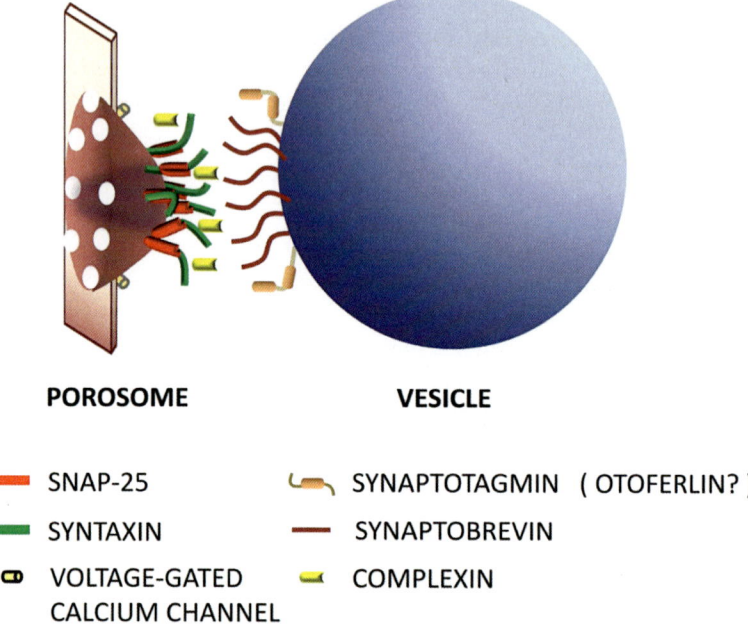

POROSOME   **VESICLE**

━━ SNAP-25   ⌐⌐ SYNAPTOTAGMIN (OTOFERLIN?)

━━ SYNTAXIN   ━━ SYNAPTOBREVIN

◨ VOLTAGE-GATED   ◁ COMPLEXIN
   CALCIUM CHANNEL

Figure 2.4   Artist's rendition of porosome and synaptic vesicle, showing involvement of multiple SNARE molecules in vesicle fusion. Porosomes, nano-scale stable membrane invaginations observed at active zones and ribbon synapses, consist of cup-shaped structures with the base of the "cup" facing the vesicle.[12] The outer rim of the cup opens towards the outside and bears eight to twelve protein knobs. The cavity of the porosome is occupied by a mobile "plug" that is thought to move in and out during neurosecretion. Multiple t-SNAREs required for vesicle fusion are shown localized at the cup's base, ready to pair with the v-SNAREs of the vesicle. Voltage-gated calcium channels are thought to be localized close to the fusion machinery via their protein-protein interaction with t-SNARE proteins. Close proximity of channels is required for the formation of calcium nano-domains that support fast exocytosis. It is hypothesized that vesicles attach to the porosome, probably via SNARE formation, and are recycled after release. (This research[23] was originally published in Molecular and Cellular Neuroscience: Ramakrishnan, N. A., Drescher, M. J., and Drescher, D. G. (2012) The SNARE complex in neuronal and sensory cells, *Mol. Cell. Neurosci.,* **50**, 58–69. © Elsevier.)

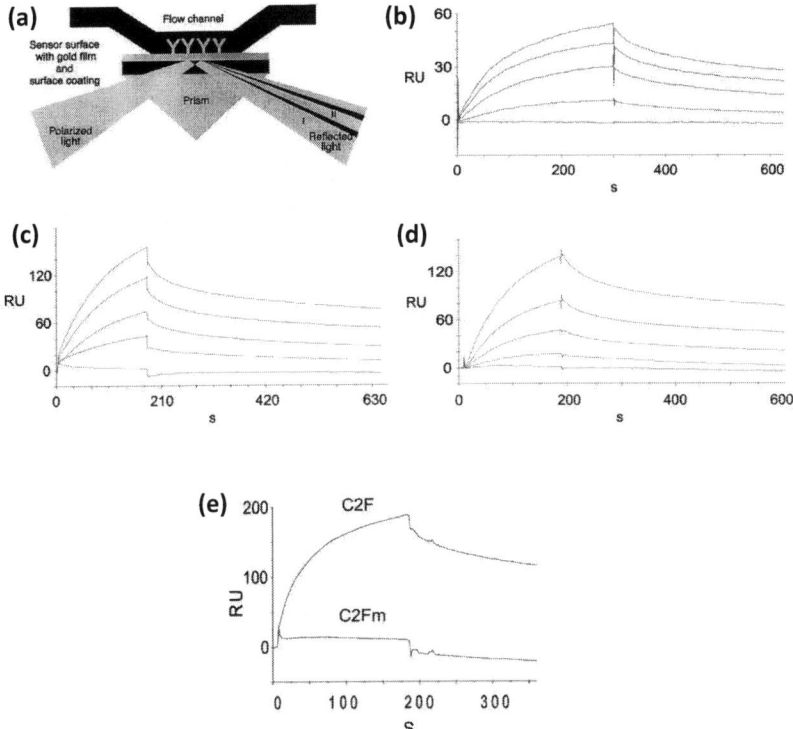

Figure 2.5   Principle of surface plasmon resonance (SPR) and quantitative SPR-interaction kinetic curves. (a) Diagram of SPR flow cell, illustrating basic mechanism of SPR analysis.[74,75] (b) Syntaxin 1A interaction with the otoferlin C2F domain, illustrating interaction of hair-cell synaptic complex components. (c) SNAP-25 interaction with the otoferlin C2F domain. (d) Cav1.3 II–III loop interaction with the otoferlin C2D domain. (e) Single amino acid substitution (Pro-Ala) in otoferlin C2F diminishes syntaxin 1A binding. SPR analysis was performed with purified native C2F (upper trace) and mutated C2Fm (lower trace) fusion peptides. (This research[17] was originally published in the Journal of Biological Chemistry: Ramakrishnan, N. A., Drescher, M. J., and Drescher, D. G. (2009) Direct interaction of otoferlin with syntaxin 1A, SNAP-25, and the L-type voltage-gated calcium channel Cav1.3, *J. Biol. Chem.*, **284**, 1364–1372. © the American Society for Biochemistry and Molecular Biology.)

Acknowledgments

Supported by NIH R01 DC000156 (DGD). Research on the porosome reported in this chapter was largely contributed by Drs. Won Jin Cho, Marian Drescher, and Neeliyath Ramakrishnan, in collaboration with the author.

References

1. Katz, B. (2003). Neural transmitter release: from quantal secretion to exocytosis and beyond, *J. Neurocyt.*, **32**, 437–446.

2. Südhof, T. C. (2004). The synaptic vesicle cycle, *Ann. Rev. Neurosci.*, **27**, 509–547.

3. Ceccarelli, B., Hurlbut, W. P., and Mauro A. (1973). Turnover of transmitter and synaptic vesicles at the frog neuromuscular junction, *J. Cell Biol.*, **57**, 499–524.

4. Dresbach, T., Qualmann, B., Kessels, M. M., Garner, C. C., and Gundelfinger, E. D. (2001). The presynaptic cytomatrix of brain synapses, *Cell. Mol. Life Sci.*, **58**, 94–116.

5. Haucke, V., Neher, E., and Sigrist, S. J. (2011). Protein scaffolds in the coupling of synaptic exocytosis and endocytosis, *Nat. Rev. Neurosci.*, **12**, 127–138.

6. Neher, E. (1998). Vesicle pools and Ca^{2+} microdomains: new tools for understanding their roles in neurotransmitter release, *Neuron*, **20**, 389–399.

7. Spassova, M. A., Avissar, M., Furman, A. C., Crumling, M. A., Saunders, J. C., and Parsons, T. D. (2004). Evidence that rapid vesicle replenishment of the synaptic ribbon mediates recovery from short-term adaptation at the hair cell afferent synapse, *J. Assoc. Res. Otolaryngol.*, **5**, 376–390.

8. Albillos, A., Dernick, G., Horstmann, H., Almers, W., Alvarez de Toledo, G., and Lindau, M. (1997). The exocytotic event in chromaffin cells revealed by patch amperometry, *Nature*, **389**, 509–512.

9. Cho, S. J., Cho, J., and Jena, B. P. (2002). The number of secretory vesicles remains unchanged following exocytosis. *Cell Biol. Int.*, **26**, 29–33.

10. Fesce, R., Grohvaz, F., Valtorta, F., and Meldolesi, J. (1994). Neurotransmitter release: fusion or 'kiss-and-run'? *Trends Cell Biol.*, **4**, 1–4.

11. Schneider, S. W., Sritharan, K. C., Geibel, J. P., Oberleithner, H., and Jena, B. P. (1997). Surface dynamics in living acinar cells imaged by atomic

force microscopy: identification of plasma membrane structures involved in exocytosis, *Proc. Natl. Acad. Sci. USA*, **94**, 316–321.

12. Jena B. P. (2009). Porosome: the secretory portal in cells, *Biochemistry*, **48**, 4009–4018.

13. Drescher, M. J., Drescher, D. G., and Hatfield, J. S. (1987). Potassium-evoked release of endogenous primary amine-containing compounds from the trout saccular macula and saccular nerve in vitro, *Brain Res.*, **417**, 39–50.

14. McDowell, E. M., and Trump, B. F. (1976). Histologic fixatives suitable for diagnostic light and electron microscopy, *Arch. Pathol. Lab Med.*, **100**, 405–414.

15. Reynolds, E. S. (1963). The use of lead citrate at high pH as an electron-opaque stain in electron microscopy, *J. Cell Biol.*, **17**, 208–212.

16. Schuck, P. (1997). Use of surface plasmon resonance to probe the equilibrium and dynamic aspects of interactions between biological macromolecules, *Annu. Rev. Biophys. Struct.*, **26**, 541–566.

17. Ramakrishnan, N. A., Drescher, M. J., and Drescher, D. G. (2009). Direct interaction of otoferlin with syntaxin 1A, SNAP-25, and the L-type voltage-gated calcium channel Cav1.3, *J. Biol. Chem.*, **284**, 1364–1372.

18. Johnsson, B., Lofas, S., and Lindquist, G. (1991). Immobilization of proteins to a carboxymethyldextran-modified gold surface for biospecific interaction analysis in surface plasmon resonance sensors, *Anal. Biochem.*, **198**, 268–277.

19. Steffner, P., and Markey, F. (1997). When the chips are down, *BIA J.*, **1**, 11–15.

20. Drescher, D. G., Ramakrishnan, N. A., and Drescher, M. J. (2009). Surface plasmon resonance (SPR) analysis of binding interactions of proteins in inner-ear sensory epithelia, *Methods Mol. Biol.*, **493**, 323–343.

21. Kretschmann, E., and Raether, H. (1968). Radiative decay of non-radiative surface plasmons excited by light, *Z. Naturforsch. Teil A*, **23**, 2135–2136.

22. Karlsson, R., Roos, H., Fägerstam, L., and Persson, B. (1994). Kinetic and concentration analysis using BIA technology, *Methods: Companion Methods Enzymol.*, **6**, 99–110.

23. Ramakrishnan, N. A., Drescher, M. J., and Drescher, D. G. (2012). The SNARE complex in neuronal and sensory cells, *Mol. Cell. Neurosci.*, **50**, 58–69.

24. Safieddine, S., and Wenthold, R. J. (1999). SNARE complex at the ribbon synapses of cochlear hair cells: analysis of synaptic vesicle- and synaptic membrane-associated protines, *Eur. J. Neurosci.*, **11**, 803–812.

25. Sollner, T., Whiteheart, S. W., Brunner, M., Erdjument-Bromage, H., Geromanos, S., Tempst, P., and Rothman, J. E. (1993). SNAP receptors implicated in vesicle targeting and fusion, *Nature*, **362**, 318–324.

26. Schoch, S., Deak, F., Konigstorfer, A., Mozhayeva, M., Sara, Y., Südhof, T. C., and Kavalali, E. T. (2001). SNARE function analyzed in synaptobrevin/VAMP knockout mice, *Science*, **294**, 1117–1122.

27. Li, F., Pincet, F., Perez, E., Eng, W. S., Melia, T. J., Rothman, J. E., and Tareste, D. (2007). Energetics and dynamics of SNAREpin folding across lipid bilayers, *Nat. Struct. Mol. Biol.*, **14**, 890–896.

28. Whyte, J. R. C., and Munro, S. (2002). Vesicle tethering complexes in membrane traffic, *J. Cell Sci.*, **115**, 2627–2637.

29. Smyth, A. M., Duncan, R. R., and Rickman, C. (2010). Munc18-1 and syntaxin1: unraveling the interactions between the dynamic duo, *Cell. Mol. Neurobiol.*, **30**, 1309–1313.

30. Gracheva, E. O., Maryon, E. B., Berthelot-Grosjean, M., and Richmond, J. E. (2010). Differential regulation of synaptic vesicle tethering and docking by UNC-18 and TOM-1, *Front. Synaptic Neurosci.*, **2**, 1–12, Art 141.

31. Dulubova, I., Sugita, S., Hill, S., Hosaka, M., Fernandez, I., Südhof, T. C., and Rizob, J. (1999). A conformational switch in syntaxin during exocytosis: role of Munc18, *EMBO J.*, **18**, 4372–4382.

32. Shi, L., Kümmel, D., Coleman, J., Melia, T. J., and Giraudo, C. G. (2011). Dual roles of Munc18-1 rely on distinct binding modes of the central cavity with Stx1A and SNARE complex, *Mol. Biol. Cell*, **22**, 4150–4160.

33. Neher, E., and Sakaba, T. (2008). Multiple roles of calcium ions in the regulation of neurotransmitter release, *Neuron*, **59**, 861–872.

34. Wadel, K., Neher, E., and Sakaba, T. (2007). The coupling between synaptic vesicles and Ca^{2+} channels determines fast neurotransmitter release, *Neuron*, **53**, 563–575.

35. Han, X., and Jackson, M. B. (2006). Structural transitions in the synaptic SNARE complex during Ca^{2+}-triggered exocytosis, *J. Cell Biol.*, **172**, 281–293.

36. Dai, H., Shen, N., Arac, D., and Rizo, J. (2007). A quaternary SNARE-synaptotagmin-Ca^{2+}-phospholipid complex in neurotransmitter release, *J. Mol. Biol.*, **367**, 848–863.

37. Giraudo, C. G., Eng, W. S., Melia, T. J., and Rothman, J. E. (2006). A clamping mechanism involved in SNARE-dependent exocytosis, *Science,* **313**, 676–680.

38. Bai, J., Wang, C. T., Richards, D. A., Jackson, M. B., and Chapman, E. R. (2004). Fusion pore dynamics are regulated by synaptotagmin t-SNARE interactions, *Neuron,* **41**, 929–942.

39. Sharma, M., Burré, J., and Südhof, T. C. (2011). CSPα promotes SNARE-complex assembly by chaperoning SNAP-25 during synaptic activity, *Nature Cell Biol.,* **13**, 30–39.

40. Weber, T., Zemelman, B. V., McNew, J. A., Westermann, B., Gmachl, M., Parlati, F., Söllner, T. H., and Rothman, J. E. (1998). SNAREpins: Minimal machinery for membrane fusion, *Cell,* **92**, 759–772.

41. Sutton, R. B., Fasshauer, D., Jahn, R., and Brünger, A. T. (1998). Crystal structure of a SNARE complex involved in synaptic exocytosis at 24 Å resolution, *Nature,* **395**, 347–353.

42. Stein, A., Weber, G., Wahl, M. C., and Jahn, R. (2009). Helical extension of the neuronal SNARE complex into the membrane, *Nature,* **460**, 525–528.

43. Südhof, T. C., and Rothman, J. E. (2009). Membrane fusion: grappling with SNARE and SM proteins, *Science,* **323**, 474–477.

44. Hanson, P. I., Roth, R., Morisaki, H., Jahn, R., and Heuser, J. E. (1997). Structure and conformational changes in NSF and its membrane receptor complexes visualized by quick-freeze/deep-etch electron microscopy, *Cell,* **90**, 523–535.

45. Stelzer, W., Poschner, B. C., Stalz, H., Heck, A. J., and Langosch, D. (2008). Sequence-specific conformational flexibility of SNARE transmemebrance helices probed by hydrogen/deuterium exchange, *Biophys. J.,* **95**, 1326–1335.

46. Hua, Y., and Scheller, R. H. (2001). Three SNARE complexes cooperate to mediate membrane fusion, *Proc. Natl. Acad. Sci. USA,* **98**, 8065–8070.

47. Han, X., Wang, C. T., Bai, J., Chapman, E. R., and Jackson, M. B. (2004). Transmembrane segments of syntaxin line the fusion pore of Ca^{2+}-triggered exocytosis, *Science,* **304**, 289–292.

48. Wong, J. L., Koppel, D. E., Cowan, A. E., and Wessel, G. M. (2007). Membrane hemifusion is a stable intermediate of exocytosis, *Dev. Cell,* **12**, 653–659.

49. Jahn, R., Lang, T., and Südhof, T. C. (2003). Membrane fusion, *Cell,* **112**, 519–533.

50. Chen, X., Tomchick, D. R., Kovrigin, E., Arac, D., Machius, M., Südhof, T. C., and Rizo, J. (2002). Three-dimensional structure of the complexin/SNARE complex, *Neuron,* **33**, 397–409.

51. Fernández-Chacón, R., Königstorfer, A., Gerber, S. H., García, J., Matos, M. F., Stevens, C. F., Brose, N., Rizo, J., Rosenmund, C., and Südhof, T. C. (2001). Synaptotagmin I functions as a calcium regulator of release probability, *Nature,* **410**, 41–49.

52. Krishnakumar, S. S., Radoff, D. T., Kümmel, D., Giraudo, C. G., Li, F., Khandan, L., Baguley, S. W., Coleman, J., Reinisch, K. M., Pincet, F., and Rothman, J. E. (2011). A conformational switch in complexin is required for synaptotagmin to trigger synaptic fusion, *Nat. Struct. Mol. Biol.,* **18**, 934–940.

53. Ngatchou, A. N., Kisler, K., Fang, Q., Walter, A. M., Zhao, Y., Bruns, D., Sørensen, J. B., and Lindau, M. (2010). Role of the synaptobrevin C terminus in fusion pore formation, *Proc. Natl. Acad. Sci. USA,* **107**, 18463–18468.

54. Shi, L., Shen, Q. T., Kiel, A., Wang, J., Wang, H. W., Melia, T. J., Rothman, J. E., and Pincet, F. (2012). SNARE proteins: one to fuse and three to keep the nascent fusion pore open, *Science,* **335**, 1355–1359.

55. Cho, S. J., Quinn, A. S., Stromer, M. H., Dash, S., Cho, J., Taatjes, D. J., and Jena, B. P. (2002). Structure and dynamics of the fusion pore in live cells, *Cell Biol. Int.,* **26**, 35–42.

56. Jena, B. P., Cho, S. J., Jeremic, A., Stromer, M. H., and Abu-Hamdah, R. (2003). Structure and composition of the fusion pore, *Biophys. J.,* **84**, 1337–1343.

57. Jeremic, A., Kelly, M., Cho, S. J., Stromer, M. H., and Jena, B. P. (2003). Reconstituted fusion pore, *Biophys. J.,* **85**, 2035–2043.

58. Elshennawy, W. W. (2011). Image processing and numerical analysis approaches of porosome in mammalian pancreatic acinar cell, *J. Am. Sci.,* **7**, 835–843.

59. Cho, W. J., Lee, J. S., and Jena, B. P. (2010). Conformation states of the neuronal porosome complex, *Cell Biol. Int.,* **34**, 1129–1132.

60. Cho, W. J., Ren, G., and Jena, B. P. (2008). EM 3D contour maps provide protein assembly at the nanoscale within the neuronal porosome complex, *J. Microsc.,* **232**, 106–111.

61. Cho, S. J., Jeftinija, K., Glavaski, A., Jeftinija, S., Jena, B. P., and Anderson, L. L. (2002). Structure and dynamics of the fusion pores in live GH-secreting cells revealed using atomic force microscopy, *Endocrinology,* **143**, 1144–1148.

62. Cho, S. J., Wakade, A., Pappas, G. D., and Jena, B. P. (2002). New structure involved in transient membrane fusion and exocytosis, *Ann. NY Acad. Sci.,* **971**, 254–256.

63. Jena, B. P. (2004). Discovery of the porosome: revealing the molecular mechanism of secretion and membrane fusion in cells, *J. Cell. Mol. Med.,* **8**, 1–21.

64. Cho, W. J., Jeremic, A., Rognlien, K. T., Zhvania, M. G., Lazrishvili, I., Tamar, B., and Jena, B. P. (2004). Structure, isolation, composition and reconstitution of the neuronal fusion pore, *Cell Biol. Int.,* **28**, 699–708.

65. Siksou, L., Rostaing, P., Lechaire, J. P., Boudier, T., Ohtsuka, T., Fejtova, A., Kao, H. T., Greengard, P., Gundelfinger, E. D., Triller, A., and Marty, S. (2007). Three-dimensional architecture of presynaptic terminal cytomatrix, *J. Neurosci.,* **27**, 6868–6877.

66. Lee, J. S., Cho, W. J., Jeftinija, K., Jeftinija, S., and Jena, B. P. (2009). Porosome in astrocytes, *J. Cell. Mol. Med.,* **13**, 365–372.

67. Drescher, D. G., Cho, W. J., and Drescher, M. J. (2011). Identification of the porosome complex in the hair cell, *Cell Biol. Int. Rep.,* **18**, 31–34. art:e00012. doi: 10.1042/CBR20110005.

68. Griesinger, C. B., Richards, C. D., and Ashmore, J. F. (2005). Fast vesicle replenishment allows indefatigable signalling at the first auditory synapse, *Nature,* **435**, 212–215.

69. Hama, K., and Saito, K. (1977). Fine structure of the afferent synapse of the hair cells in the saccular macula of the goldfish, with special reference to the anastomosing tubules, *J Neurocytol.,* **6**, 361–373.

70. Parsons, T. D., and Sterling, P. (2003). Synaptic ribbon: conveyor belt or safety belt? *Neuron,* **37**, 379–382.

71. Drescher, M. J., Drescher, D. G., Hatfield, J. S., and Seitz, C. M. (1987). Synaptic morphology and vesicle morphometry of hair cells in the teleost saccular macula, *Soc. Neurosci. Abstr.,* **13**, 43.

72. Lee, J. S., Jeremic, A., Shin, L., Cho, W. J., Chen, X., and Jena, B. P. (2012). Neuronal porosome proteome: molecular dynamics and architecture, *J. Proteomics,* **75**, 3952–2962.

73. Jena, B. P. (2009). Membrane fusion: Role of SNAREs and calcium, *Prot. Pept. Res.,* **16**, 712–717.

74. Markey F. (1995). What is SPR anyway? *BIA J.,* **1**, 14–17.

75. Stenberg, E., Persson, B., Roos, H., and Urbaniczky, C. (1991). Quantitative determination of surface concentration of protein with surface plasmon resonance using radiolabeled proteins, *J. Colloid Interface Sci.,* **143**, 513–526.

Chapter 3

The Neuronal Porosome Complex in Mammalian Brain: A Study Using Electron Microscopy

Mzia G. Zhvania,[a,b] Nadezhda J. Japaridze,[b] Mariam G. Qsovreli,[a] Vera G. Okuneva,[a,b] Arkadi G. Surmava,[b] and Tamar G. Lordkipanidze[a,b]

[a]Institute of Chemical Biology, Ilia State University,
3/5 K. Cholokhashvili Avenue, 0162 Tbilisi, Georgia
[b]Department of Neuroanatomy, I. Beriitashvili Center of Experimental BioMedicine,
14, Gotua Street, 0160 Tbilisi, Georgia

mzia_zhvania@iliauni.edu.ge

Porosomes are the universal secretory machinery in cells where membrane-bound secretory vesicles dock and transiently fuse (kiss-and-run) to release intravesicular contents to the outside of the cell during secretion. Earlier it was shown that in rat neurons 12–17 nm cup-shaped lipoprotein porosomes, possessing a central plug are present at the presynaptic membrane sometimes with 35–50 nm in diameter docked synaptic vesicles. In the current study, neuronal porosome structures following hypokinetic stress were evaluated using electron microscopy. Experiments were carried out to identify and evaluate the porosome structure at the presynaptic membrane in the rat and cat brain in control and experimental

NanoCellBiology: Multimodal Imaging in Biology and Medicine
Edited by Bhanu P. Jena and Douglas J. Taatjes
Copyright © 2014 Pan Stanford Publishing Pte. Ltd.
ISBN 978-981-4411-79-0 (Hardcover), 978-981-4411-80-6 (eBook)
www.panstanford.com

animals and to demonstrate the presence of porosomes in the dog brain. The results reveal for the first time the presence of neuronal porosomes in dog brain and further confirm their existence at the presynaptic membrane in rat and cat brain. Furthermore, the results demonstrate neuronal porosomes to possess a cup-shaped morphology in all the three mammalian species examined, i.e., the rat, cat, and dog. The next series of experiments were designed to evaluate morphological changes in the porosome structure as a consequence of pathological condition—chronic hypokinetic stress. This condition is known to produce structural alterations in the synapses, including the presynaptic regions of limbic region. The depth and diameter of porosome in the central nucleus of amygdale of normal rat and rat subjected to 90 day hypokinetic stress were measured. Morphometric analysis point out the heterogeneity of porosome dimensions that remain unchanged in pathological states. These studies demonstrate for the first time that despite alterations in the presynaptic terminal structure and synaptic transmission provoked by chronic hypokinetic stress in the limbic region, the gross morphology of porosome is unaffected. These results do not, however, rule out possible changes in the composition of the porosome complex following stress. Furthermore, longer period of stress may elicit changes in the neuronal porosome complex, which remains to be established.

3.1 Introduction

In all cells, cellular cargo destined for secretion is packaged and stored within membranous vesicles that transiently dock and establish continuity at the base of cup or flask-shaped plasma membrane structures called "porosomes"[1-5] and neurons are no exception.[6-16] Therefore, "porosomes" are the universal secretory machinery in cells where vesicles transiently dock and fuse to release intravesicular contents to the outside of the cell during secretion.

It is suggested that in each type of secretory cell special content of secretory vesicles, different speed of release and different volume of content release dictates specific size of porosomes. In neurons and astrocytes, representing fast secretory cells, porosomes range in size from 10 to 17 nm. In an earlier study, using the atomic force microscope (AFM) and the electron microscope (EM), it was demonstrated that in the neurons 40–50 nm synaptic vesicles are docked at roughly 10 nm in diameter neuronal porosomes.[4] Recent EM 3D tomography in rat brain also reveals the presence of 12–17 nm permanent presynaptic densities to which 35–50 nm synaptic vesicles are found docked.[15] Moreover, the inside-out ultrahigh-resolution AFM study of presynaptic membrane preparations of isolated synaptosomes clearly displays the presence of the inverted cup-shaped 10–17 nm neuronal porosomes. In contrast, in slow secretory cells, such as in acinar cells of the exocrine pancreas, secretory granules measuring approximately 1000 nm in diameter, expel their content following transient fusion at the porosome base measuring 100–180 nm.[1,6] Furthermore, the results from more recent studies using AFM, EM, electron density, and 3D contour mapping provided additional nanoscale information on the structure and assembly of proteins within the neuronal porosome complex.[8] Particularly, it has become clear that neuronal porosomes possess a central plug that is absent in porosomes in other kinds of secretory cells. This central plug interacts with proteins at the periphery of the structure, conforming to an eightfold symmetry; each of them is connected with spoke-like elements to the central plug that is involved in the rapid opening and closing of the neuronal porosome to the outside.[9] The neuronal central plug has been further examined at various conformational states, providing its gatekeeping role in neurotransmitter release during neurotransmission. Thus, the central plug at various conformations–fully pushed outward, halfway retracted, and completely retracted into porosome cup– has been elegantly demonstrated.[17] Although it is easy to observe porosomes in intact synaptosomes and in inside-out synaptosome preparations using the AFM, in fixed cells it becomes difficult, owing to artifacts due to fixation, dehydration, and tissue processing for EM, and compounded with the fact that the presynaptic membrane contain a high density of plasma membrane proteins resulting in heavy metal staining, rendering it difficult to separate porosomes from the other structures. Nonetheless, porosomes are clearly identifiable in electron micrographs in numerous reported studies.[4,7–9,11,13–15]

In the present research, we continue our EM studies of the porosome morphology in healthy and disease conditions. Two main goals are identified: (1) to further evaluate the porosome structure at the presynaptic membrane in the rat and cat brains and to demonstrate the presence of porosome in the dog brain and to further determine if porosome morphology (diameter and depth) is species dependent; (2) to establish if porosome structure/size is altered as a result of pathological conditions, specifically, as a result of chronic hypokinetic stress.

3.2 Neuronal Porosome in Rat, Cat, and the Dog Brains

It is now well established that porosome represents universal secretory machinery in cells. However, to further determine its universality, the description of structure and other characteristics of porosome in various types of secretory cells as well in various animal species are of great importance. For the first time, the neuronal porosome was studied in the rat brain using the AFM and the EM and the size of neuroporosome was measured.[4] More recently, additional structural parameters of neuroporosome in rat brain, specifically, its depth and diameter, were further assessed and clarified and the same parameters in the cat brain were determined.[14] However, no information is available regarding the neuroporosome structure in the dog brain. In a series of studies presented in this chapter, the EM morphometric analysis of the porosome at the presynaptic membrane in rat, cat, and dog brains was undertaken. Specifically, the porosome opening diameter and depth of the porosome cup were carefully measured and documented.

3.2.1 Material and Methods

3.2.1.1 Animals and experimental design

Adult male Wistar rats (90–95 days old), adult male cats (110–120 days old) and adult male dogs (8 years old) from ordinary vivarial conditions (temperature 20–22°C, humidity 55–60%, light on 07.30–19.30) were used in this study (n = 4 for each specie). The procedures for handling and caring for the animals were approved by the Animal Studies Committee of Georgian Life Science Research Center and are in accordance with current international laws and policies.

3.2.1.2 Electron microscopic examination

Following pentobarbital injection (100 mg/kg), animals to have EM examination of their brains underwent transcardiac perfusion with heparinized 0.9% NaCl, followed by 500 mL of 4% paraphormaldehyde and 2.5% glutaraldehyde in 0, 1 M phosphate buffer (PB), pH 7.4, at a perfusion pressure 120 mm Hg. The brains were removed from skull and placed in the same fixative overnight. The right hemispheric tissue blocks containing hippocampi were cut into 400 micron-thick coronal slices. Slices were washed in cold 0.1 M PB and kept in 2.5% glutaraldehyde in 0.1 M PB until processing; When processing, the slices were washed in cold PB, post-fixed in 1% osmium tetroxide in cold PB for 2 h and again washed in 0.1 M PB. The hippocampus was identified with an optical microscope Leica MM AF, cut out from the coronal slices, dehydrated in graded series of ethanol and acetone, and embedded in araldite. Blocks were trimmed and 70–75 nm-thick sections were cut with an ultra-microtome Reichert, picked up on 200-mesh copper grids, double-stained with uranyl-acetate and lead-citrate, and examined with a JEM 100 C (JEOL, Japan), HF 3300 (Hitachi, Japan) and Tesla BS 500, Czechoslovakia) transmission electron microscopes. For each case, 115 sections were observed.

3.2.1.3 Morphometric analysis of porosome diameter and depth

In order to identify any differences with regard to diameter and depth of porosome complexes in the brain tissue of the various mammals (rat, cat, and dog), a morphometric analysis was performed. The following abbreviations are used: DiPR, diameter of porosome in rat brain; DePR, depth of porosome in rat brain; DiPC, diameter of porosome in cat brain; DeRC, depth of porosome in cat brain; DiPD, diameter of porosome in dog brain; DePD, depth of porosome in dog brain. A total of 274 synaptic terminals were studied in the hippocampus of the rat ($n = 91$), cat ($n = 93$), and dog brains ($n = 90$) and 99 neuronal porosomes (33/specie) were identified. The depth and diameter of these porosomes were carefully measured with Image J software (version 1.41) and the data are represented in Tables 3.1 and 3.2.

Table 3.1 The dimensions of porosomal complex in the rat, cat, and dog brains (in nanometers)

#	DiPR	DePR	DiPC	DePC	DiPD	DePD
1	17	15	11	8	13	7
2	16	15	11	7	13	7
3	12	12	16	7	15	14
4	14	11	13	8	21	13
5	18	12	15	12	12	10
6	13	9	16	5	15	10
7	9	7	13	9	17	18
8	10	8	13	5	16	16
9	13	14	18	10	19	10
10	14	8	10	5	15	17
11	19	19	13	10	17	11
12	19	15	17	7	10	8
13	17	14	16	5	8	6
14	9	9	14	8	19	16
15	15	11	17	10	18	7
16	15	14	10	10	11	15
17	17	11	11	12	11	6
18	12	7	18	13	14	10
19	14	21	15	12	11	8
20	17	19	15	12	11	7
21	15	16	18	18	12	13
22	15	15	15	10	17	10
23	18	21	18	18	14	11
24	17	14	15	16	16	16
25	17	10	13	10	13	19
26	17	9	15	11	13	11
27	19	11	13	15	14	11
28	15	14	20	11	16	16
29	19	11	17	14	11	15
30	16	9	14	16	13	19
31	15	16	20	16	13	11
32	15	12	17	11	15	14
33	19	14	19	10	21	13

Table 3.2 Dimensions of neuronal porosome complex (in nanometers)

#	Diameter (control)	Diameter (experiment)	Depth (control)	Depth (experiment)
1	17	16	15	9
2	16	17	15	9
3	12	18	12	6
4	14	16	11	6
5	18	12	12	6
6	13	11	9	6
7	9	10	7	6
8	10	11	8	9
9	13	13	14	9
10	14	12	8	8
11	19	13	19	9
12	19	14	15	9
13	7	14	14	15
14	9	14	9	10
15	15	14	11	10
16	15	10	14	17
17	17	16	11	11
18	12	17	7	14
19	14	18	21	14
20	17	16	19	13
21	15	16	16	13
22	15	17	15	18
23	18	18	21	11
24	17	16	14	10
25	17	16	10	14
26	17	17	9	14
27	19	18	11	13
28	15	16	14	14
29		14		20

3.2.1.4 Statistical analyses

To determine whether specie identity affects the parameters of neuronal porosomes—diameter of opening and depth of cup-shaped structure at the presynaptic membrane—the one-way ANOVA was performed separately on the diameter and depth of rat, cat, and dog brains. In the case of statistical significance of F-value, comparison between the two groups of measurements would be made using two sample t-tests with a p-value threshold ≤ 0.05.

3.2.2 Results

In Fig. 3.1, axo-dendritic synapse with well-defined presynaptic membrane, postsynaptic density, and porosome-docked vesicles are presented.

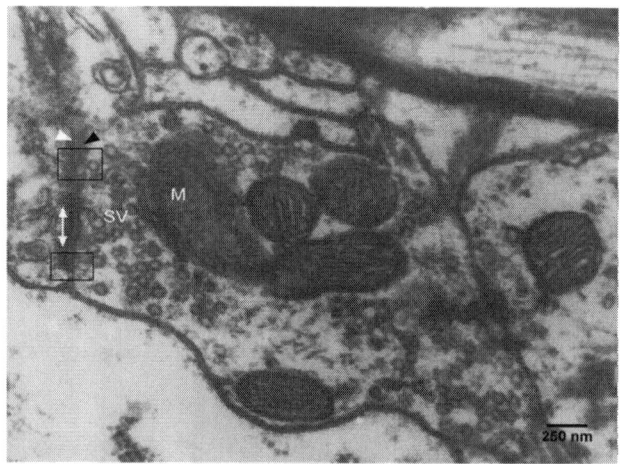

Figure 3.1   Electron micrograph of axo-dendritic synapse in the cat brain. Note the 10–17 nm cup-shaped porosomes at the presynaptic membrane, where 50 nm in diameter synaptic vesicles are found docked (rectangles). (Black arrow)–presynaptic membrane, (white arrow)–postsynaptic membrane, (double arrowhead)–postsynaptic density, (M)–mitochondria, (SV)–synaptic vesicles.

In several micrographs, clearly visible cup-shaped profiles of neuronal porosomes are seen with (Figs. 3.2a–c,e,f, 3.3a, 3.4b,c) and without (Figs. 3.2d, 3.3b,c, 3.4d) docked synaptic vesicle in the rat, cat, and dog brains. The existence of porosomes away from synaptic vesicles further demonstrates that porosome are permanent structures at the presynaptic membrane and not formed as a consequence of synaptic vesicle docking. The one-way ANOVA results presented in Fig. 3.5 revealed for the first time that the type of secretory cell rather than the species influence porosome diameter ($F_{(3,131)} = 1.02$, $p > 0.05$) and depth ($F_{(3,131)} = 2.86$, $p > 0.05$). According to our measurements, the parameters of the porosome complex in the brain tissue are found to be heterogeneous in all species. The DiPR fluctuates from 9 nm up to 19 nm, DiPC from 10 nm up to 18 nm, and DiPD from 8 up to 21 nm. Similarly, the DePR fluctuates from 7 nm up to 18 nm, DePC from 5 nm to 18 nm, and DePD from 6 nm up to 19 nm.

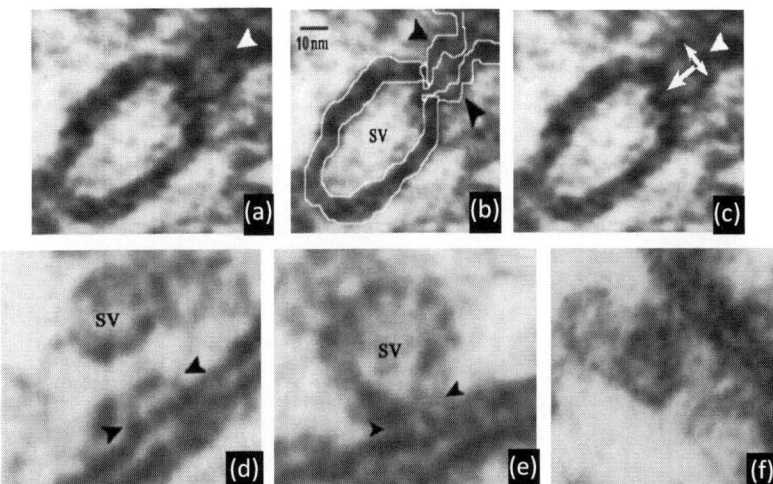

Figure 3.2   Electron micrographs of neuronal porosome complexes in the rat brain. (a) Neuronal porosome complex with docked vesicle; the central plug at the porosome opening (white arrowhead) is clearly identified; (b) porosome (black arrowhead) and synaptic vesicle (SV) are allocated with a white contour; (c) porosome parameters: diameter of opening (double head arrow) and depth (white arrow); (d) the porosome structure (black arrowheads) away from synaptic vesicles; (e) synaptic vesicles docked to a cup-shaped porosome (black arrowhead), the central plug of porosome is clearly identified; (f)—a small approximately 25–30 nm clear vesicle docked at the base of the porosome complex at the presynaptic membrane. The small size of the synaptic vesicle suggests that the vesicle may have discharged some of its contents.

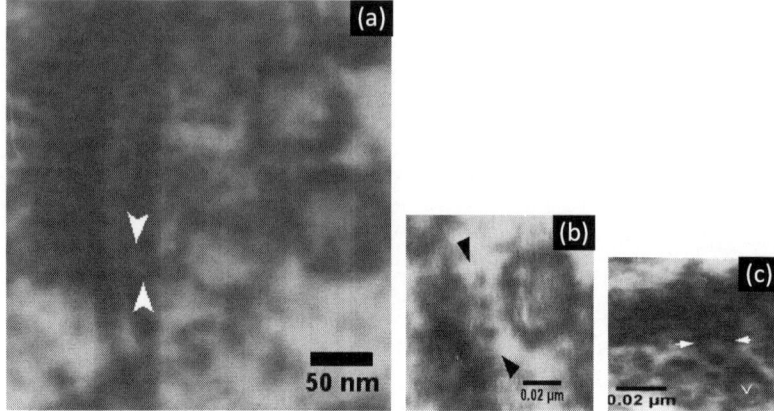

Figure 3.3   Electron micrograph of neuronal porosome complexes in the cat brain. (a) Magnified synaptic active zone of an axo-dendritic synapse with 17–19 nm cup-shaped porosome at the presynaptic membrane (white arrowheads) with docked synaptic vesicles (SV); (b) vesicle-free porosome structures (black arrowheads); (c) porosome complex (white arrows) and small approximately 25–30 nm vesicles near the porosome complex.

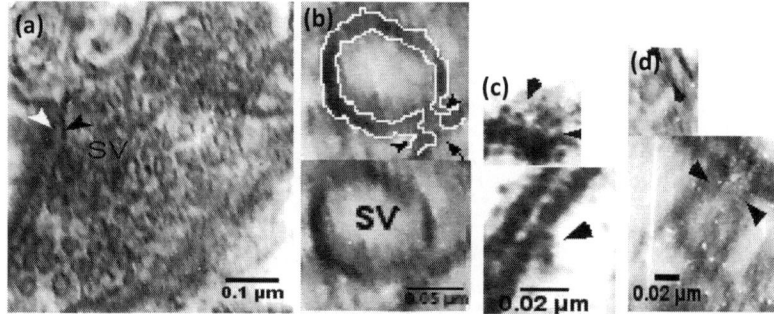

Figure 3.4   Electron micrograph of neuronal porosome complex in the dog brain. (a) The axo-dendritic synapse in the dog brain, with clearly defined pre- (black arrow) and postsynaptic membrane (white arrow), SV—synaptic vesicles; (b) 15–17 nm cup-shaped porosome (black arrowhead), 50–70 nm diameter synaptic vesicle (SV) and opening of porosome (black arrow) allocated with a white contour; (c) vesicle-free porosomes (black arrows); (d) vesicle-bound porosomes (black arrows).

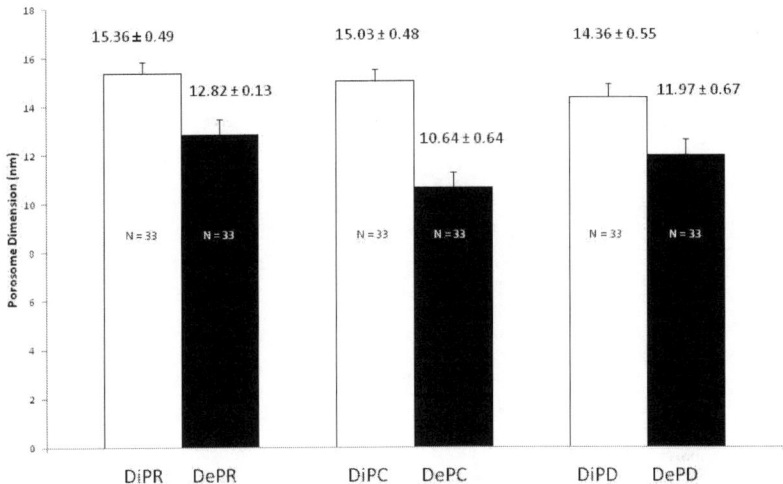

Figure 3.5   Depth and diameter of the neuronal porosome complex measured in nanometers in the rat, cat, and dog brains. DiPR, diameter of porosome in the rat brain; DePR, depth of porosome in rat brain; DiPC, diameter of porosome in the cat brain; DePC, depth of porosome in the rat brain; DiPD, diameter of porosome in the dog brain; DePD, depth of porosome in the rat brain. Note the similarity in the dimension of the porosomes in all three mammalian species. Data are presented as average ± SEM.

In our prior work, statistically significant difference in the porosome depth was revealed between rat and cat brains.[14] In particular, the first set of measurements (I) ranges in rats (DePR I) from 7 nm up to 21 nm and in cats (DePC I) from 5 nm up to 18 nm; the mean value of DePR I goes above DePC I by 26% (12.82 vs. 9.48, $p < 0.001$)[14]. In the present study, the data range (5–18 nm) and mean values of DePR and DePC are the same (12.23 vs. 9.03 $p > 0.05$), but the comparison of depth distribution histograms of recent and present studies (Fig. 3.6) reveals that in the present set the porosomes exhibit depth size from 10 to 18 nm. This difference between the two sets of measurements may reflect the highly dynamic nature of the neuronal porosome complex, in addition to minor changes in fixation conditions between experiments, and the number of measurements taken. Regardless, the size parameters of the porosome complex measured (diameter and depth values) fluctuate in the same ranges and their mean value do not display statistically significant difference between various species of mammals.

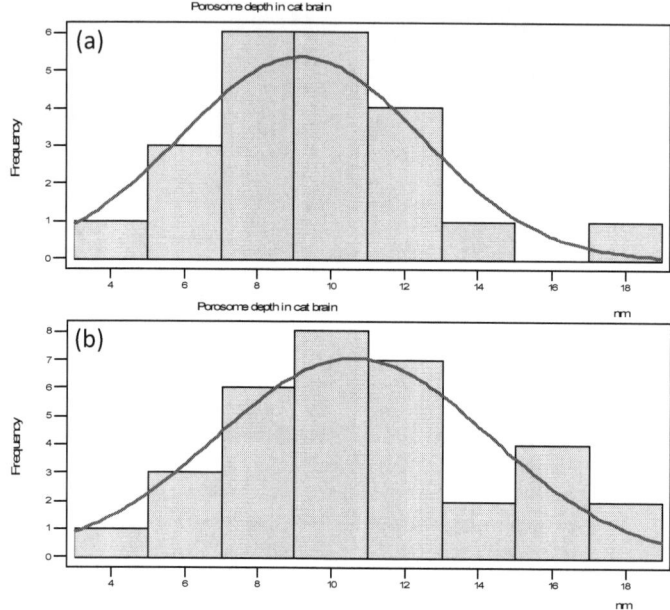

Figure 3.6   Histograms showing the frequency distribution of neuronal porosome based on depth value in synaptic terminals in the cat brain: (a) previous and (b) present study. All measured porosomes are separated by the depth's size and binned by area in 2 nm natural log increment.

3.3   Porosome in the Rat Brain: Effect of Hypokinetic Stress

In further evaluation of the porosome structure at the presynaptic membrane in the rat brain and to determine if the porosome morphology is altered in disease states, the current study was undertaken for the first time using high-resolution EM. One of pathologies that produce subtle structural alteration at the presynaptic terminals and synaptic transmission has been demonstrated to be stress.[18–22] Specifically, clearly defined pathological changes have been observed following chronic hypokinetic stress.[22-25] One of the brain structures to which the hypokinetic stress-response "sends a quick message" is the central nucleus of amygdale. Various ultrastructural alterations in the presynaptic terminals of the central nucleus as a result of 90 day hypokinetic stress are well documented.[22-24] Thus, the aggregation of synaptic vesicles, destruction or vacuolization of presynaptic mitochondria, clearly visible polymorphism of synaptic vesicles, significant increase of synaptic vesicles with granular core, and several other alterations were detected.[23,24] However, it remains uncertain whether such alterations are reflected on the structure of porosome. The present experiments were designed to show if such ultrastructural alterations are associated with the changes in the porosome structure. Therefore, comparative study of porosome diameter and depth was made in the brain from normal rats and rats subjected to 90 day hypokinetic stress.

3.3.1 Material and Methods

3.3.1.1 Animals and experimental design: simulation of hypokinesia

Studies were carried out using 90–95 day-old male Wistar rats, weighting 280–300 g. The conditions of animals' housing and their handling and care were identical to what is described earlier in Section 3.2.1.1. Experiments were conducted to minimize the number of animals used and the suffering caused by the procedures used in the study.

For the simulation of chronic hypokinetic stress, experimental rats were kept for 90 days in individual Plexiglas cages. Cage dimensions of 195 cm × 80 cm × 95 cm allowed movements to be restricted in all directions without hindering food and water consumption. The cages were housed under normal controlled environment (temperature 20–22°C, humidity 55–60%, light on from 07:30 to 19:30). Five age-matching control rats were kept in ordinary vivarium conditions.

3.3.1.2 EM examination

Procedures related to the preparation of brain tissue for EM examination and EM study were identical to as described in Section 3.2.1.2.

3.3.1.3 Morphometric analysis of porosome diameter and depth

A morphometric analysis of the porosomes in presynaptic membrane of central nucleus of amygdale in normal rats (control group of animals) and rats subjected to hypokinetic stress (experimental group of animals) was carried out in order to identify any differences in size with regard to porosome diameter and depth. A total of 196 synaptic terminals (100 control and 96 experimental) were observed, and 57 porosome measured (control 28, experiment 29). The neuronal porosomes in each case was identified and measured with Image J software (version 1.41).

3.3.1.4 Statistical analyses

To determine whether a 90 day hypokinetic stress in rats provokes changes on the sizes of neuronal porosome diameter and depth, the one-way ANOVA of quantitative data was performed in control and experimental groups of animals. A *p*-value less than 0.05 was considered statistically significant. In the case of significant effect, planned comparisons were carried out using the *t*-test.

3.3.2 Results

According to our results, the EM micrographs demonstrate the obviously defined difference in the presynaptic terminals of axo-dendrtic synapses in control and experimental group of animals. Particularly in the synaptic terminals of control rats, synaptic vesicles are distributed in the whole presynaptic area; mitochondria have moderate osmiophility and well-preserved structure (Fig. 3.7a). In the synaptic terminals of experimental rats, however, the aggregation of synaptic vesicles, presence of osmiophilic mitochondria (Fig. 3.7b), and polymorphism of vesicles—ellipsoid and dark-cored granular vesicles—are observed (Fig. 3.8). Regardless, in several micrographs, clearly visible cup-shaped profiles of the neuronal porosomes are observed with (Fig.s 3.9a–c,f) and without (Figs. 3.9d,e) docked synaptic vesicles.

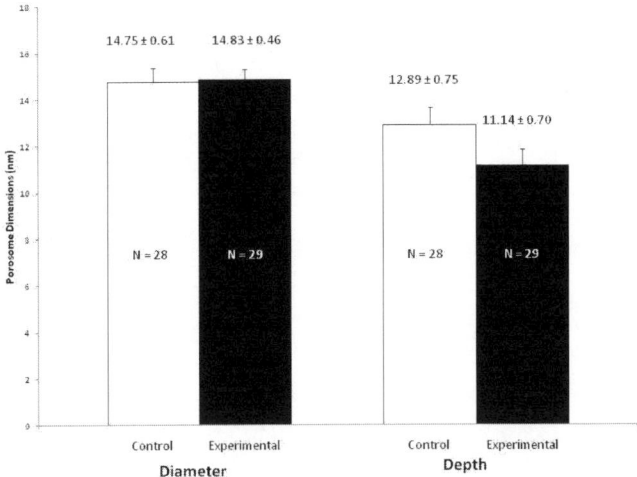

Figure 3.7 Depth and diameter of the neuronal porosome complex in normal and pathological rat brain (90 days of hypokinetic stress). Data are presented as average ± SEM. Note that there was no significant change in the experimental porosome complex over control, in both diameter and depth measurements.

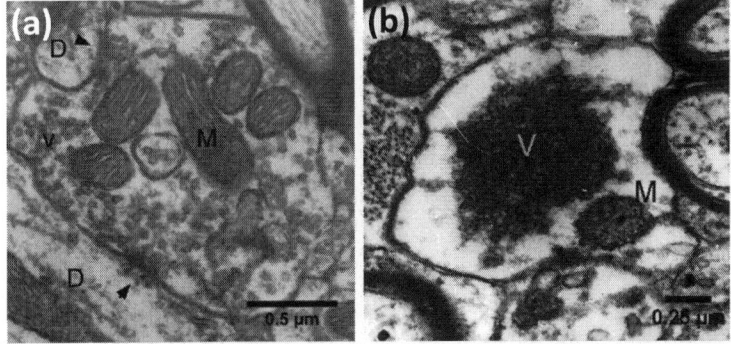

Figure 3.8 Electron micrograph of the synaptic terminal in the central nucleus of rat brain amygdale. (a) axo-dendritic synapse with well preserved mitochondria (M), synaptic vesicles (v) and active zones (black arrowheads), in the normal rat brain, D—dendrites; (b) profile of presynaptic terminal in rat brain after 90 days of hypokinetic stress with vesicle aggregation (v), mitochondria (M) with moderately osmiophilic matrix.

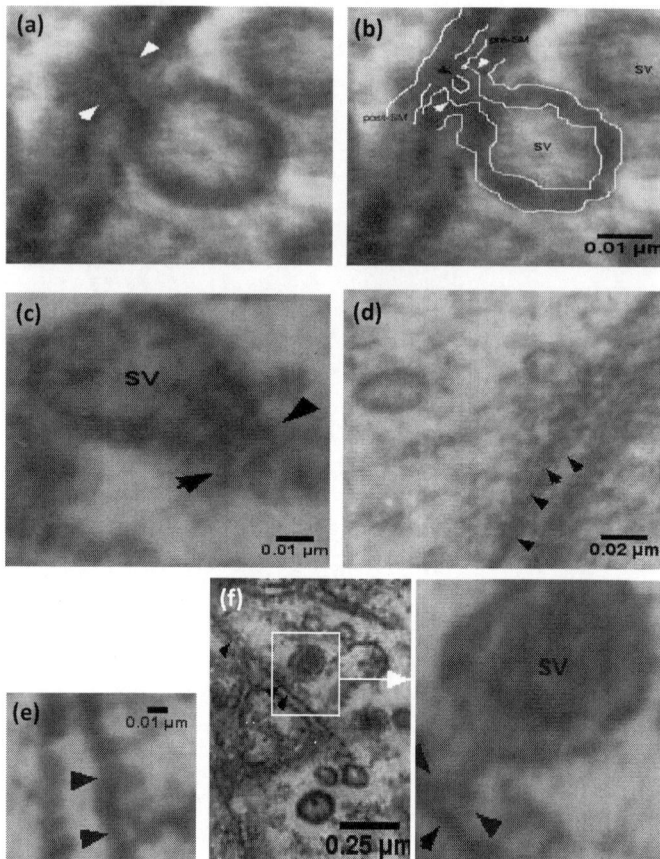

Figure 3.9   Electron micrographs of neuronal porosome complexes in the central nucleus of the amygdale and in the normal rat brain (a, b, c, d) following 90 day hypokinetic stress (e, f). (a) Synaptic vesicle docked neuronal porosome complex (white arrowhead); (b) the porosome structure (white arrowhead) and the central plug at the porosome opening (black arrowhead), sv—synaptic vesicle; (c) synaptic vesicle (sv) docked neuronal porosome complex (black arrowhead); (d) synaptic vesicle–free neuronal porosome complex in the normal rat brain (black arrowhead); (e) presynaptic membrane with synaptic vesicle-free porosomes (black arrowheads); (f) left—presynaptic terminal, right—magnified porosome (black arrowhead) with docked dark-cored granular synaptic vesicle (sv), which is several times larger (80–100 nm) than spherical synaptic vesicles (50–60 nm).

As previously discussed, imaging of neuronal porosomes in EM micrographs is difficult due in part to their size (10–18 nm) and the presence of high concentration of proteins at the presynaptic membrane. That is why in the normal brain, a total of 196 axo-dendritic synapses were observed and only 28 porosomes were clearly identified. In rats subjected to hypokinetic stress, 96 axo-dendritic synapses were examined and 29 porosomes were clearly identifiable. The data from measurements of diameter and depth are represented in Fig. 3.7. Statistical analysis of the porosome complex in the central nucleus of the amygdala using ANOVA demonstrate no change in the diameter [$F_{(3,252)}$ = 0.02 ($p > 0.05$)] and depth [$F_{(3,252)}$ = 2.92 ($p > 0.05$)] of the complex following hypokinetic stress. The range of diameter is 12–16 nm and that of depth is 5–20 nm.

The diameter of porosome in normal and experimental rat brain is almost the same, as presented in Fig. 3.7. These data are consistent with the evidence obtained with AFM and photon correlation spectroscopic studies, according to which the mean value of porosome diameter is 12–17 nm.[7] Thus, our results demonstrate that parameters of porosomes and especially diameter are very heterogeneous. However, despite the highly dynamic nature of the neuronal porosome complex, the ranges of dimension (diameter 12–16 nm, depth 5–20 nm) remain the same in the normal rat brain and in the brain of rats subjected to chronic hypokinetic stress.

In summary, our results obtained with the morphometric analysis of neuronal porosome complex in the brain of different species of mammals (rat, cat, and dog) and in normal and pathological condition revealed the following:

- existence of porosome complex at the presynaptic membrane of synapses in the brain of different mammals
- presence of porosome complex at the presynaptic membrane, either free or with docked synaptic vesicles
- intra- and inter-species heterogeneity of porosome complex parameters
- range of fluctuation in the neuronal porosome diameter and depth remain similar between species

3.4 Discussion

Porosomes have been identified as the universal secretory machinery in cells, and their presence has been well documented in various secretory cells, such as acinar cells of the exocrine pancreas, endocrine cell, astrocyte, and neuron.[1,4,13,26] Also, it has become clear that despite many structural and compositional similarities of porosomes in different secretory cells, their size differs in different cell types. Among other reasons, this fact could be nature's adaptation to the different size of secretory vesicles present in different cell types; e.g., in fast and slow secretory cells as well as diverse porosomes for special cargo containing vesicles in the same cell.

Therefore, because of different size of fusing vesicles (smaller vesicles fuse more efficiently than larger ones),[27-29] the curvature of both secretory vesicles and the porosome base would dictate the efficacy of vesicle fusion at the cell plasma membrane.[27-29] In agreement with earlier studies,[27-29] neuronal porosomes of various sizes are observed in our study. In an earlier collaborative study, using AFM and EM, we demonstrated in the rat brain 35–50 nm synaptic vesicles docked with near 10 nm in diameter neuronal porosomes.[4] In contrast, in slow secretory cells, such as acinar cells of exocrine pancreas, secretory granules measuring >1000 nm in diameter, expel their content following transient fusion at the porosome base measuring 100–180 nm.[1,6] Based on these findings, it is suggested that secretory vesicle size dictates the size of porosomes, which in turn dictates the speed and volume of content release. In the current study, we have extended our observation of natural design of the neuronal porosome complex, demonstrating heterogeneity in size both within and between species using high-resolution EM morphometry. In the study, we were able to determine the size of neuronal porosome in the hippocampus of three mammalian species, namely the rat, cat, and dog, demonstrating porosome heterogeneity. To obtain a more integrated and complete evaluation of neuronal porosome, we measured not only porosome diameter, as in our earlier study,[4] but also the porosome depth.[14]

Our results demonstrate for the first time the existence of porosome in the dog brain and confirm the cup-shaped morphology of porosomes in the rat[4,14] and cat brains.[14] These results further display parameters, especially depth heterogeneity in mammalian neuronal porosomes. The structure and assembly of proteins within the neuronal porosome complex obtained by different approaches such as AFM, EM, electron density, and 3D contour mapping,[8,17] and the functional reconstitution into artificial lipid membrane provide new insights into the porosome assembly and structure function. Using AFM, the ultrahigh-resolution imaging of the presynaptic membrane of isolated synaptosomes preparations demonstrates the presence of porosome plug at various conformations (open, partially open, and closed) in situ and allowed an understanding of the involvement of the central plug in the rapid opening and closing of the neuronal porosome structure to the outside during neurotransmission via its vertical movement.[17] The different positions of the central plug signify the neuroporosome to be a highly dynamic structure. Although the porosome is a permanent structure at the presynaptic plasma membrane and appears similar in size in all three mammalian species examined, the wide range of porosome depth identified in EM micrographs supports the dynamic nature of the structure. Future studies will provide an explanation regarding the size heterogeneity of the porosome.

Similar heterogeneity of above-mentioned parameters was demonstrated in the disabled brain, specifically in the central nucleus of amygdala of rat, subjected to 90 day hypokinetic stress. It is well known that amygdala plays a crucial role in the orchestration and modulation of the organism response to aversive, stressful events.[30-32] Earlier studies report significant alterations in neurons, glial elements, and synapses of the central nucleus of amygdala provoked by 90 day hypokinetic stress.[22-25] Besides pathological changes at the presynaptic terminals, aggregation of synaptic vesicles, the presence of large osmyophilic bodies, destruction or vacuolization of mitochondria, existence of large irregular forms or relatively large dense-core synaptic vesicles, significant increase of symmetric synapses on dendritic shafts,[22,23] an altered synaptic vesicle distribution pattern and their depletion, have been reported.[18-21] Despite alterations in the presynaptic terminal structure and synaptic transmission provoked by chronic hypokinetic stress in the limbic region, the present EM study demonstrates for the first time that the gross morphology of the porosome is unaffected in such pathological conditions. Specifically, the morphometric analysis of the diameter and depth of the neuronal porosome is unchanged. These results do not, however, rule out the possible changes that may have occurred in the biochemical profile of the neuronal porosome complex following such stress. Furthermore, longer period of stress may elicit changes in the neuronal porosome complex, which remains to be examined. Therefore, we plan to carry out stress studies for longer duration and determine both morphometric and biochemical changes that may occur in neuronal porosomes following stress. Biochemical assays involving both proteomics and lipidomics will be carried out using immunoisolated neuronal porosome complexes obtained from control and experimental animals following stress.

3.5 Conclusion and Future Studies

The electron microscopic study of porosome structure in the hippocampus of different mammalian species was performed. For the first time, the neuroporosome in the dog brain was described and neuroporosome existence at the presynaptic membrane in rat and cat brains was further confirmed. The results prove the cup-shaped morphology of porosome in the rat and cat brains and reveal their similar shape and size in the dog brain. Moreover, the results show the heterogeneity of porosome diameter and depth not only within but also between species. Such data point to the fine-tuning of porosome size to the size of the corresponding secretory vesicle. This is also true not only in various secretory cells but also within the same region where secretion is occurring. Consequent evaluation of neuroporosome structure in other regions of brain and in other mammals, are under way. Furthermore, the biochemical evaluation of the neuronal porosome complex in hypokinetic-stressed rats is being carried out. It is also being examined whether a longer period of stress may elicit alterations in the structure of porosome complex.

References

1. Cho, S. J., Jeftinija, K., Glavaski, A., Jeftinija, S., Jena, B. P., and Anderson, L. L. (2002). Structure and dynamics of the fusion pores in live GH-secreting cells revealed using atomic force microscopy, *Endocrinology,* **143,** 1144–1148.
2. Cho, S. J., Kelly, M., Rognlien, K. T., Cho, J. A., Horber, J. K., and Jena, B. P. (2002). SNAREs in opposing bilayers interact n a circular array to form conducting pores. *Biophys. J.* **83,** 2522–2527.

3. Cho, S. J., Quinn, A. S., Stromer, M. H., Dash, S., Cho, J., Taatjes, D. J., and Jena, B. P. (2002). Structure and dynamics of the fusion pore in live cells, *Cell Biol. Int.*, **26**, 35–42.

4. Cho, W.-J., Jeremic A., Rognlien, K. T., Zhvania, M. G., Lazrishvili, I. L., Tamar, B., and Jena, B. P. (2004). Structure, isolation, composition and reconstitution of the neuronal fusion pore, *Cell Biol. Int.,* **2**, 699–708.

5. Cracium, C. (2004). Elucidation of cell secretion: pancreas led the way, *Pancreatology*, **4**, 487–489.

6. Schneider, S. W., Sritharan, K. C, Geibel, J. P., Oberleithner, H., and Jena, B. P. (1997). Surface dynamics in living acinar cells imaged by atomic force microscopy: identification of plasma membrane structures involved in exocytosis, *Proc. Natl. Acad. Sci. USA*, **94**, 316–321.

7. Cho, W.-J., Jeremic, A., Jin, H., Ren, G., and Jena, B. P. (2007). Neuronal fusion pore assembly requires membrane cholesterol, *Cell Biol. Int.,* **31**, 1301–1308.

8. Cho, W.-J., Ren, G., and Jena, B. P. (2008). 3D contour maps provide protein assembly at the nanoscale within the neuronal porosome complex, *J. Microscopy*, **232**, 106–111.

9. Cho, W.-J., Ren, G., Lee, J.-S., Jeftinija, K., Jeftinija, S., and Jena, B. P. (2009). Nanoscale three-dimensional contour map of protein assembly within the astrocyte porosome complex, *Cell Biol. Int.*, **33**, 224–229.

10. Cho, W.-J., Lee, J.-S., Zhang, L., Ren, G., Shin, L., Manke, C. W., Potoff, J., Kotaria, N., Zhvania, M. G., and Jena B. P. (2011). Membrane-directed molecular assembly of the neuronal SNARE complex, *J. Cell. Mol. Med.*, **15**(1), 31–37.

11. Drescher, D. G., Cho, W. J., and Drescher, M. J. (2011). Identification of the porosome complex in the hair cell, *Cell Biol. Int.,* Rep. Published 07 Oct 2011 as manuscript CBR20110005.

12. Elshenawy, W. W. (2011). Image processing and numerical analysis approaches of porosome in mammalian pancreatic acinar cells, *J. Am. Sci.*, **7**(6), 835–843.

13. Lee, J.-S., Cho, W.-J., Jeftinija, K., Jeftinija, S., and Jena, B. P. (2009). Porosome in astrocytes, *J. Cell Mol. Med.*, **13**, 365–372.

14. Okuneva, V. G., Japaridze, N. J., Kotaria, N. T., and Zhvania, M. G. (2012). Neuronal porosome in the rat and cat brain, *Tsitologiya*, **54**(3), 210–215.

15. Siksou, L., Rostaing, P., Lechaire, J. P., Boudier, T., Ohtsuka, T., Fejtova, A., Kao, H. T., Greengard, P., Gundelfinger, E. D., Triller, A., and Marty, S. (2007). Three-dimensional architecture of presynaptic terminal cytomatrix, *J. Neurosci.*, **27**, 6868–6877.

16. Zhao, D., Lulevich, V., Liu, F., and Liu, G. (2010). Applications of atomic force microscopy in biophysical chemistry, *J. Phys. Chem. B.*, **114**, 5971–5982.

17. Cho, W. J., Lee, J. S., and Jena, B. P. (2010). Conformation states of the neuronal Porosome complex, *Cell Biol. Int.*, **34**, 1129–1232.

18. Magarinos, A. M., Lo, C. J., Gal Toth, J., Bath, K. G., Jing, D., Lee, F. S., and McEven B. S. (2011). Effect of brain-derived neurotrophic factor haploin sufficiency on stress-induced remodeling of hippocampal neurons, *Hippocampus*, **3**, 253–264.

19. Magarinos, A. M., Verdugo, J. M., and McEwen, B. S. (1997). Chronic stress alters synaptic terminal structure in hippocampus, *Proc. Natl. Acad. Sci., USA*, **94**(25), 14002–14008.

20. McEwen, B. S., and Magarinos, A. M. (2011). Stress effects on morphology and function of the hippocampus, *Ann N Y Acad. Sci.*, **832**(1), 271–284.

21. Mo, B., Feng, N., Renner, K., and Forster, G. (2008). Restaint stress increases serotnin release in the central nucleus of amygdale via activation of corticotrophin-releasinf facto receptors, *Brain Res. Bull.*, **76**(5), 493–498.

22. Zhvania, M. G. (1996). Influence of hypokinesia on the synaptoarchitectonical futures of rat's limbic and extrapyramidal structures, in *Neurochemistry: Cellular, Molecular and Clinical Aspects* (Teelken, A., and Jaap Korp ed.), New York, 491–496.

23. Zhvaniia, M. G. (1996). Ultrastructural reorganizations in the formations of the rat forebrain in decreased motor activity not evoking stress, *Morfologiia*, **109**(3), 10–13.

24. Zhvaniya, M. G., and Bliadze, M. G. (1991). Influence of hypokinesia on the ultrastructure of the emotional structures of the rat cerebrum, *Neurosci. Behav. Physiol.*, **1**, 59–64.

25. Zhvaniya, M. G., and Kakabadze, I. M. (1996). Ultrastructure of telencephalic myelinated fibers of the hypokinetic rat, *Neurosci. Behav. Physiol.*, **26**(3), 201–206.

26. Jena, B. P. (2009). Porosome: the secretory portal in cells, *Biochemistry*, **49**, 4009–4018.

27. Jena, B. P. (2007). Secretion machinery at the cell plasma membrane, *Curr. Opin. Struct. Biol.*, **17**, 437–443.

28. Jena, B. P. (2008). Porosome: the universal molecular machinery for cell secretion, *Mol. Cells.*, **26**, 517–529.

29. Jena, B. P. (2009). Porosome: the secretory portal in cells, *Biochemistry,* **49**, 4009–4018.

30. Beretta, S., (2005). Cortico-amygdala circuits: role in the conditioned stress response, *Stress,* **8**, 4, 221–232.

31. Carrasco, G. A., and Van de Kar, L. D. (2003). Neuroendocrine pharmacology of stress, *Eur J Pharamcol.,* **453**, 235–272.

32. Carter, R. N., Pinnock, S. B., and Herbert, J. (2004). Does the amygdala modulate adaptation to repeated stress? *Neurosci.,* **126**(9), 9–19.

Chapter 4

Granule Size Distribution Suggests Mechanism: The Case for Granule Growth and Elimination as a Fusion Nano-Machine

Ilan Hammel[a] and Isaac Meilijson[b]

[a]Sackler Faculty of Medicine, Department of Pathology,
Tel Aviv University, Tel Aviv 6997801, Israel
[b]Raymond and Beverly Sackler Faculty of Exact Sciences,
School of Mathematical Sciences, Department of Statistics and Operations Research,
Tel Aviv University, Tel Aviv 6997801, Israel

ilanh@patholog.tau.ac.il, isaco@math.tau.ac.il

4.1 Introduction

Secretory cells store a notable array of preformed molecules of various sizes in their secretory granules. When these cells are activated to degranulate, granule content is secreted through a fusion pore to the extracellular space. Analysis of secretory granule size (or content) distributions in various cells documented a periodic multimodal behavior, as would be the case under homotypic fusion of granules (G_n) with unit granules of quantal size (G_1), a mechanism we proposed[1] for granule growth. We first documented quantal size characteristics of mast cell secretory granules,[1] whose cross-

NanoCellBiology: Multimodal Imaging in Biology and Medicine
Edited by Bhanu P. Jena and Douglas J. Taatjes
Copyright © 2014 Pan Stanford Publishing Pte. Ltd.
ISBN 978-981-4411-79-0 (Hardcover), 978-981-4411-80-6 (eBook)
www.panstanford.com

sectional areas, measured by digitized planimetry, were recorded in transmission electron micrographs (TEM) of sections of rat peritoneal mast cells. A histogram of equivalent volumes calculated from the measured areas (assuming spherical granules) showed a periodic multimodal distribution in which the modes fell at volumes that were successive integral multiples of the volume at the first mode. Application of a moving-bin technique to the data confirmed the presence of these periodic modes of discrete steps.[2] Biochemical analysis[3] and pulse and chase autoradiography,[4] as well as immunohistochemical approaches correlated with quantitative electron microscopy,[5] assisted to establish the notion of granule quantal size. The quantal model of granule size and content gained acceptance after it was confirmed in different cells by other methodologies, e.g., the patch clamp technique,[6] intra-granule content estimation[3] and in vitro biochemical studies correlated with granule morphology.[2] For instance, follow-up of single granule degranulation of eosinophils by time-resolved patch-clamp capacitance measurements discloses that the plasma membrane increases in discrete steps. The capacitance step size distributions in promyelocytes and myelocytes[7-8] confirm that mature large specific granules are formed by homotypic fusion of unit granules of similar size. Homotypic fusion is facilitated during early stages of differentiation associated with granulogenesis, thus confirming our earlier work on maturation of rat bone marrow and peritoneal eosinophils.[9] The morphometric study demonstrated a periodic, multimodal granule volume distribution, in mature and immature eosinophils. Since the basic volume quantum was observed to be equal in both cases, we suggested that turning the vesiculated young eosinophil granule into a mature dense one depends on intra-granule processes rather than significant volume change. Molecular players coupled with secretion processes are progressively becoming identified. Although these players appear to be highly conserved in all non-virus-mediated membrane fusion processes, the dynamic mechanisms of granule homotypic fusion and exocytosis remain to be resolved.[9-12]

One of the major cornerstones for the establishment of granule growth emerged from quantitative follow-up of granule reconstitution (in various cells) following massive secretion. Subsequent to secretion, the earliest identifiable activated cell showed an extensive diminution in cell volume associated with granule loss (90–95%). Cell volume then increased almost to the original level over a period of a month (mast cells) or a day (pancreas). The size of the Golgi apparatus increased markedly during the granule packaging period and then returned to its original size.[13] The reconstitution of mast cells and pancreatic acinar cells after secretion is a prolonged process with several phases, resulting in cells of varying appearance and content. Granule periodicity remained in evidence throughout the re-granulation time, with constant G_1 volume but increasing mean granule size, stabilizing at a constant mean value.[13-16] Pulse and chase autoradiography supported the conclusion that granule growth is the result of homotypic fusion.[4] In addition, we were able to document TEM micrograph of dumb-bell shaped granules, suggesting homotypic fusion.[1] On the other hand, VAMP8-knockout mice result in lack of homotypic fusion.[17]

Intracellular membrane fusion in eukaryotes requires SNARE (an acronym derived from "SNAp [Soluble NSF Attachment Protein] Receptor") proteins, whose principal role is to mediate vesicle fusion by forming complexes bridging the two membranes (Fig. 4.1). SNARE proteins constitute large eukaryotic protein super-families. For example, the family consists of more than 65 members in yeast and mammalian cells. SNARE protein aggregation in evoked secretion and homotypic growth of granules are at the center of secretory cell research. SNAREs are divided into two classes: v-SNAREs, integrated into the membranes of transport vesicles during budding, and t-SNAREs, located in the membranes of target compartments/membranes. Binding of v- and t-SNAREs induces a conformational change and leads to membrane fusion. Elimination of one of the SNARE-associated proteins results in significant inhibition of the fusion process of the vesicle with the target compartment. A mature granule fuses with a unit granule (to grow into a bigger granule) or with the cell's membrane (to exit the cell) via the formation of a circular rosette composed of SNARE proteins.[18-28] The self-aggregation capacity of the SNARE protein complex has been reported,[17,22-24,26] whereby each unit granule has a limited number of "hooks" and the target membranes have a limited number of "loops." For example, in neurosecretory vesicles the hook is one t-SNARE (SNAP-25+syntaxin complex), the loop is one v-SNARE (VAMP protein), and a SNARE pair is a complex of one t-SNARE and one v-SNARE.[23] The resulted core

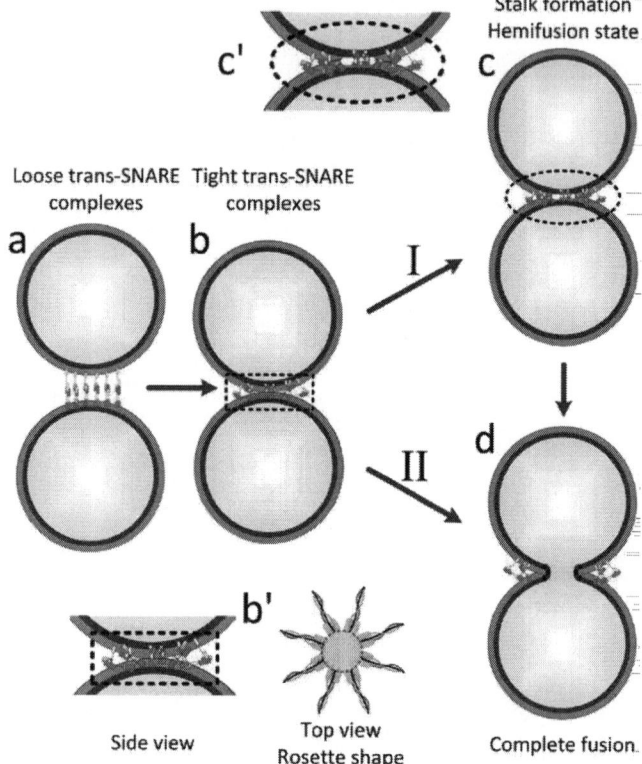

Figure 4.1 Granule life events during homotypic fusion. Membrane fusion requires transient structural reorganization of at least some lipids, with two optional pathways (I and II).[59-61] During membrane fusion, the outer membrane leaflets are loosely connected by SNAREs (step a) and brought to close proximity (step b), whereas the distal membrane leaflets remain separate (step c) until the opening of a fusion pore (Fig. 4.1d). The current perception correlates the initiation of hemifusion (Fig. 4.1c') with the formation of a contact stalk-like zone between the membranes in which the two proximal monolayers are connected by a SNARE rosette-like aggregate (Fig. 4.1b'), where each rosette ring is a *K*-SNARE complex. The stalk forms a hydrophobic "narrow bridge" between contacting vesicle outer leaflets.[59-61] The initial local stalk may evolve to a fusion pore (Pathway II), or expand to hemifusion (Pathway I). This transitional hemifusion stage is a critical metastable intermediate for membrane fusion. Interactions (Fig. 4.1b) can be homotypic (granule–granule fusion) or heterotypic (granule–plasma membrane fusion).

SNARE complex is a four-α-helix bundle, where one α-helix is contributed by syntaxin-1, one α-helix by synaptobrevin and two α-helices are contributed by SNAP-25. It has been well established that for the formation of a circular arrangement between two fusing membranes there must be a formation of an aggregate of SNARE-pairs.[23] The circular arrangement of SNARE-pair complexes is historically denoted as rosette[18,23,29–30] and the pairing of the rosette with the membrane at cup-shaped structures is designated as porosome. Such a membrane fusion structure has been investigated by various methods by the Jena group[19,22–23] leading to the conclusion that vesicle curvature dictates more generally the size of the SNARE ring complex formed.

4.2 Intracellular Management of the Inventory of Granules

Granule stock management is a good term for understanding the cytoplasmic granule mix and the different demands on granule stock. At steady state, demands are influenced by both extracellular and cytoplasmic factors.[31] Thus, at homeostatic state granule growth can be modeled as a steady-state machine in which there is a constant production of granules of unit size (G_1), with independent life cycles that follow a (size-dependent) constant rate of granule size increase and a (size-dependent) constant rate of granule elimination (Figs. 4.2a,b). Large granules (G_n, $n \geq 2$), lacking the capacity to fuse with each other, display fusion capacity with newly generated vesicles only. In other words, all granules enclosed with a smooth line in Fig. 4.2b have terminated the capacity to fuse with other mature granules. This *unit addition* model represents a fusogenic apparatus in which only the newly generated G_1 granule has the capacity to fuse with other granules. These unit building granules G_1 (Fig. 4.2b, enclosed by broken-line ellipsoid) have two options: to develop into a mature unit granule and lose the fusogenic capacity (Fig. 4.2b, G_1 enclosed by smooth-line ellipsoid), or to fuse with another granule of any size.[31–32] Mature granules also have two options: to fuse with a newly generated unit granule or to go through an elimination process (Figs. 4.2a,b, G_n enclosed by smooth-line ellipsoid). Under the Markovian model to be presented, the content distribution of basally eliminated granules may differ from the steady-state content distribution of the cytoplasmic granule population (Figs. 4.2a,c).

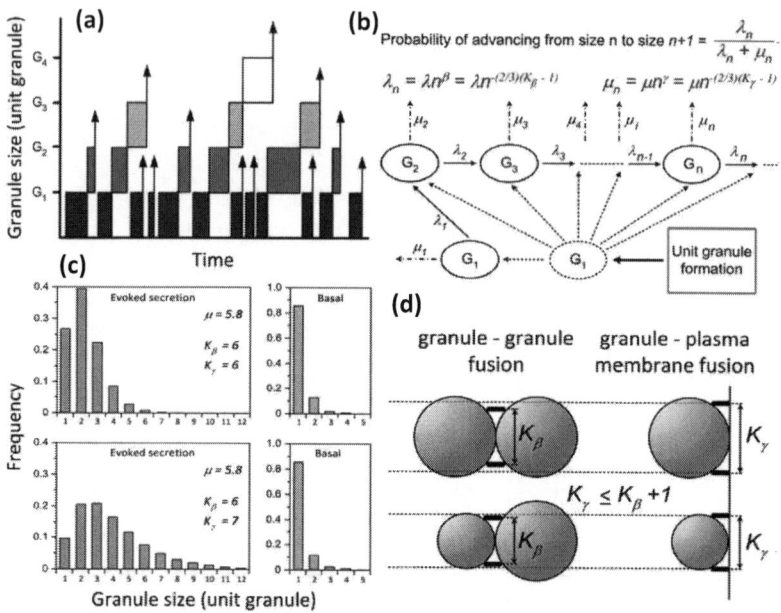

Figure 4.2   (a, b) Single steps in the G&E Markov model, based (via four global parameters β, γ, λ, μ) on two transition rates $\lambda_n = \lambda n^\beta$ and $\mu_n = \mu n^\gamma$ that describe the probability rate of a granule of size G_n to move one level up and become a granule of size G_{n+1} (i.e., $G_n + G_1 \rightarrow G_{n+1}$) or to move out of the system, respectively. The transition, which occurs after an exponentially distributed time with mean $1/(\lambda_n + \mu_n)$, is an increase in size with probability $\lambda_n/(\lambda_n + \mu_n)$ and exit with the complementary probability $\mu_n/(\lambda_n + \mu_n)$. "Birth" (Fig. 4.2b) follows a Poisson process of formation of mature unit granules, each of which undergoes, independently, a Markov history as in Fig. 4.2a. Theoretical stationary (similar to fast evoked state) and exit (similar to basal secretion) size distributions are presented in Fig. 4.2c. Interactions (Fig. 4.2d) can be homotypic (granule–granule fusion, on the left) or heterotypic (granule–plasma membrane fusion, on the right). The top left figure represents homotypic fusion between large unit granules, while the bottom left figure represents fusion between granules of sizes G_1 and G_3.

A Markov process is a memory-efficient statistical representation of the progression of a random system, in the sense that conditional on the current state of the system, future and past states are independent.[31-32] In our proposed growth and elimination (G&E) model of granule life cycle, schematically illustrated in Fig. 4.1b, growth is quantified by the transition rate λ_n from level n to level $n + 1$ and Elimination, by the rate μ_n at which granules at level n exit the system. The Markov property holds if these two transition rates act on a granule at the current level regardless of how this level was accomplished. By modeling the corresponding transition rates λ_n and μ_n as $\lambda_n = \lambda n^\beta$ and $\mu_n = \mu n^\gamma$, this model can parsimoniously describe the steady-state and exit size distributions of the granule population in a resting cell as deterministic functions of three parameters: the *effective kinetics factor* μ/λ, the *effective surface factor* γ-β and the *surface tension* γ (Fig. 4.2b). Thus, prediction of normal behavior and pathological perturbations are workable. While emphasis is placed on a resting cell, the model has implications on evoked secretion as well.[31-32]

This particular parametric model was introduced and empirically validated[31-32] in terms of statistical parameter values that afforded excellent fit to data but involved surface tension values that lacked proper classical biophysical interpretation (e.g., via van der Waals interaction forces).[33-35] Resorting to a statistical mechanics approach in which aggregates of SNARE components are viewed as interacting particles, we were able to solve in terms of γ and β for a simple "nano-machine" behind granule growth and secretion.[32] The results reveal two linear size relations $\gamma = -(2/3)(K_\gamma - 1)$ and $\beta = -(2/3)(K_\beta - 1)$, where K_γ and K_β are integer and define the rosette space sizes, i.e., the number of SNARE units needed for granule fusion. Since $\gamma - \beta > -1$ is obligatory for the existence of a steady state,[31] it follows that $K_\gamma \leq K_\beta + 1$ (graphical display in Fig. 4.2d).

4.3 Does Granule Inventory Represent "New Old Stock"?

Studies using pulse-labeling techniques have suggested that secretory proteins may be stored and secreted heterogeneously.[36-39] Although most of these studies found that newly synthesized proteins are secreted selectively, little is known about how non-parallel secretion results from the integration of synthesis with the sequential cellular processing, storage, and mobilization of this protein. If the composition of the secretory proteins following stimulation approaches the enzyme composition of pancreatic tissue, we speak of parallel secretion. In contrast, nonparallel secretion can be defined as an asynchrony[40-41] in the ratio of the output content of a secretory protein pair or of a single enzyme to overall protein secretion, as correlated with time or a stimulus. Thus, non-parallel secretion will lead to a variable mixture of enzymes, dependent on the secretagogue type or rate of exocytosis. Granules constitute a heterogeneous vesicular population even within a single cell type, which can differ in size,[42-43] content,[3,42-44] and kinetics of content release.[39,45-50] Several of these differences are correlated with granule age.[15,37-39] Aging can also affect the probability of secretory granules undergoing exocytosis, with newly generated secretory granules being preferentially released.[36-37] For example, by tagging secretory granules with the fluorescent protein dsRed-E5, which changes its emission from green to red over time, Duncan et al. investigated the age-dependent distribution of secretory vesicles within chromaffin cells.[51] These authors point out that granule age is a critical factor that segregates granules with respect to their localization and mobility, and affects their probability of undergoing exocytosis in response to different stimuli. Adelson et al. followed pancreatic acinar cell secretions at various states and concluded[36-37] that "under conditions of constant stimulation with different secretagogues or basal conditions, pancreatic exocrine secretion differed with respect to the percent of each enzyme secreted as a proportion of the total enzyme species, with respect to each other, and with respect to the overall variation from the mean in both enzyme percentage and ratio. The fact that nonparallel secretion occurred among and between basal conditions and the two classical types of stimulant, hormonal (CCK) and neurotransmitter (cholinergic stimulation), indicates that nonparallel secretion of digestive enzymes is likely the rule, not the exception." These observations raise the question of how do secretory cells distinguish between granules "sorted" for regulated or constitutive pathways so they can be correctly targeted for secretion.

Regulated secretion is protein secretion that is monitored by a secretagogue. Regulated secretory cells have an alternative pathway for protein secretion, than the constitutive pathway.[36-38,46] Wherever both constitutive and regulated pathways for granule secretion exist in the same cell, the various proteins have a preference for one pathway over another.[46,52] Such phenomena suggest a preferential release of granules from the total granule inventory. Already at an early state of investigation, it was realized that there must exist two discrete mechanisms for granule turnover—a slow basal release and a faster evoked state.[53] Based on this dichotomy and motivated by the proposed dynamic life cycle of granules (Fig. 4.1a), the G&E model sets apart the granule inventory into two major granule "reserve pools" that co-exist in the cytoplasm. The first is a group of typically small granule size with a high chance to be secreted and the second group is composed of larger granules with low turnover (Fig. 4.3).

There are two essential goals for keeping granule stock: uncertainty of demand and lead time for production. Granule stock is maintained as a buffer to meet insecurity in demand by the extracellular environment, and as a source of supply during the lead time to produce granules of adaptive content. Figure. 4.1c demonstrates that the typical constitutively secreted granule is smaller than the typical granule in the stock. It is well documented that the constitutive pathway for secretion is favored by granules which, following packaging in the Golgi apparatus, bud off and directly move towards the plasma membrane, where they fuse and secrete their contents into the extracellular environment.[37,54] Our model suggests that these granules reflect the granule group secreted at the basal state.

Inherent in some studies and explicit in others, is the assumption that the "correct" updated response to environmental change is to embrace its predictability. Such expectation-oriented response is a pre-requisite for cell adaptation to meet new environmental demands. The G&E model offers an alternative option for response, by identifying basal secretion with the exit/elimination distribution and active/evoked secretion with the stationary distribution. The phenomenon of size discrepancy (whose effect on the shape of these distributions is displayed in Fig. 4.4) will result from the reality that smaller granules tend to remain within the cell for a shorter time as compared to granules that reached larger size (Fig. 4.3), inducing an almost "last-in, first out" queue management.[31] This empirically validated fact may assist cell adaptation to new environmental changes by establishing a mechanism that enables the newly formed granules with the "adaptive content" to be sorted for secretion before the "older granules," adapted to previous states. Thus, favored exocytosis of fresher granules might allow cells to

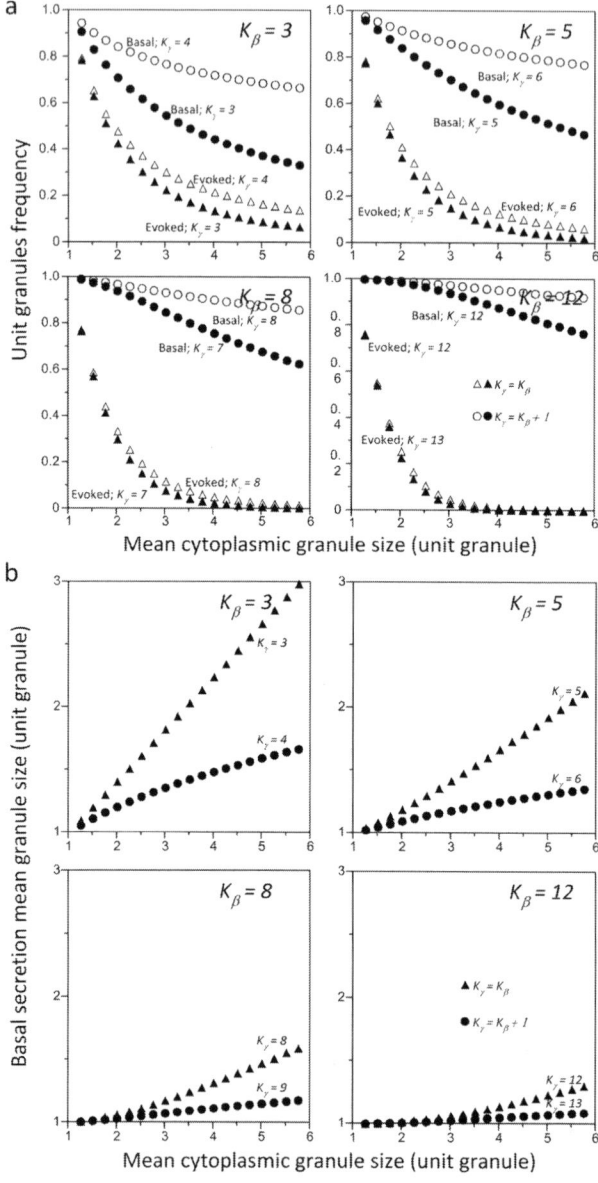

Figure 4.3   Mean cytoplasmic granule size correlated with basal secretion granule size, as a function of rosette size. Preferential basal secretion for smaller granules (a) induces stationary or evoked mean granule size that is higher than basally secreted mean granule size (b).

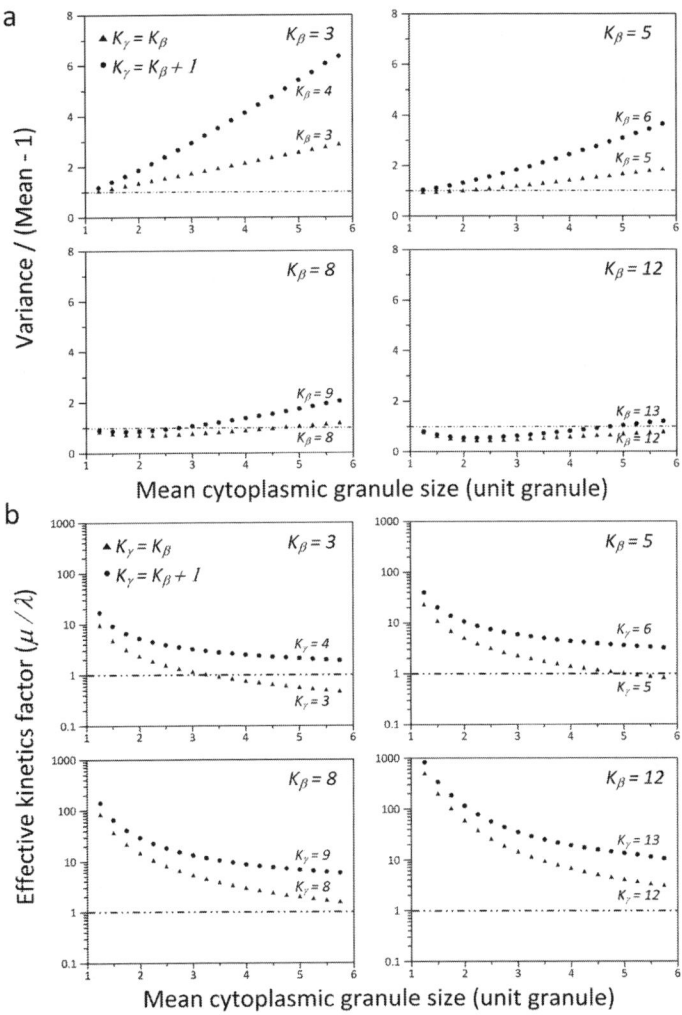

Figure 4.4   Mean cytoplasmic granule size correlated with granule dispersion (Fig. 4.4a) and with effective kinetics factor (Fig. 4.4b). For small granules, requiring few SNAREs (e.g., neurosecretory vesicles, $K_\beta = 5$), quantal granule size is Poisson-like distributed only for mean quantal granule size below 2.5 unit granules. In contrast, for larger granules requiring rosette sizes above 8 SNAREs, almost the entire range of cytoplasmic quantal granule size leads to a Poisson-like distribution. The effective kinetics factor (Fig. 4.4b) is significantly regulated by rosette size. Benchmark simulations suggest that rosette size should be below 20 for most biologically feasible cases.

adjust their secretory response to changing physiological needs quickly by swiftly modifying the granule composition.[54] Such favored basal secretion of recently synthesized secretory material will result in secretion of various combinations of soluble and granular factors as a direct adaptive response to the extracellular demands, reflecting distinctive challenges and organizational dynamics that arise during the cell life-cycle. Thus, the "new old stock" inventory management reflects adaptation based on "cell experience" learned from extracellular communication with the cell. Namely, "stationary" granules, similar to original stock parts that remained in inventory for a use that rarely came, undergo evoked secretion as exocytosis of "adapted granules" based on cell long time experience, while basal secretion reflects "tailor made" granules. Thus, the evoked state represents the term "new old stock," which refers to granule stock being available for secretion and which was "manufactured" long ago but that has not been spent.

4.4   Cellular Communication: Evoked State Yields Information

Communication is an emergent reality, a state of associations unique in its characteristics. In our context, it is the exchange and flow of information from one cell to another or to the extracellular environment, about the cell's granule exocytosis *mode*, basal or evoked. The G&E model suggests that under basal secretion, secreted granule size should follow the exit distribution. Assuming that under evoked secretion, granules in the vicinity of the membrane exit with their typical content, secreted granule size should follow the steady-state distribution. Thus, the G&E model of intra-cell processes has direct and explicit implications on the extra-cellular gradient of secretion modes. As such, it provides a numerical tool for capacity assessment, the evaluation of the evoked burst size needed for reliable transmission of a yes/no answer, the disclosure of the secretion mode.

The two main statistical disciplines we call upon for this analysis are information theory and detection of a change-point in distribution. Information theory provides numerical quantization of the uncertainty of a set of consequences (entropy), of the transmission quality of a communication channel (mutual information and capacity), and of the distance between two probability assignments (Kullback–Leibler divergence). The change-point detection problem has received great notoriety over the last century, in particular in the biostatistical literature. It deals with quick detection/recognition, with a low false alarm rate, of a change in a system from one stochastic behavior to another, based on observations sampled sequentially from the system. In our context, the timely recognition that basal secretion has been switched into a short-lived evoked event enables the "receiver" (e.g., synapse) to identify and react to a potential "new event" more successfully. The CUSUM (cumulative sum control chart) sequential analysis technique[55] has been proved optimal for detecting such a shift and is commonly used in quality control to detect deviations from benchmark values.

As explained above, the G&E model suggests that there are two options for granule elimination, basal and evoked (fast) secretion of granules, of which the latter reflects the intracellular granule population. The steady-state and exit equations that we have developed[31-32] provide a tool for the assessment of the number of evoked granules (burst size) needed to generate a reliable yes/no answer for communication between a cell and another (Fig. 4.5) or with the environment. For fixed granule–granule rosette size K_β (typical values in the range $3 \le K_\beta \le 12$) and fast evoked secretion (basal secretion rate × 10), the number of granules needed for detection decreases as either mean granule size or granule-membrane rosette size K_γ increases. For example, if mean granule size is in the range of 3–5 unit granules and $K_\beta \le 5$, at least 4 granules are needed to detect the evoked state, while 3 granules might be enough for $K_\beta \ge 8$. Recent observations in neurosynaptic communications suggest that basal secretion and action potential-evoked secretion preferentially activate distinct subsets of postsynaptic receptors, supporting the proposal that synapses use physically segregated pathways to decode basal and evoked neurotransmission.[56-58] Under physiological as well as pathophysiological circumstances, spontaneous fusion events can set the concentration of ambient levels of secretory mediators within extracellular milieu or the synaptic cleft. Therefore, elucidation of information content underlying basal secretion may help better understand the functional significance of this form of secretion and provide tools for its selective manipulation. Since most of the informative communication is done at the synapse ($K_\beta \leq 5$; mean vesicle size $\approx$ 40 nm, about 2 ± 0.5 unit granules[32]) an evoked burst consisting of more granules than the above is needed to confirm a yes/no answer.

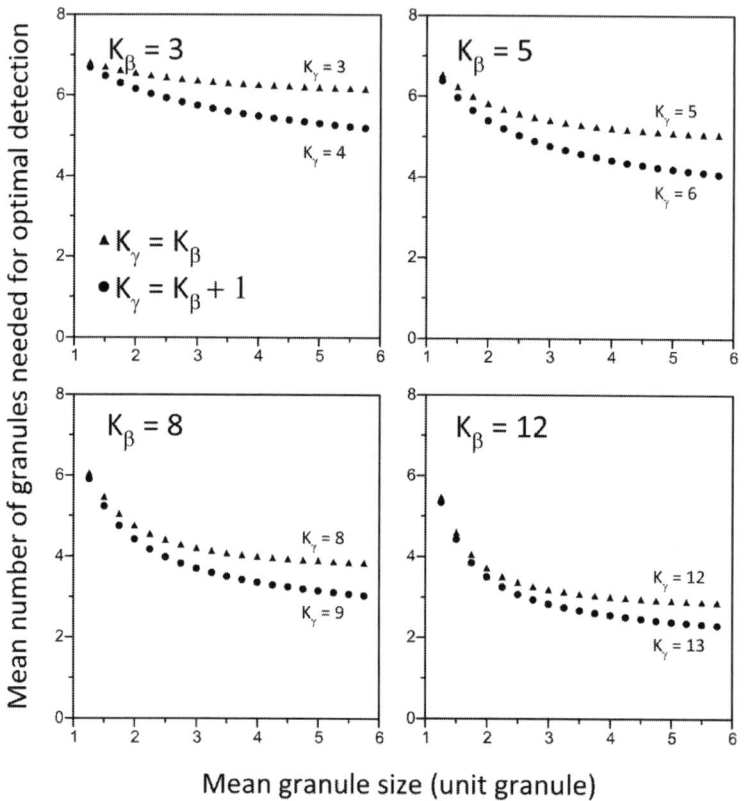

Figure 4.5 Mean number of evoked granules needed for detection of evoked mode, in terms of mean cytoplasmic quantal granule size and rosette size. Benchmark simulations suggest that secretion magnification rates from 8 to 32 yield similar results and that in case the mean number of granules needed for optimal secretion is above 8, a yes/no confirmation will be achieved with one mistake per day at basal frequency of 1 Hz.

It has been recognized for at least half a century that two distinct modes of secretion are operative in various cells: active degranulation of secretory mediators triggered by cell specific activators (i.e., evoked state) and spontaneous secretion. Secretory cells communicate with their environment by exocytosis of the granule content. The significant information processing capacity of the cell environment depends on the receiver, which has to differentiate between the two secretion modes. In connective tissue there is less demand for accurate communication as compared to the information processing needed in the mammalian brain. Namely, information about change point is most important at the synapse level, the nerve terminal, as it is in hormonal secretion.

The rate of single granule "elimination" events varies between the two secretion modes (let η_{EVOKED} and η_{BASAL} denote the two secretion rates) and so does the distribution of secreted granule size (the *steady-state* and *exit-size* distributions mentioned earlier). To efficiently detect a mode switch, the CUSUM method evaluates an action potential AP(t) of the integrate-and-fire type, affected by both elements of change, and declares a detection whenever this potential exceeds a threshold. The action potential AP($t + \Delta$) at time $t + \Delta$ is obtained from AP(t) at time t by STEP (i): subtracting $const_1$ × Δ and adding $const_2$ × S_Δ and ($const_3$ – $const_4$) × N_Δ, where $const_3$, $const_4$, $const_1 = \eta_{EVOKED} - \eta_{BASAL}$ and $const_2 = -\gamma$ are positive constants, N_Δ is the number of granules secreted from the cell in the time interval ($t, t + \Delta$) and S_Δ is the sum of the logarithms of their quantal volumes. Then STEP (ii): whenever the action potential becomes negative, it is instantly redefined as zero. The parameter $const_1$, tantamount to the membrane time constant (except that it acts additively rather than multiplicatively), ideally equal to the difference in secretion rates between the two modes, reacts as $const_3$ does, only to changes in secretion rate, while the parameters $const_2 = -\gamma$ and $const_4$ react only to changes in the granule volume distribution. The sum of the two last summands constitutes the input to the action potential, much like in the neuronal integrate-and-fire model. While a summand of the type $const_3$ × N_Δ appears explicitly in ref. [58], the summands $(-\gamma)$ × S_Δ and $-const_4$ × N_Δ are intrinsic to a homotypic fusion model such as G&E. Summarizing the composition of the action potential, the time clock is inhibitory, the volume counter is excitatory and the granule counter can be either, depending on whether the mode gradient is primarily due to magnified secretion rate or to a sharp change in granule-size distribution. As for the comparison with integrate-and-fire, the usual multiplicative time constant model under which the action potential loses a fixed *proportion* of its size per unit time and never reaches zero, is very similar to the CUSUM version above, in which the action potential loses a fixed *amount* per unit time and is not allowed to go negative. The two models are so interchangeable in practice, that it should be up to evolution to determine which is more feasible biologically, and within this choice, set the secretion rate gradient contingent on the feasible choice of time constant.

We evaluated numerically the optimal mean number of granules (MGCD) needed for correct detection of the evoked mode while keeping the mean number of granules between false alarms (MGFA) under control.[32] The evaluation was performed on a wide benchmark of parameters: K_β from 3 to 12, stationary mean granule size (MGS, in unit granules) from 1.25 to 5.75 at 0.25 intervals, MGFA = 10^d for d from 2 to 7, and secretion rate magnified by a factor 2^f for f from 0 to 6. For MGFA = 10^5, MGCD is displayed (Fig. 4.5) as a function of MGS. It is evident that for small K_β (4 to 5) at least 8 granules (on the average) are needed for detection, while for $K_\beta > 8$ much fewer granules are needed. However, since the total volume of 10 granules with $K_\beta = 4$ is about 37% of the volume of one granule with $K_\beta = 12$, the drive for miniaturization of granules is apparent. Differentiation of cells by secretory activity (e.g., neuro-secretion, hormones, immune activity) is coupled with differentiation by granule diameter size (i.e., by rosette size K), where small K corresponds to high granule turnover.[32]

For $K_\gamma = K_\beta$, MGS decreases if the common value of K_γ and K_β is increased. However, MGS decreases further if K_β is reduced by one rosette petal with respect to K_γ. Thus, according to the G&E model, the granule-membrane porosome size K_γ should optimally exceed the granule–granule nano-machine size K_β by one rosette petal, although equality between the two is close to optimal. It may be biologically safer to keep them equal, since $K_\gamma = K_\beta + 1$ is borderline stable in the sense that $K_\gamma > K_\beta + 1$ would not admit a steady nano-machine.[32] In most examples analyzed, the two were estimated to be indeed equal (supporting information figures 1–3 in Hammel and Meilijson[32]). Figure 4.5 demonstrates that secretion magnification factors above 8 achieve MGCD quite insensitive to granule size distribution or further magnification factor. Indeed, the literature reports that evoked secretion rates are commonly roughly 10 times above basal secretion rates.

References

1. Hammel, I., Lagunoff, D., Bauza, M., and Chi, E. (1983). Periodic, multimodal distribution of mast cell granule volumes. *Cell Tissue Res.,* **228**, 51–59.

2. Hammel, I., Lagunoff, D., and Galli, S. J. (2010). Regulation of secretory granule size by the precise generation and fusion of unit granules. *J. Cell. Mol. Med.,* **14**, 1904–1916.

3. Mroz, E. A., and Lechêne C. (1986). Pancreatic zymogen granules differ markedly in protein composition. *Science,* **232**, 871–873.

4. Hammel, I., Dvorak, A. M., Fox, P., Shimoni, E., and Galli, S. J. (1998). Defective cytoplasmic granule formation. II. Differences in patterns of radiolabeling of secretory granules in beige versus normal mouse pancreatic acinar cells after [^{3}H]-glycine administration in vivo. *Cell Tissue Res.,* **293**, 445–452.

5. Bialik, G. M., Abassi, Z. A., Hammel, I., Winaver, J., and Lewinson, D. (2001). Evaluation of atrial natriuretic peptide and brain natriuretic peptide in atrial granules of rats with experimental congestive heart failure. *J. Histochem. Cytochem.,* **49**, 1293–1300.

6. Alvarez de Toledo, G., and Fernandez, J. M. (1990). Patch-clamp measurements reveal multimodal distribution of granule sizes in rat mast cells. *J. Cell Biol.,* **110**, 1033–1039.

7. Hartmann, J., Scepek, S., and Lindau, M. (1995). Regulation of granule size in human and horse eosinophils by number of fusion events among unit granules. *J. Physiol. (London),* **483**, 201–209.

8. Scepek, S., and Lindau, M. (1997). Exocytotic competence and intergranular fusion in cord blood-derived eosinophils during differentiation. *Blood,* **89**, 510–517.

9. Elmalek, M., and Hammel, I. (1987). Morphometric evidence that the maturation of the eosinophil granules is independent of volume change. *J. Submicrosc. Cytol.,* **19**, 265–268.

10. Guo, Z., Turner, C., and Castle, D. (1998). Relocation of the t-SNARE SNAP-23 from lamellipodia-like cell surface projections regulates compound exocytosis in mast cells. *Cell,* **94**, 537–548.

11. Kögel, T., and Gerdes H. H. (2010). Maturation of secretory granules. *Results Probl. Cell Differ.,* **50**, 1–20.

12. Wickner, W. (2010). Membrane fusion: five lipids, four SNAREs, three chaperones, two nucleotides, and a Rab, all dancing in a ring on yeast vacuoles. *Annu. Rev. Cell Dev. Biol.,* **26**, 115–136.

13. Lew, S., Hammel, I., and Galli, S. J. (1994). Cytoplasmic granule formation in mouse pancreatic acinar cells. Evidence for formation of immature granules (condensing vacuoles) by aggregation and fusion of progranules of unit size, and for reductions in membrane surface area and immature granule volume during granule maturation. *Cell Tissue Res.,* **278**, 327–336.

14. Hammel, I., Lagunoff, D., and Krüger, P.-G. (1988). Studies on the growth of mast cells in rats. Changes in granule size between one and six months. *Lab. Invest.,* **59**, 549–554.

15. Hammel, I., Lagunoff, D., and Krüger, P. G. (1989). Recovery of rat mast cells after secretion: a morphometric study. *Exp. Cell. Res.,* **184**, 518–523.

16. Hammel, I., Lagunoff, D., and Wysolmerski, R. (1993). Theoretical considerations on the formation of secretory granules in the rat pancreas. *Exp. Cell Res.,* **204**, 1–5.

17. Hammel I., Wang C. C., Hong W., Amihai D. (2012). VAMP8/endobrevin is a critical factor for the homotypic granule growth in pancreatic acinar cells. *Cell Tissue Res.,* **348**, 485–490.

18. Bowen, M. E., Weninger, K., Brunger, A. T., and Chu, S. (2004). Single molecule observation of liposome-bilayer fusion thermally induced by SNAREs. *Biophys. J.,* **87**, 3569–3584.

19. Cho, W. J., Jeremic, A., and Jena, B. P. (2005). Size of supramolecular SNARE complex: membrane-directed self-assembly. *J. Am. Chem. Soc.,* **127**, 10156–10157.

20. Domanska, M. K., Kiessling, V., Stein, A., Fasshauer, D., and Tamm, L. K. (2009). Single vesicle millisecond fusion kinetics reveals number of SNARE complexes optimal for fast SNARE-mediated membrane fusion. *J. Biol. Chem.,* **284**, 32158–66.

21. Hua, Y., and Scheller, R. H. (2001). Three SNARE complexes cooperate to mediate membrane fusion. *Proc. Natl. Acad. Sci. USA,* **98**, 8065–8070.

22. Jena, B. P. (2009). Functional organization of the porosome complex and associated structures facilitating cellular secretion. *Physiology (Bethesda),* **24**, 367–376.

23. Jena, B. P. (2011). Role of SNAREs in Membrane Fusion. *Adv. Exp. Med. Biol.,* **713**, 13–32.

24. Megighian, A., Scorzeto, M., Zanini, D., Pantano, S., Rigoni, M., Benna, C., Rossetto, O., Montecucco, C., and Zordan, M. (2010). Arg206 of SNAP-25 is essential for neuroexocytosis at the Drosophila melanogaster neuromuscular junction. *J. Cell Sci.,* **123**, 3276–3283.

25. Mohrmann, R., de Wit, H., Verhage, M, Neher, E., and Sorensen, J. B. (2010). Fast vesicle fusion in living cells requires at least three SNARE complexes. *Science,* **330**, 502–505.

26. Montecucco, C., Schiavo, G., and Pantano, S. (2005). SNARE complexes and neuroexocytosis: how many, how close? *Trends Biochem. Sci.,* **30**, 367–372.

27. van den Bogaart, G., and Jahn, R. (2011). Counting the SNAREs needed for membrane fusion. *J. Mol. Cell Biol.,* **3**, 204–205.

28. Jahn, R. (2004). Principles of exocytosis and membrane fusion. *Ann. N. Y. Acad. Sci.,* **1014**, 170–178.

29. Burwen, S. J., and Satir B. H. (1977). A freeze-fracture study of early membrane events during mast cell secretion. *J. Cell Biol.,* **73**, 660–671.

30. Weber, T., Zemelman, B. V., McNew, J. A., Westermann, B., Gmachl, M., Parlati, F., Söllner, T. H., and Rothman, J. E. (1998). SNAREpins: minimal machinery for membrane fusion. *Cell,* **926**, 759–772.

31. Nitzany, E., Hammel, I., and Meilijson, I. (2010). Quantal basis of vesicle growth and information content, a unified approach. *J. Theor. Biol.,* **266**, 202–209.

32. Hammel, I., and Meilijson, I (2012). Function suggests nano-structure: electrophysiology supports that granule membranes play dice. *J. R. Soc. Interface,* **9**, 2516–2526.

33. Nir, S., and Nieva, J. L. (2000). Interactions of peptides with liposomes: pore formation and fusion. *Prog. Lipid Res.,* **39**, 181–206.

34. Papahadjopoulos, D., Nir, S., and Düzgünes, N. (1990). Molecular mechanisms of calcium-induced membrane fusion. *J. Bioenerg. Biomembr.,* **22**, 157–179.

35. Rädler, J. O., Feder, T. J., Strey, H. H., and Sackmann, E. (1995). Fluctuation analysis of tension-controlled undulation forces between giant vesicles and solid substrates. *Phys. Rev. E. Stat. Phys. Plasmas Fluids Relat. Interdiscip. Topics,* **51**, 4526–4536.

36. Adelson, J. W., Clarizio, R., and Coutu, J. A. (1995). Pancreatic digestive enzyme secretion in the rabbit: rapid cyclic variations in enzyme composition. *Proc. Natl. Acad. Sci. USA,* **92**, 2553–2557.

37. Adelson, J. W., and Miller, P. E. (1989). Heterogeneity of the exocrine pancreas. *Am. J. Physiol.,* **256**, G817–G825.

38. Case, R. M. (1978). Synthesis, intracellular transport and discharge of exportable proteins in the pancreatic acinar cell and other cells. *Biol. Rev. Camb. Philos. Soc.,* **53**, 211–354.

39. Rothman, S., Liebow, C., and Grendell, J. (1991). Nonparallel transport and mechanisms of secretion. *Biochim. Biophys. Acta,* **1071**, 159–173.

40. Oron, U., Kinamon, S., and Bdolah, A. (1978). Asynchrony in the synthesis of secretory proteins in the venom gland of the snake Vipera palaestinae. *Biochem J.,* **174**, 733–739.

41. Dagorn, J. C. (1978). Non-parallel enzyme secretion from rat pancreas: in vivo studies. *J. Physiol. (London),* **280**, 435–448.

42. Hammel, I., Elmalek, M., Castel, M., and Kalina, M. (1989). Variability in gold bead density in cells. Quantitative immunocytochemistry. *Histochemistry,* **91**, 527–530.

43. Ermak, T. H., and Rothman S. S. (1986). Zymogen granule size in pancreas of nursing rats. *J. Morphol.,* **187**, 289–299.

44. Kalina, M., Elmalek, M., and Hammel, I. (1988). Intragranular processing of pro-opiomelanocortin in the intermediate lobe of the rat pituitary glands. A quantitative immunocytochemical approach. *Histochemistry,* **89**, 193–198.

45. Satir, B. (1974). Membrane events during the secretory process. *Symp. Soc. Exp. Biol.,* **28**, 399–418.

46. de Dios, I., Garcia-Montero, A. C., Orfao, A., and Manso, M. A. (1999). Selective exocytosis of zymogen granules induces non-parallel secretion in short-term cholecystokinin stimulated rats. *J. Endocrinol.,* **163**, 199–206.

47. Hammel, I., Shilo-Rabinovich, H., and Nir I. (1988). Two populations of mast cells on fibroblast monolayers: correlation of quantitative microscopy and functional activity. *J. Cell Sci.,* **91**, 13–19.

48. Pollard, H. B., Pazoles, C. J., Creutz, C. E., and Zinder, O. (1979). The chromaffin granule and possible mechanisms of exocytosis. *Int. Rev. Cytol.,* **58**, 159–197.

49. Wingren, U., Wasteson. A., and Enerbäck, L. (1983). Storage and turnover of histamine, 5-hydroxytryptamine and heparin in rat peritoneal mast cells in vivo. *Int. Arch. Allergy Appl. Immunol.,* **70**, 193–199.

50. Zinder, O., and Pollard, H. B. (1980). The chromaffin granule: recent studies leading to a functional model for exocytosis. *Essays Neurochem. Neuropharmacol.,* **4**, 125–162.

51. Duncan, R. R., Greaves, J., Wiegand, U. K., Matskevich, I., Bodammer, G., Apps, D. K., Shipston, M. J., and Chow, R. H. (2003). Functional and spatial segregation of secretory vesicle pools according to vesicle age. *Nature,* **422**, 176–180.

52. Arvan, P., Castle, D. (1998). Sorting and storage during secretory granule biogenesis: looking backward and looking forward. *Biochem. J.,* **332**, 593–610.

53. Ginsburg, H., Nir, I., Hammel, I., Eren, R., Weissman, B. A., and Naot, Y. (1978). Differentiation and activity of mast cells following immunization in cultures of lymph-node cells. *Immunology,* **35**, 485–502.

54. Solimena, M., and Gerdes, H. H. (2003). Secretory granules: and the last shall be first. *Trends Cell Biol.,* **13**, 399–402.

55. Page, E. S. (1954). Continuous Inspection Scheme. *Biometrika,* **41**, 100–115.

56. Sutton, M. A., and Schuman, E. M. (2009). Partitioning the synaptic landscape: distinct microdomains for spontaneous and spike-triggered neurotransmission. *Sci. Signal.,* **2**, pe19.

57. Wasser, C. R., and Kavalali E. T. (2009). Leaky synapses: regulation of spontaneous neurotransmission in central synapses. *Neuroscience,* **158**, 177–188.

58. Kavalali, E. T., Chung, C., Khvotchev, M., Leitz, J., Nosyreva, E., Raingo, J., and Ramirez, D. M. (2011). Spontaneous neurotransmission: an independent pathway for neuronal signaling? *Physiology (Bethesda),* **26**, 45–53.

59. Chernomordik, L. V., and Kozlov, M. M. (2008). Mechanics of membrane fusion. *Nat. Struct. Mol. Biol.,* **15**, 675–683.

60. Kasson, P. M., and Pande, V. S. (2007). Control of membrane fusion mechanism by lipid composition: predictions from ensemble molecular dynamics. *PLoS Comput. Biol.,* **3**, 2228–2238.

61. Kunding, A. H., Mortensen, M. W., Christensen, S. M., Bhatia, V. K., Makarov, I., Metzler, R., and Stamou, D. (2011). Intermembrane docking reactions are regulated by membrane curvature. *Biophys. J.,* **101**, 2693–2703.

Chapter 5

Probing Protein Assembly, Biomineralization, and Biomolecular Interactions by Atomic Force Microscopy

Kang Rae Cho[a] and James J. De Yoreo[b]

[a]The Molecular Foundry, Lawrence Berkeley National Laboratory,
Berkeley, CA 94720, USA
[b]Physical Sciences Division, Pacific Northwest National Laboratory,
Richland, WA 99352, USA

kangraecho@lbl.gov, james.deyoreo@pnnl.gov

Over the past 20 years, the atomic force microscopy (AFM) has been playing a pivotal role in new discoveries and understanding in biological fields through its capability to image surfaces with nanometer-scale spatial resolution in liquid environments. In this chapter, we describe the basic operation of AFM, interpretation of AFM images, and experimental design and discuss three topics in biology investigated by the AFM: recombinant human prion protein aggregation, calcium oxalate monohydrate crystal growth, and molecular interactions between amelogenin and hydroxyapatite. Then, we point out future directions in AFM imaging that will impact research in biology on topics such as these.

NanoCellBiology: Multimodal Imaging in Biology and Medicine
Edited by Bhanu P. Jena and Douglas J. Taatjes
Copyright © 2014 Pan Stanford Publishing Pte. Ltd.
ISBN 978-981-4411-79-0 (Hardcover), 978-981-4411-80-6 (eBook)
www.panstanford.com

5.1   Introduction

The atomic force microscope (AFM) measures surface properties of matter such as topographical,[1-4] mechanical,[5-8] electrical,[9,10] and magnetic[11] properties at the nanometer and micrometer scale by sensing the interaction between the sample surface and a probe that is raster-scanned across the surface. Since its invention by Binnig, Quate, and Gerber in 1986,[1] the AFM has significantly advanced scientific understanding in materials science,[2,3,12-15] biology,[16-28] medicine,[7,29,30] and other areas.[31,32] In particular, its ability to operate on electrically insulating surfaces and in liquid environments has opened new chapters in the histories of these fields. Researchers have used AFM to both image and measure the bond strength,[22] binding energy,[33] and stiffness[5,7,8,29,30] of biological specimens such as proteins, cells, and bacteria in solution conditions that mimic their natural environments. The growth of crystals[15,34-43] has been observed in real time in solution, bringing a new understanding of phenomena at the nanometer and micrometer scale, which was not possible prior to the invention of the AFM.

In this chapter, we introduce the reader to the application of the AFM to investigations of protein assembly,[12,23,24,26,27] biomineralization,[15,35-37,39,40-43] and biomolecular interactions.[14,22,33,44-45] For these phenomena, we will discuss three topics, recombinant human prion protein (HurPrP^C) aggregation,[27] calcium oxalate monohydrate (COM) crystal growth,[40-42] and molecular interactions[33] between amelogenin protein (Amel) and hydroxyapatite (HAP). Human prion protein (HuPrP^C) aggregation is an essential step leading to the occurrence of fatal human neurodegenerative prion diseases; COM is the primary crystal phase comprising human kidney stones; and the interaction between Amel and HAP plays an essential role in the formation of tooth enamel. Relevant background necessary for the reader to understand these topics will be given in each section. Prior to the introduction of these topics, we will also address issues related to instrumentation, interpretation of AFM images, and experimental design. This will give the reader familiarity with the AFM and the data it produces, as well as a better understanding of the topics presented in this chapter.

5.2 Materials and Methods

5.2.1 Operation of AFM

In order to collect an image by AFM, either the scanning tip moves across the surface of the sample or the sample is moved while the tip is kept stationary. In either case, the movement is extremely precise and controlled by a piezoelectric controller. The design of the latter is shown in Fig. 5.1.[46-48] The sample sits on top of a piezo-scanner tube and interacts with a sharp tip at the end of a cantilever, which is fixed to the stationary body of the AFM. Scanning of the sample in X–Y plane is driven by AC voltages applied to the piezo-scanner tube. A laser beam reflects off the cantilever and hits a position-sensitive photo diode detector. As the piezo-scanner tube moves the sample laterally, variations in the interaction force between the tip and the sample surface are detected by displacement of the beam position on the photo diode detector. (Forces used in AFM imaging include contact force, van der Waals force, electrostatic force, capillary force, magnetic force, friction force, surface adhesion and viscoelasticity, etc. The specific forces sensed by the AFM depend on the type of imaging mode.[46-48]) These changes in beam position on the photo diode are recorded as variations in detector output voltage, which in turn are used by the feedback system to move the beam position back to its initial value. This is accomplished by applying a voltage to the Z-piezo element to precisely adjust the distance between the tip and the sample surface. In this way, the interaction force is kept constant during scanning.

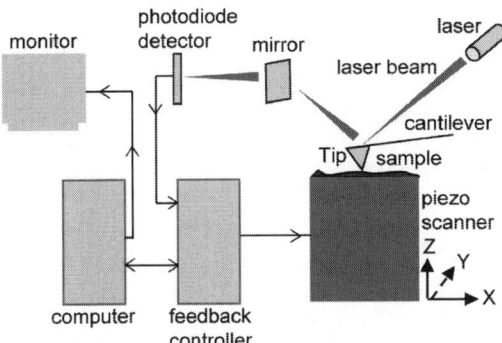

Figure 5.1   Schematic of the AFM setup.

Among various AFM imaging modes, the most widely used are the contact and the tapping modes, which use cantilever deflection and the oscillation amplitude as the feedback signal, respectively. The working principles of contact and tapping mode are explained below.

5.2.1.1 Contact mode

To get the topography of the sample, the tip is gently touched to the surface and raster-scanned across the surface with a chosen deflection (d) of the cantilever, i.e., a chosen beam position on the photodiode detector. The chosen deflection or so-called set point value (voltage) of the deflection corresponds to a fixed contact force (F), because $F = kd$, where k is the spring constant of the cantilever. When the tip encounters a surface feature, an instant change in the beam position on the detector occurs due to the change in the deflection of the cantilever. This deflection is the source of the voltage change used in the feedback loop described above. When the image is made directly from these incoming voltage signals to the feedback system, it is called a deflection image. When the image is obtained from the display of the applied voltage to the Z-piezo versus position in the X–Y plane, it is called a height image and records the topography of the surface.

Deflection and height images each have their utility. Deflection images essentially show the derivative of height with lateral position. Thus, two regions of an image that have the same slope but different heights have the same intensity in a deflection image and gradual changes in height are difficult to observe. On the other hand, sudden changes in height between two regions that are, nonetheless, nearly at the same height are easily seen. Thus, deflection images are especially useful for imaging the surfaces of biominerals where atomic-height steps separate otherwise flat regions. Though the overall height may vary quite a bit from one side of the image to the other, the deflection images bring out the atomic steps in stark relief.

5.2.1.2 Tapping mode

In the tapping mode, a cantilever oscillating near its resonance frequency is scanned across the sample surface with a fixed amplitude of oscillation—the so-called set point amplitude. Typically, at the maximum of the oscillatory cycle, the tip just comes into contact with the surface, though this is not a requirement. Just as the extent of the cantilever deflection in the contact mode is manifest as a voltage value on the detector, so is the cantilever amplitude in the tapping mode. Because the laser beam reflects off the cantilever before hitting the detector, the oscillating cantilever results in an oscillating signal at the detector output. When the tip encounters a surface feature, the resulting change in the amplitude of the cantilever oscillation is sent to the feedback system as an error in the amplitude of the voltage signal, just as the displacement of the beam position in contact mode is sent to the feedback system as an error in the deflection of the voltage signal. The image made directly from these incoming voltage signals to the feedback system is called an amplitude image. The feedback system recovers the signals to the set point voltage or amplitude by using Z-piezo element to adjust the tip–sample separation. The Z-piezo voltages needed to do these processes during the scanning are recorded and used, as in the case in the contact mode, to produce an image of the topography of the sample. This is called a height image.

Tapping mode and contact mode each have their advantages and disadvantages. Tapping mode is typically carried out at a much slower scan rate than contact mode (~1 Hz vs. ~10 Hz), thus reducing its utility for imaging dynamic processes. In addition, on relatively hard materials, contact mode generally gives much higher lateral resolution and thus is often used in studies of biominerals such as COM[40–42] and calcite.[35,36] On the other hand, tapping mode reduces the magnitude of the lateral forces present during contact mode which can cause smearing of the image and even surface damage when soft materials are investigated.[17,49] Indeed, tapping mode has proved to be very successful on soft materials including single adsorbed protein molecules as shown in earlier works of Radmacher et al.[17] and Bezanilla et al.[49] Another advantage of the tapping mode is that it is not sensitive to drift of the cantilever. The cantilevers most often used in the AFM are made of silicon nitride and they bend as a result of temperature changes.[50] This bending is reflected in the cantilever deflection. As explained in Section 5.2.1.1, in contact mode the feedback system tries to keep the deflection constant during the operation; thus, any thermally induced bending will lead to drift in the loading force applied during the imaging.

5.2.1.3 Force spectroscopy

In addition to its the most popular function of producing topographical images, the AFM is also widely used to measure quantitative forces between the sample surface and tip or mechanical properties of samples such as material elasticity[5,7,29,30] and adhesion[14,33,44,45,51] using data from the AFM force curves and theoretical force models.[52-54] This function of the AFM has been widely used to investigate interactions in biological samples[51] such as protein–ligand interactions[22] and molecule–crystal surface interactions.[33,44,45] Here the AFM tip is often modified to have targeted chemical functionalities.

Figure 5.2a shows an example of simple contact AFM force plot using a soft cantilever such as a silicon nitride probe, which is very sensitive to attractive and repulsive forces.[46] The horizontal and vertical axes indicate the Z-piezo movement (i.e., sample position) relative to the cantilever deflection, respectively. The latter is measured by the laser beam position on the photodiode detector. Again, note that contact force can be obtained from the measured cantilever deflection by $F = kd$. Beginning at location 1 in Fig. 5.2a, the Z-piezo starts to extend, however, the sample will not contact the tip until location 3. As soon as the Z-piezo starts to extend beyond position 2, the tip is pulled down by attractive forces near the surface, in the region between locations 2 and 3. This is because the separation of the tip–sample surface is within the attractive region of the intermolecular force curve as indicated in Fig. 5.2b. This phenomenon is often called the "jump to contact" and is usually due to electrostatic attraction and/or capillary forces during operation in air. When operated in liquid, these capillary forces are eliminated or dramatically reduced. As the Z-piezo continues to extend, the tip contacts the sample surface at location 3 and the cantilever bends or deflects upward, as shown by the region between 3 and 4. After reaching the targeted cantilever deflection, the Z-piezo retracts. During the retraction between 4 and 5, the cantilever relaxes downward until the tip forces are in equilibrium with the surface forces. As the Z-piezo continues to retract, the cantilever bends downward as the surface attraction holds onto the tip. This is represented by the region between 5 and 6. This distance is typically larger than that between 2 and 3 because the tip binds to the sample surface due to attractive forces. When the difference in cantilever deflection between 5 and 6 is multiplied by spring constant of the cantilever, the attractive force of the tip–sample interactions is measured. As the Z-piezo continues retracting further, the tip finally breaks free of the surface attraction and the cantilever rebounds sharply upward as shown by the region between 6 and 7. The cantilever then no longer experiences any deflection because the tip is off the sample surface, as shown by the region between 7 and 8. The work (W) associated with breaking the tip–sample bond is obtained by integrating over the shaded area. As may be inferred from the force curve (Fig. 5.2a), during contact mode imaging with constant force or cantilever deflection, operation within the region of "jump to contact" is not stable; thus, it usually requires a set point voltage above the line indicated by "Set point voltage" in Fig. 5.2a.

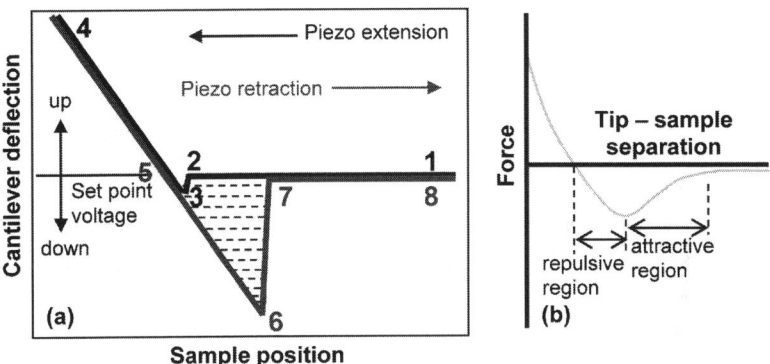

Figure 5.2   Schematic of the contact AFM force plot (a) and potential experienced by a tip approaching the sample surface (b).

5.2.2   Artifacts Observed in AFM Images: Tip Convolution

The morphology of a feature in AFM images is formed by convolution of the shape of the feature and the shape of the tip.[46–48] When the scanned feature is much larger than the radius of the tip, the true morphology of the feature is obtained, except in a few special cases such as deep vertical-wall trenches where the interaction between the vertical wall and the tip edge generates artifacts in the AFM image.[46] However, when the feature is smaller or not much larger than the tip radius, the AFM image is dominated by the shape of the tip and can be quite different from the true morphology. This artifact is called a tip convolution effect.

Figure 5.3a shows how this artifact is produced.[46–48] Let us assume that a tip has a pyramidal shape. When the tip scans a surface feature such as that shown by the filled blue rectangle, the AFM image of the feature is that shown by the filled red object at the bottom of Fig. 5.3a. The shape and lateral dimension of the feature in the image differs from its true shape and lateral dimension and is strongly affected by the pyramidal shape of the tip. This is because the profile is formed by adopting the shape of the tip as shown in the red dotted line at the bottom of Fig. 5.3a. When the feature's true shape is compared with that of the image, one sees that the lateral dimension in the image is larger than that of the feature. However, their heights are the same, because height is free from tip convolution.

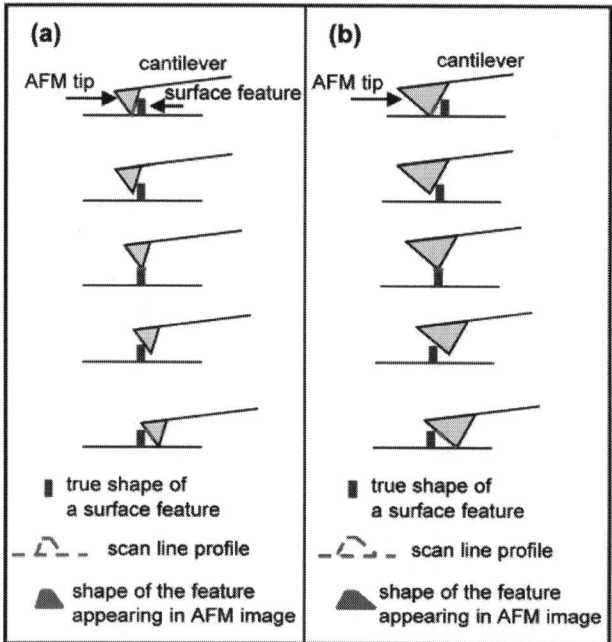

Figure 5.3 Tip convolution effect. A surface feature (blue) is scanned by a tip with a small radius (a) and a larger radius (b). The apparent morphologies of the object (red) in the collected AFM images are compared.

As can be surmised from the above discussion, the observed lateral dimension of any feature is influenced by the size of the tip radius. This is illustrated by Fig. 5.3b. When the same blue rectangle is scanned by a tip with a larger radius than that of the tip used in Fig. 5.3a, the AFM image of the object (drawn as a filled red object in Fig. 5.3b) has a larger lateral dimension. However, regardless of which tip is used, the height of the feature appearing in the AFM image represents the feature.

This tip convolution effect has to be considered when one characterizes small objects such as protein particles adsorbed on a surface such as mica, because they are typically much smaller than the AFM tip radius. As explained above, the lateral dimension of the protein particles obtained from images cannot be used to characterize their size due to tip convolution, but the heights can be. As an example, let us consider Figs. 5.4a,b, which are ex situ AFM images showing HurPrP^C particles adsorbed on mica.[27] (AFM images obtained in air and solution conditions are called ex situ and in situ images.) The tips used for collecting Figs. 5.4a,b were of the same brand from Veeco Probes with a nominal tip radius of less than 10 nm. However, the tip used for collecting Fig. 5.4b had become blunt after extensive use, resulting in a tip radius much larger than that of the fresh tip to collect Fig. 5.4a. Both HurPrP^C monomers and oligomers indicated by green and yellow arrowheads respectively, in Fig. 5.4a have very similar heights to those indicated by the arrowheads in Fig. 5.4b. However, the protein particles in Fig. 5.4b have larger lateral dimensions than those in Fig. 5.4a. For examples, the monomer with a height of 1.4 nm in Fig. 5.4b has a diameter of 26 nm, whereas the monomer with a height of 1.5 nm in Fig. 5.4a has a diameter of only 11 nm. Similarly the diameter with a height of 2.3 nm in Fig. 5.4b has a diameter of 31 nm, whereas the oligomer with the same height of 2.3 nm in Fig. 5.4a has a diameter of only 15 nm. Thus, while the lateral dimension of protein particles in AFM images are highly influenced by the tip radius and cannot be used to characterize particle size, the heights are free of tip convolution and enable one to compare sizes of adsorbed protein particles. In the example shown here, the heights provide information about the extent of HurPrP^C oligomerization described in Section 5.3.

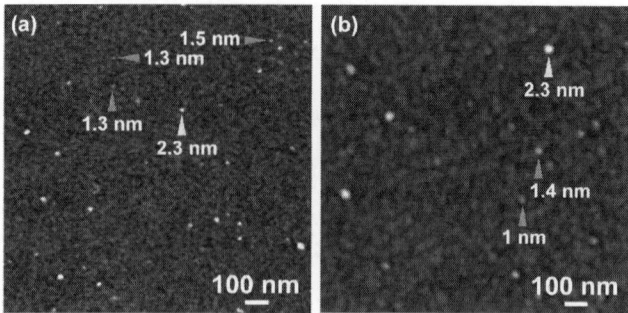

Figure 5.4 Ex situ AFM images of an insertion mutant HurPrP^C (10OR) with additional five octapeptide (PHGGGWGQ) repeats imaged by the tapping mode. The radius of the tip used to obtain the image (a) was smaller that that to obtain the image (b) due to the tip blunting.

5.2.3 Protein Imaging

To obtain ex situ AFM images of proteins such as those in Fig. 5.4, a small aliquot (~10 µL) of protein solution is typically deposited on a prepared substrate such as freshly cleaved mica, left for a few minutes, rinsed with distilled or Milli-Q water, and dried with compressed nitrogen gas. The sample is then imaged with the AFM. Sometimes, it is necessary to alter the surface charge state of the mica from the negative to positive in order to adsorb a negatively charged protein like apoferritin. In this case, ~10 µL Poly-L-lysine is typically deposited on freshly cleaved mica for a few minutes, rinsed with distilled or Milli-Q water, and dried gently with compressed nitrogen gas. Imaging is then carried out using the procedure described.

To obtain in situ (solution) AFM images of proteins, a fluid cell designed specifically for the AFM by the manufacturer is utilized to contain the solution (schematic of the fluid cell is shown in Section 5.2.4). A ~50 µL aliquot of protein solution is introduced into the pre-installed fluid cell using a syringe or pipette. The proteins then adsorb to a substrate such as mica placed inside the fluid cell. Alternatively, an aliquot can be directly deposited on the substrate first and the fluid cell then installed. Occasionally air bubbles can form in the fluid cell during these processes and block the laser light, thus preventing imaging.

5.2.3.1 Determining protein particle size by using calibration curves of protein heights versus molecular weights

AFM measurements are typically conducted with sharp AFM tips having radii of ~5–10 nm. However, as explained above, AFM tip convolution effects still prevent accurate measurements of lateral dimensions of particles below about 10 nm.[27] During early stages of protein aggregation, which are the most important stage for probing underlying mechanisms, oligomers and/or aggregates contain small numbers of monomers.[27] These small sizes prevent lateral dimensions from being used for estimation of their true size. However, the AFM height measurements, which do not suffer from these tip convolution effects and are accurate to ~1 Å, can be used for size estimation.

By measuring heights of proteins with known molecular weights on mica surfaces in air as well as solution environments that mimicked physiological conditions and by combining these measurements with the hydrodynamic diameters of various proteins,[55–59] Cho et al. established a calibration curve of protein height versus molecular weight.[27] This approach was adopted from gel electrophoresis where marker proteins are used to estimate the mass of unknown proteins and protein fragments, as well as the work of Wilkins et al.[57] showing that hydrodynamic diameters of proteins scale with the number of amino acid residues for proteins ranging in size from 58 to 760 amino acids. Figures 5.5 and 5.6 show AFM height images of five marker proteins with different molecular weights obtained in air (ex situ) and solutions (in situ), respectively. They include an insertion mutant of HurPrP^C (referred to as 10OR as discussed in Section 5.3, Figs. 5.5a, and 5.6a, molecular weight (MW): 26.8 KDa) in the non-denatured monomeric state, bovine serum albumin (BSA, Figs. 5.5b and 5.6b, MW: 66 KDa), immunoglobulin G (IgG, Figs. 5.5c and 5.6c, MW: 150 KDa), mycocerosic acid synthase (MAS, Figs. 5.5d and 5.6d, MW: 224 KDa) and apoferritin (Figs. 5.5e,f and 5.6e,f, MW: 481.2 KDa).[27] Note that high-resolution AFM amplitude images, Figs. 5.5f and 5.6f show the typical structure of the apoferritin monomer with its internal cavity (see particles indicated by circles). For each protein, histograms of particle heights were constructed (for example, see

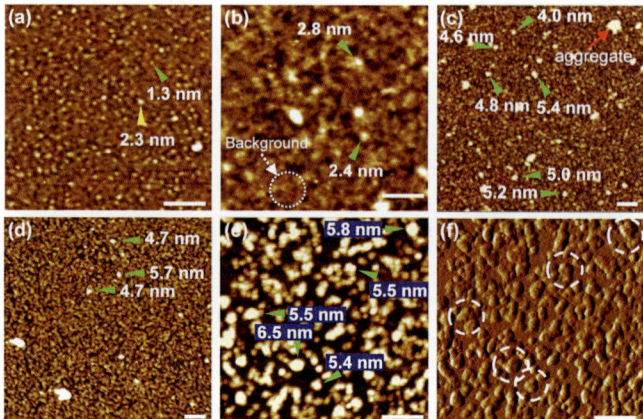

Figure 5.5 Ex situ AFM images of marker proteins with known molecular weights. (a) 10OR (26.8 KDa), (b) bovine serum albumin (BSA, 66 KDa), (c) human immunoglobulin G (IgG, 150 KDa), (d) mycocerosic acid synthase (MAS, 224 KDa), and (e, f) apoferritin monomers (481.2 KDa). Numbers next to arrowheads in (a–e) give heights of adjacent protein particles where green = monomers and yellow = oligomers. Images (a–e) are height images and image (f) is an amplitude image. Scale bars, 100 nm.

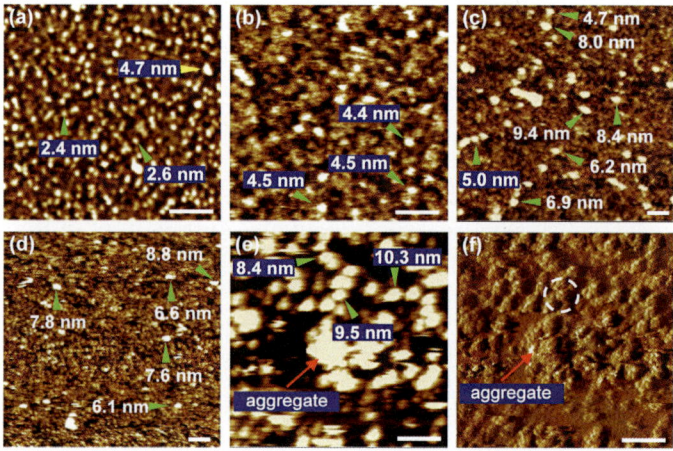

Figure 5.6 In situ AFM images of marker proteins shown in Fig. 5.5 (a) 10OR, (b) BSA, (c) IgG, (d) MAS, and (e, f) apoferritin. Images (a–e) are height images and image (f) is an amplitude image. Numbers next to arrowheads are heights of adjacent particles with color-coding as in Fig. 5.5. Scale bars, 100 nm.

Fig. 5.7a) and the average height was determined. Figure 5.8 shows the dependence of measured protein height on molecular weight for both the in situ and ex situ measurements, as well as the hydrodynamic diameter versus molecular weight.[27] Over the investigated range of ~25–150 kDa, the protein height calibration curves correlate well with the known dependence of hydrodynamic diameter on molecular weight. Thus, these curves can be used to directly estimate these two quantities for globular proteins of unknown weight in the range of this calibration. In addition, for protein whose molecular weight is known, the curves can be used to approximately estimate the number of protein monomers in its oligomer even over the range beyond ~150 KDa.

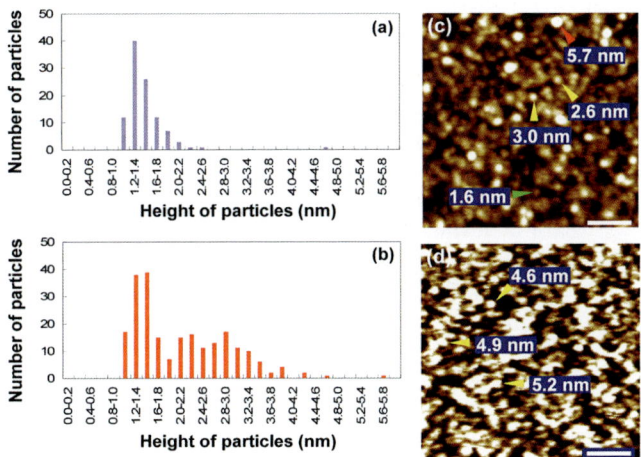

Figure 5.7 The height distribution of 10OR in native-like solution (20 mM NaOAc) and partially denaturing solution conditions (~50 mM NaOAc, ~150 mM NaCl, 0.5 M guanidine hydrochloride (GdnHCl), pH ~4.0). Histogram (a) was obtained from an ex situ image from a native-like condition (Fig. 5.5a) and (b) was from two ex situ images obtained immediately after preparation of partially denaturing solution (at $t = 0^+$ min). Upon partial denaturation, 10OR oligomers with height of ~2–3 nm begin to form in majority, making a bimodal distribution, as shown in (b). Image (c) is an ex situ image used for (b) and (d) is an in situ image at $t = 0^+$ min. Numbers next to arrowheads in (c, d) are heights of adjacent particles where green = monomers, yellow = oligomers and red = β-oligomer. Height measurements were carried out with section analysis of standard Veeco Nanoscope image analysis software. Scale bars, 100 nm.

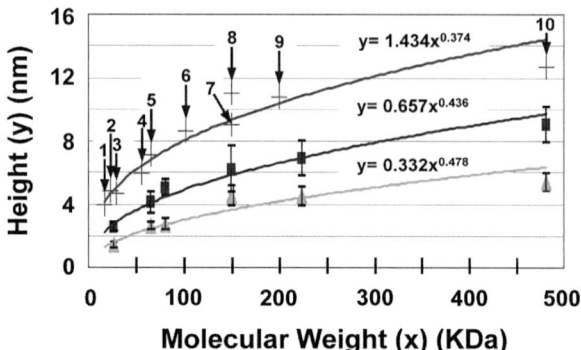

Figure 5.8   Calibration curves showing heights of marker proteins versus their molecular weights. Heights were measured from the surface of mica to the top of protein particles. Light blue triangles—ex situ curve where, from left to right, points are for: 10OR monomers (26.8 KDa), BSA monomers (66 KDa), majority of newly formed 10OR oligomers (seeds) at $t = 0^+$ min such as those marked with yellow arrowheads in Figs. 5.7c,d assuming they are trimers (80.4 KDa), IgG monomers (150 KDa), MAS monomers (224 KDa) and Apoferritin monomers (481.2 KDa). Dark blue squares—in situ curve where order left to right is the same as for ex situ curve. Green crosses—hydrodynamic diameters. Numbers indicate: 1, myoglobin monomer (16.9 KDa)[55]; 2, WT monomer (22.8 KDa)[56]; 3, carbonic anhydrase monomer (29 KDa)[55]; 4, yeast triosephosphate isomerase dimer (56 KDa)[57]; 5, BSA monomer[55]; 6, Hexokinase monomer (102 KDa) (Malvern Instruments); 7, alcohol dehydrogenase dimer (150 KDa)[55]; 8, IgG monomer[58]; 9, β-amylase monomer (200 KDa)[55]; 10, Apoferritin monomer.[59] Error bars represent ± one standard deviation. Reprinted with permission from ref. 27. Copyright (2011) American Chemical Society.

5.2.4   Experimental Design of in situ AFM Investigation of Macromolecular and Biomineral Crystal Growth

Figure 5.9 shows a schematic of an experimental setup for in situ AFM investigation of solution crystal growth. A seed crystal is either glued on a glass cover-slip or is directly nucleated and immobilized on a substrate that has been glued to glass cover-slip. Alternatively, when large, high-quality crystals (e.g., calcite) that cleave well are available, the cleaved crystal is directly glued to the specimen disk, eliminating the need for the cover-slip. Over the years we have found that urethane glues work well with aqueous solutions. Next, this seed crystal is placed within the fluid cell, which has both an inlet and outlet port and makes a seal with the glass cover-slip via a silicone O-ring (see schematic of the fluid cell in Fig. 5.9). To obtain quantitative information from these experiments, important variables such as supersaturation, temperature, and pH must be controlled. To achieve this control, the growth solution is held at fixed composition within a well-mixed and temperature-controlled container with active control over pH or CO_2 content as shown in Fig. 5.9a (if an experiment is conducted at room temperature, the temperature controller is not used). The flow rate of the solution is controlled by a peristaltic pump that is placed between the container and the AFM. Alternatively, a syringe pump can be used instead of using both the peristaltic pump and solution container, in cases where the continuous control over pH or CO_2 content is not needed and the solution can simply be kept at room temperature. A chamber containing an air pocket inserted in the flow line serves to damp out any vibrations from the peristaltic pump, and an in-line filter (not shown) can also be placed to remove any particles introduced during preparation of the solution or transfer to the container. (When a syringe pump is used, a vibration damper may not be needed.) Before passing into the fluid cell, the solution flows through the meander of a heat exchanger where temperature is controlled by using a thermoelectric element. The heat is taken away from or added to the back-side of the Peltier device by flowing chilled or heated water through the copper block of the device. The control thermocouple is inserted into the fluid tubing within the heat exchanger and a measurement thermocouple is placed into the outlet port of the fluid cell. The resulting system has good temperature control of about 0.1°C at the cell.

To obtain quantitative data on growth rates, flow rates are increased until the growth rate of the crystal surface is no longer a function of flow rate. However, for very slowly growing crystals, it may not be necessary to flow solution because the supersaturation drop inside the fluid cell is negligible over the time scale of the experiment. Control over supersaturation is achieved in one of two ways: holding the temperature fixed and changing the solute content of solution, or varying the temperature at fixed solute content. The latter is applicable only to systems in which the solubility depends moderately on temperature.

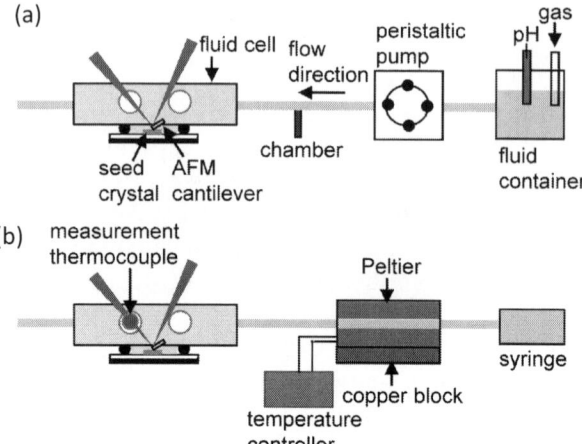

Figure 5.9 Schematic of an experimental setup for in situ AFM investigation of solution crystal growth using flow systems. Either peristaltic pump (a) or syringe (b) can be utilized to flow solution and Peltier device can be used to control temperature in either system.

Most AFM studies of solution crystal growth have been performed in contact mode, which generally provides higher resolution than tapping mode. However, tapping mode has also been used, primarily for imaging biological materials such as protein and virus crystals.[60,61] As described in Section 5.2.1.2, tapping mode reduces the lateral forces usually present in contact mode; thus, it can prevent or reduce alterations in the details of surface structure and small adsorbates.

5.3 In vitro Investigation of the Pathway for Recombinant Human Prion Protein Aggregation

An essential phenomenon associated with prion diseases including Creutzfeldt–Jakob disease (CJD) is the conversion of the normal monomeric α-helix–rich prion protein (PrP^C) into the infectious and pathogenic β-sheet–rich oligomeric scrapie prion protein (PrP^{Sc}) and subsequent aggregation.[62–69] Prion diseases can occur as infectious, sporadic, and familial diseases, all sharing the same pathogenic mechanism which involves the conversion of PrP^C into

PrP^{Sc}.[64,66] A number of recent results suggest that smaller oligomeric aggregates, rather than the much larger amyloid fibrils, are critical in the pathogenesis.[69] Recombinant prion protein ($rPrP^{C}$) has been extensively used to investigate the aggregation process in vitro in order to obtain insight into the in vivo formation of β-sheet–rich PrP^{Sc} oligomers and aggregates.[27,70–78] Oligomers formed from the aggregation of initial α-helix–rich $rPrP^{C}$ in partially denaturing solution conditions have been shown by Fourier-transform infrared (FTIR) spectroscopy to have an extended anti-parallel β-sheet structure and termed "β-oligomers."[74] These are octameric or larger and the most representative β-sheet–rich isoforms produced during in vitro experiments with $rPrP^{C}$.

Wild-type (WT) human recombinant prion protein ($HurPrP^{C}$) contains four octapeptide (PHGGGWGQ) repeat sequences. Two insertion $HurPrP^{C}$ mutants, 8OR and 10OR, which contain three and five additional repeats, respectively, are associated with familial prion disease. A recent AFM study[27] by Cho et al. revealed identical formation pathways for these β-oligomers of WT, 8OR and 10OR and subsequent non-fibrillar aggregates at the molecular level by using the calibration curves of protein height versus molecular weight obtained from the AFM height measurements as described in Section 5.2.3.1 (see refs. 70 and 71 for details of generation and purification of bacterially produced WT, 8OR and 10OR).

5.3.1   Seed Formation of HurPrP^{C}

When $HurPrP^{C}$ is introduced into partially denaturing acidic solution conditions with moderate additions of denaturants such as GdnHCl and Urea, oligomers and aggregates form through conversion of their α-helix–rich into β-sheet–rich structure.[27,72–76] Figure 5.10 shows that oligomer and aggregate formation of WT and 10OR occurred when they were introduced into GdnHCl-containing acidic solution.[27] Prior to aggregation ($t = 0^{-}$ min), WT and 10OR are shown to exist in the non-denaturing (20 mM NaOAc) solution mostly as monomers with heights of around 1.5 nm ex situ (Figs. 5.10a,b) and 2–3 nm in situ (Fig. 5.6a). As soon as they are introduced into the GdnHCl solution ($t = 0^{+}$ min), small oligomers begin to form with highly uniform heights in the range of ~2–3 nm ex situ (Figs. 5.10d,e) and ~4.5–5.5 nm in situ (Fig. 5.7d). As can be seen from the calibration curves in Fig. 5.8 (third points from left), the oligomers are found to be most likely trimers or possibly tetramers (MWs are as follows: WT trimer, 68.4 KDa; 10OR trimer, 80.4 KDa; WT tetramer, 91.2 KDa; 10OR tetramer, 107.2 KDa).

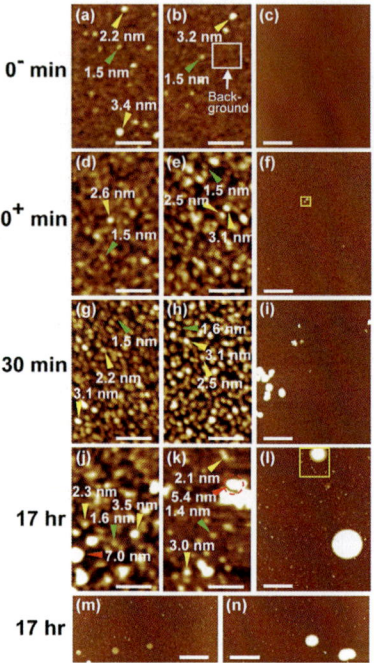

Figure 5.10 The formation of oligomers and non-fibrillar aggregates. (a–l) Ex situ images of WT and 10OR in native like solution (a–c) and partially denaturing solution conditions (d–l). (a) WT and (b, c) 10OR. (d) WT and (e, f) 10OR at $t = 0^+$ min. (g) WT and (h, i) 10OR at $t = 30$ min. (j) WT and (k, l) 10OR at $t = 17$ h. (m, n) Ex situ images of WT (m) and 8OR (n) at $t = 17$ h in the partially denaturing solution condition. Numbers next to arrowheads are heights of adjacent particles with color-coding as in Fig. 5.7. Scale bars are 100 nm for (a–b, d–e, g–h, j–k) and 2 μm for (c, f, i, l, m–n). Reprinted with permission from ref. 27. Copyright (2011) American Chemical Society.

These trimers and/or tetramers represent what have been interpreted as seeds or nuclei for prion aggregation.[79,80] They are the smallest oligomer that has overcome the energy barrier associated with entropy loss[79–82] due to oligomerizaton, thus being able to proceed to further growth rather than dissociation. However, they are not nuclei as defined in the classical sense. Unlike the critical nucleus whose size depends on concentration (i.e., supersaturation) and undergoes further growth by subsequent monomer attachment, the size of these seeds does not depend on concentration and, as will be shown in Section 5.3.2, they do not undergo further growth by monomer attachment.[27] This AFM result proving a seed size of trimer and/or tetramer is consistent with other experimental studies. The analysis from the turbidity curves of rPrP^C aggregation indicates trimers and/or tetramers as the seed size.[27,75] Electron microscopy and simulation studies on two-dimensional crystals of truncated PrP^{Sc} (PrP$^{Sc\ 27\text{-}30}$) formed during the in vitro purification of infected tissues show a trimeric assembly of PrP$^{Sc\ 27\text{-}30}$ as the basic unit.[65,82] Another study suggested that the smallest infectious PrP^{Sc} particle contains three or less prion protein monomers.[62] These results all correlate well with the AFM observation of trimeric or tetrameric seeds as the smallest stable unit of the β-sheet form of HurPrP^C.

5.3.2   Growth of β-Oligomers and Non-Fibrillar Aggregates

Immediately after preparation of partially denaturing solution conditions (at $t = 0^+$ min), larger oligomers—so-called β-oligomers[72-76] —form in addition to the seeds, in both WT and 10OR solutions (Fig. 5.7c, red arrowhead). They also increase in number with longer incubation time (Figs. 5.10j,k, red arrowheads).[27] These β-oligomers have heights mostly in the range of ~5 to ~7 nm in ex situ AFM images and seldom grow beyond 8 nm at 20 μM HurPrP^C concentration, with their average size scaling with HurPrP^C concentration. According to the calibration curve (Fig. 5.8), this height corresponds to oligomers containing ~11–22 monomers, or ~3–7 seeds. The size of these β-oligomers is very similar to that of the most infectious PrP^{Sc} particles,[68] which were reported to contain about 14–28 prion protein monomers, corresponding to about 4 to 9 of the seeds described here.

While fibrillar protein aggregates have usually been considered to grow by monomer attachment to seeds,[73,75,76,79,80] the AFM observations including the analysis of the number of oligomers of different sizes versus incubation time showed that β-oligomers do not form through addition of monomers to seeds.[27] Rather, as documented in the series of AFM images in Figs. 5.11a–e and shown

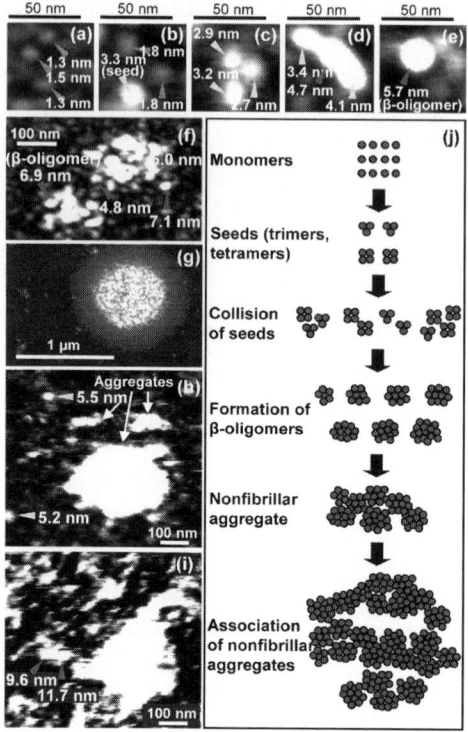

Figure 5.11 Pathway leading to formation of β-oligomers and subsequent non-fibrillar aggregates. (a–g) Ex situ images of 10OR showing: (a) monomers; (b) monomers and a small oligomer (seed); (c) collision of three seeds; (e) a β-oligomer; (f) non-fibrillar aggregates at early stage of formation showing that they are comprised of β-oligomers (higher resolution image of yellow rectangle in Fig. 5.10f); and (g) large aggregate formed by direct association of small, irregular non-fibrillar aggregates like those in f (higher resolution image of yellow rectangle in Fig. 5.10l). (h and i) Sequential in situ images captured in the same spot on mica showing small oligomers (seeds) and aggregates composed of β-oligomers. Following collection of (h), freshly mixed 10OR solution (partially denaturing) was added and image (i) was collected. It shows growth of the aggregates in (h) as well as a newly formed aggregate composed of two β-oligomers with heights 9.6 nm and 11.7 nm. (j) Schematic of observed pathway to formation of β-oligomers and subsequent non-fibrillar aggregates illustrating multiple stages of association by stable oligomeric intermediates. Reprinted with permission from ref. 27. Copyright (2011) American Chemical Society.

schematically in Fig. 5.11j, they are created primarily by direct interaction and coalescence of trimeric or tetrameric seeds with the formation pathway described as follows: Monomers (Fig. 5.11a) combine to form trimeric or tetrameric seeds (Fig. 5.11b). Then, instead of monomer attachment to the seeds for the subsequent seed growth, the seeds collide (Fig. 5.11c) and coalesce to create the β-oligomers (Fig. 5.11e).

A range of β-oligomer sizes are created depending on how many seeds coalesce. However, as mentioned, nearly all of the β-oligomers are comprised of approximately 11 to 22 monomeric units or ~3 to 7 seeds regardless of incubation time at the 20 μM HurPrP^C concentration used by Cho et al.[27] This suggests that their size is thermodynamically limited, perhaps by strain[83] which should be generated when a number of seeds coalesce into one larger oligomer.

The β-oligomer formation mechanism revealed by the AFM work of Cho et al. is consistent with that proposed by Sokolowski et al.[74] based on their FTIR measurements. Moreover, this mechanism is expected to extend beyond the growth of the β-oligomers, because Grégoire et al.[84] also proposed that protofibrils of Lithostathine grow by tetramer attachment to the ends (see Fig. 5.8 in ref. 84). Once β-oligomers are formed, they have strong tendency to be closely spaced to make non-fibrillar aggregate structures as shown in Fig. 5.11f. These then directly associate to form the largest aggregates (Fig. 5.11g). Figures 5.11h–i show sequential in situ AFM images documenting growth of β-oligomers and aggregates captured in the same location on mica. Following collection of Fig. 5.11h, freshly prepared partially denaturing 10OR solution was added and image Fig. 5.11i was obtained. The images show growth of the aggregates, as well as a newly formed aggregate composed of two β-oligomers with heights 9.6 nm and 11.7 nm.

This work highlights the utility of the AFM-based approach to probing aggregate formation. As shown in Fig. 5.11f, the β-oligomers are closely spaced within the aggregates. In this case, using more common light scattering methods to obtain hydrodynamic diameters of protein aggregates cannot give a true representation of aggregate structure because they can not resolve individual oligomers within the aggregates. In this situation, AFM imaging with the calibration curve of Fig. 5.8 is more useful because it enables one to resolve the aggregate structure and provide the approximate hydrodynamic diameters of individual oligomers comprising the aggregates.

5.4 Inhibition of the Calcium Oxalate Monohydrate Crystal Growth by Biomolecules

Human kidney stones are typically aggregates of small crystals containing matrices of proteins, carbohydrates, and lipids and often found attached to epithelial cells.[85-89] The primary crystal phase comprising human kidney stones is calcium oxalate monohydrate (COM). Biomolecules with acidic functional groups such as citrate[90-93] and osteopontin protein (OPN)[94-100] contained in normal urine are found to inhibit COM crystal formation from numerous in vitro experiments, suggesting that they may play a pivotal role in human kidney stone formation as potent urinary inhibitors. Others, such as Tamm–Horsfall protein (THP)[91,92,101-103] exist in high concentrations in urine but show little inhibition in vitro at physiological levels. The AFM has been successfully used to observe the way that these biomolecules modify COM crystal growth at the molecular level, producing new insight into their specific interactions that regulate kidney stone formation.

In this section, first the bulk crystal morphology of the in vitro grown COM crystal and the structure of the growth sources on its different faces observed by in situ AFM will be introduced. Then in situ AFM results showing the modulation of the COM growth at the molecular level by biomolecules such as citrate, OPN, and THP will be discussed.

5.4.1 Equilibrium Shape of Pure COM Crystal and Its Growth

As Fig. 5.12a shows, COM crystals typically have hexagonal crystal shapes with mirror symmetry (space group: $P2_1/n$[104]) and express three major families of crystallographically distinct faces, i.e., the {-101}, {010}, and {120}, although under certain conditions others are also possible.[105-108] (The miller indices (hkl) denote a single plane and {hkl} represent equivalent planes by symmetry. [UVW] specify a unique vector direction, and <UVW> indicate a set of vectors equivalent by symmetry.)

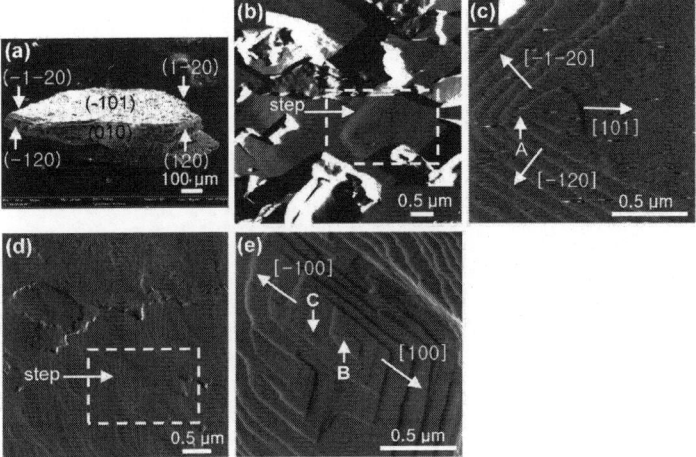

Figure 5.12 COM crystal morphology and dislocation hillock shapes. (a) A scanning electron microscopy (SEM) image of a synthetic COM crystal grown in vitro by gel method.[105] The in vitro grown COM often has a twinned structure, as shown here. (b–e) In situ AFM images of the surfaces of COM on (–101) (b, c) and (010) faces (d, e).

At low to moderate superstations, the source of growth on each COM face consists of a spiral of single molecule-high steps called a growth hillock, which is generated by defects known as screw dislocations inherent in the crystals.[39,108] Growth of each face proceeds by the advance of these steps. As Figs. 5.12b–e show, the continuous generation and growth of steps produces the hillock structures on COM (see dotted boxes in Figs. 5.12b,d). Figures 5.12c,e are higher resolution images of hillocks such as those seen in Figs. 5.12b,d and show approximate locations (A–C) of dislocation outcrops from which steps are generated. Steps on the (–101) face (Figs. 5.12b–c) form hillocks with triangular shapes and the [101] step has been observed by AFM measurements[40] to move ~12 times faster than the [–1–20] and [–120] steps, which are structurally identical and possess mirror symmetry. ([101] lateral step speed is (~23 nm/s) at supersaturation (σ) of 0.91 and its speed as a function of σ is given in ref. 41.) On the (010) face, as shown in Figs. 5.12d–e, steps form hillocks in the form of parallelograms. Steps on this face grow slowly compared to the [101] step on the (–101) face. For example, the lateral step speed along <100> was 1.2 nm/s at σ = 0.82.[109] Further details of steps in relation to the equilibrium hexagonal crystal shape are described in refs. 40 and 109.

5.4.2 Inhibition of Calcium Oxalate Monohydrate Crystal Growth by Biomolecules at the Nanometer and Micrometer Scales

5.4.2.1 Citrate

Many clinical studies have shown that urinary citrate plays an important role in renal stone disease.[93] For example, a deficiency of citrate secreted into urine caused by renal and gastrointestinal disorders is often observed in patients with stone disease and oral citrate therapy is an effective treatment for preventing recurrence of stone disease.[93,108] Citrate has direct inhibitory effects on COM crystallization phenomena that include nucleation, growth, and aggregation of COM crystals.[90,92]

Now let us see the effect of citrate on the growth of COM crystals observed in real time by in situ AFM. Figure 5.13 shows the effect of citrate (12 μM) on growth hillocks of the (−101) face, as well as the bulk crystal habit.[40] The growth hillock morphology markedly changes with increasing time during growth in the presence of citrate, as shown in Figs. 5.13a–d. This is because citrate binds to the step edges and blocks incorporation of solute ions, thus pinning the steps.[41] As the citrate concentration is increased, step-pinning occurs more rapidly and extensively, leading to more inhibition of the step growth and greater effects on hillock morphology.[40,41] As shown in Figs. 5.13a–d, the effect is especially pronounced on the fast-growing [101] step, with its step speed being reduced by a factor of 25. Although the two <120> steps are also roughened and slowed, the effect is relatively minor with their speed dropping only by a factor of 2. As a result, at this citrate level, the step speed becomes nearly independent of orientation, resulting in a disk-shaped hillock morphology (Figs. 5.13c,d).

Weaver et al. examined inhibition of the growth of the fast-growing [101] step as a function of citrate concentration at various calcium activities, i.e., supersaturations using in situ AFM.[41] Their results are shown in Fig. 5.14. Here, the effect of citrate concentration on inhibition is demonstrated by using relative step velocity (V/V_0), which is given by the step velocity (V) in solution containing citrate over that from pure solution (V_0). As shown,

V/V_0 decreases with increasing citrate concentration, dropping rapidly at low concentrations and leveling off starting at ~2 μM concentration, which is far below physiological values[110] for all investigated supersaturations. This demonstrates citrate's great effect on [101] step growth. In contrast to its effect on the steps of the (−101) face, citrate showed no significant effect on either morphology or step speed on the (010) face, indicating that the citrate-step interactions are much weaker on this face.[40] Molecular modeling conducted to understand the source of this highly specific interaction showed that binding of citrate to steps on the (010) face is energetically much less favorable than those on the (−101) face due to the details of the stereochemical relationships between citrate and the steps on the two faces.[40] The effect of citrate on the growth hillocks observed by in situ AFM is directly reflected in the macroscopic growth habit of COM as shown in Fig. 5.13f.[111,112]

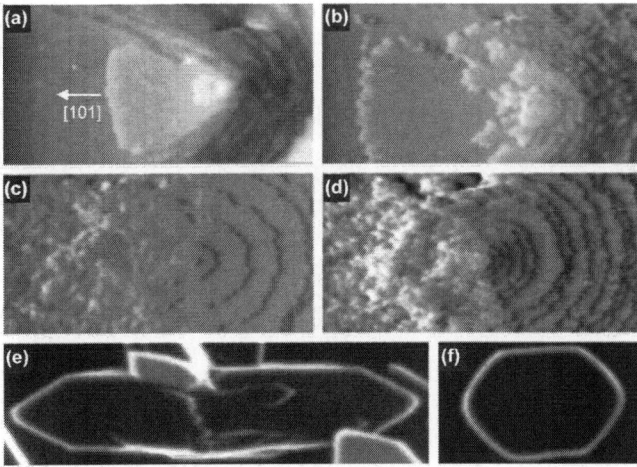

Figure 5.13 Representative images (deflection mode) showing the effect of citrate on COM morphology. (a–d) Sequential AFM images during growth in citrate (12 μM)-bearing solution (supersaturation: 0.7; accounted for the citrate effect, citrate-to calcium ratio, 0.1) at t = 3.5 min (a), t =11 min (b), t = 79 min (c), and t = 103 min (d). Shape of hillock in pure solution is shown in Fig. 5.12c. SEM images of COM crystals grown in the absence (e) and presence (f) of citrate in solution. Citrate-to-calcium ratio is 0.05. Image horizontal dimensions are as follows: 1.5 μm (a and d), 1 μm (b and c), 100 μm (e), and 25 μm (f). Images reproduced, with permission from ref. 40. Copyright (2004) National Academy of Sciences, U.S.A.

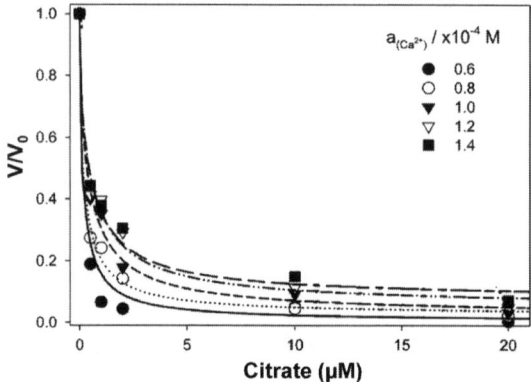

Figure 5.14 Dependence of V/V_0 on citrate concentration at various calcium activities. Reprinted from ref. 41, Copyright (2007), with permission from Elsevier.

5.4.2.2 Osteopontin

Osteopontin (OPN) is a urinary protein abundant in glutamic- and aspartic acid. It has a peptide sequence identical to that of OPN from human bone.[113] OPN has posttranslational modifications such as phosphorylation,[114–116] glycosylation,[113–115] and sulfation.[116] These modifications are considered to play an important role in regulating stone formation, adding to the role of OPN's inherent acidic residues. Human urinary OPN inhibits COM nucleation,[95] growth,[97] and attachment of COM crystals to renal epithelial cells.[86] It shifts the predominant crystalline form of calcium oxalate from COM to calcium oxalate dihydrate (COD), which is considered to be protective since COD crystals show less binding than COM crystals to renal cells.[87] Strong evidence for a pivotal role for urinary proteins such as OPN and THP in stone disease has been provided by recent gene ablation investigations.[98,103,108] In vivo studies conducted in mice with disruptions of the genes encoding OPN[98] and THP[103] showed that renal calcium oxalate deposited in the mice with deficiency of these proteins, but not in normal mice, demonstrating the important role of urinary acidic proteins in the regulation of stone formation.

Now let us see in situ AFM results showing the effect of OPN on the growth of COM. Figure 5.15 shows the effect of OPN on growth of hillocks on the (010) and (–101) faces of COM.[40] As shown in Figs. 5.15a–d, modification of the growth hillocks on the (010) face proceeds with time in the presence of a mere 5 nM OPN. All steps become strongly pinned and lose lateral stability with increasing time, with step speeds dropping by an order of magnitude. Similar effects were observed to occur at other OPN levels (1–25 nM), but developed more rapidly and were greater at higher OPN concentrations. In contrast to its effect on the (010) face, OPN did not alter either step speeds or the step morphology of hillocks on the (–101) face, as shown in Figs. 5.15e,f. Discrete adsorbates are seen on the (–101) terraces at all OPN concentrations (1–25 nM) with their number increasing but their dimensions not changing over time. Sequential images (Figs. 5.15g–i) show that steps freely pass under the adsorbates (circled) and once the steps pass by them, they again adsorb on the terrace. This indicates that OPN molecules adsorb weakly to the terraces and interact weakly with the steps on the (–101) face, resulting in no changes in step kinetics or morphology. Thus, these AFM findings show that OPN has highly face- and step-specific interactions, just like citrate.

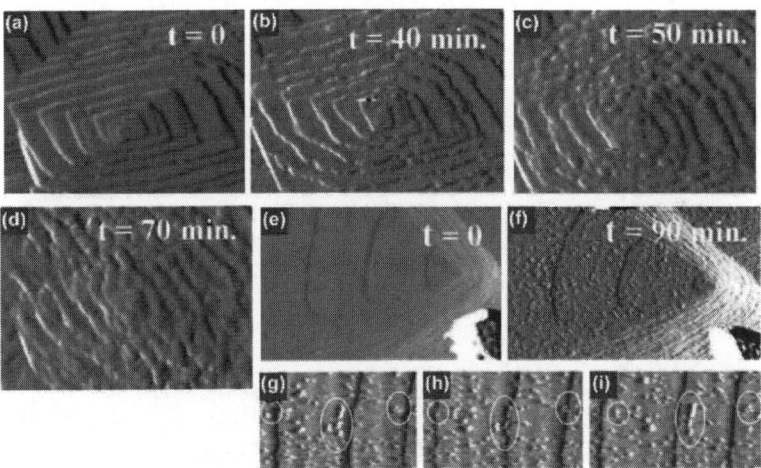

Figure 5.15   AFM images showing the effect of OPN on COM growth hillock morphology. (a–d) Sequential images (horizontal dimension, 1.7 μm) showing OPN modification of the growth hillocks on the (010) face by 5 nM OPN. (e and f) Sequential images (horizontal dimension, 6.0 and 4.5 μm) showing the effect of OPN on growth hillock morphology of the (–101) face. (g–i) Temporal evolution of step train and protein adsorbates on the (–101) face (horizontal dimension, 2 μm). Images reproduced, with permission from ref. 40. Copyright (2004) National Academy of Sciences, U.S.A.

5.4.2.3 Tamm–Horsfall Protein

Tamm–Horsfall Protein (THP) is the most commonly found protein in human urine. It has been shown to be a potent inhibitor of COM crystal aggregation, while having much less effect on crystal growth.[102,108,117] THP, along with OPN is found to readily attach to the COM surface[85] and this affinity is considered to play a role in inhibiting COM aggregation or attachment to cell membranes.

Now let us look at in situ AFM images showing COM growth in THP-containing supersaturated solution. Figure 5.16 shows the evolution of hillock morphology on the (–101) face of COM after exposure to 100 nM immunoaffinity-purified THP.[42] As clearly seen in Figs. 5.16c,d, THP adsorbs as aggregates onto the surface. Typically, they have an irregular filamentary shape with a dominant orientation along the <010> direction. Although the surface is fully covered with THP, the triangular hillock shape is still preserved. Steps also advance with no obvious change in speed, up until the time when the steps are no longer visible due to the adsorption, suggesting that THP has little effect on the growth of this face.

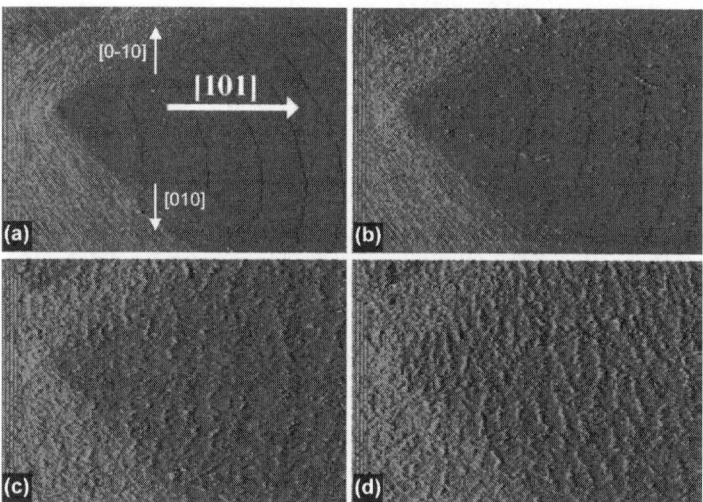

Figure 5.16 Growth hillock on the (–101) face of COM after (a) 0 min, (b) 1.5 min, (c) 3 min, and (d) 4.5 min of exposure to 100 nM immunoaffinity-purified THP. Horizontal dimension of images = 3.5 μm. Images reproduced from ref. 42, Copyright (2009), with kind permission from Springer Science + Business Media.

Figure 5.17 shows THP's effect on the growth of hillocks on the (010) face.[42] The (010) face was exposed to THP at various concentrations (0–75 nM THP) for 1 h. As shown, aggregates like those seen on the (–101) face do not form on this surface. However, as THP concentration increases, the hillock exhibits changes in its overall morphology, with individual steps becoming less recognizable due to the formation of a thin film of THP, and step spacing in <100> directions becoming narrower due to a slight decrease of lateral growth speed along these directions (Figs. 5.17c,d). This AFM result on the micrometer scale is commensurate with observations from bulk experiments that THP has less effect on COM crystal growth.[102] However, although the film of THP has little effect on growth, it may be responsible for the observed inhibitory effect on aggregation[117] and prevention of COM attachment to cell surfaces.[88,89,102]

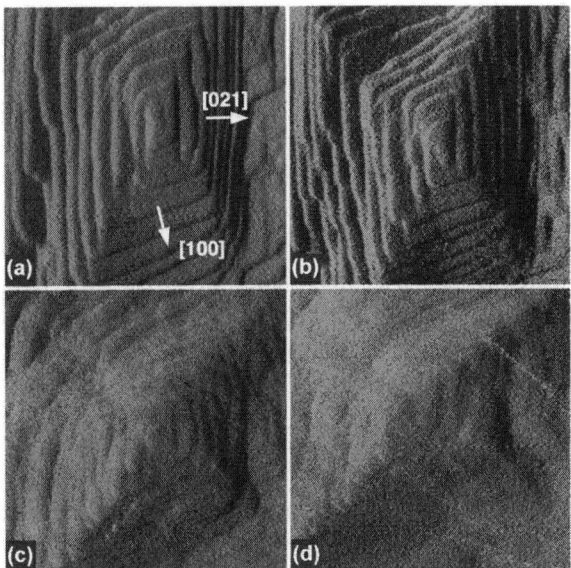

Figure 5. 17 Growth hillock on the (010) face of COM after 1 h of exposure to 0 nM (a), 10 nM (b), 25 nM (c), and 75 nM immunoaffinity-purified THP (d). As THP concentration increases, step edges become difficult to resolve. The loss of resolution does not prevent measurement of step speed. Horizontal dimension of images = 1 μm. Images reproduced from ref. 42, Copyright (2009), with kind permission from Springer Science+Business Media.

5.5 Free Energy of Amelogenin Protein C-Terminal Fragment Binding to Hydroxyapatite

During biomineralization, such as bone and tooth formation, organized protein matrices and non-matrix proteins or other molecules direct formation of mineral components.[118-122] Although interactions between proteins and the mineral components are still not known in great detail, the adhesion force and binding energy between the specific functional group of molecules and crystal surfaces have been informed by AFM force measurements for biominerals such as COM,[44,45] introduced in Section 5.4, and HAP,[33] the mineral phase in tooth enamel and bone. During formation of HAP in tooth enamel, amelogenin (Amel), especially its C-terminal region, is believed to play an essential role in directing mineral growth through the interaction between Amel and the growing HAP crystallites.[123-127]

In the last part of this chapter, we will describe an application of the AFM force measurements to investigation of face-specific binding energy of porcine Amel C-terminal fragment (AmelC) to the HAP crystal surface.[33]

Gold-coated Si_3N_4 AFM tips were functionalized with AmelC (sequence: W PATDKTKREE VD) by using a heterobifunctional cross-linker LC-SPDP as shown in Fig. 5.18.[33] LC-SPDP contains a pyridyl disulfide, which adsorbs to gold, and a Nhydroxysuccinimide (NHS) ester that reacts with the N-terminal amine or lysine residues of the AmelC to form an amide bond. In this way the AmelC was attached to the AFM tip. This functionalized tip was brought to the (100) face of HAP in calcium phosphate solution at approximately equilibrium saturation with the crystal in order to obtain force–distance curves associated with the AmelC binding to the (100) face (see Fig. 5.19a for SEM image of the HAP crystal).

Figure 5.18b shows representative force–distance curves obtained during retraction of the Z-piezo for each level of chemical functionalization. As shown in the graphs, AmelC-functionalized tips have much stronger adhesion force (~200 pN) to the (100) HAP than the other two, indicating a great affinity of AmelC for the crystal surface. The hatched region is the work, W, required to move the cantilever tip from the HAP surface to the minimum of the pulling potential. This corresponds to the work associated with breaking the AmelC-HAP bond. Histograms in Fig. 5.18c show the frequency of these W values produced at different loading rates (r_f) and the graph to the left gives the mean of W versus r_f. As shown, as r_f increases, W increases. This phenomenological picture is well explained by Evans and Ritchie[53] who have described molecular unbinding process occurring in AFM force measurements as a first order kinetic process assisted by an external loading force, whose magnitude changes with time at a given constant r_f. The breaking or "pull-off" force $(f_{\text{pull-off}})$ is expressed by the following equation[14,53]:

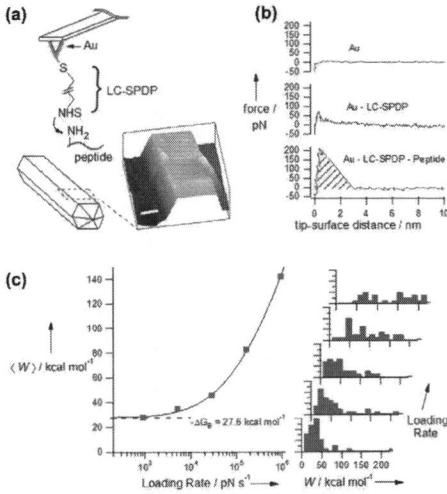

Figure 5.18 Determination of the free energy of binding of AmelC to the (100) face of HAP. (a) The AmelC is linked to a gold-coated AFM cantilever by way of a bifunctional linker. The tip is placed directly on the (100) face of individual HAP crystals, which were visible under bright-field optics. Inset: typical AFM scan of an HAP crystal used to characterize the surface before force measurements (scale bar 4 mm). (b) Representative force–distance curves for each level of functionalization (Au coating, LC-SPDP linker, and peptide) at approximately equivalent loading rates (4.7, 3.6, and 5.3 nNs⁻¹, respectively). (c) Means (solid squares) and corresponding histograms (solid bars) of work measured from repeated force–distance trajectories for a peptide-functionalized tip (spring constant 93 pNnm⁻¹) from the (100) face of HAP. Solid curve is a fit to a two-state theoretical model by Friddle et al.[33] The mean work tends asymptotically to a finite value given by the free-energy difference (dashed line) of $\Delta_{\text{GB}} = -27.6$ kcal mol⁻¹. The normalized histograms for increasing loading rate (0.93, 5.25, 29.5, 165.9, and 933 nNs⁻¹) are offset for clarity using identical ranges along the two axes. Copyright (2011) Wiley. Used with permission from ref. 33.

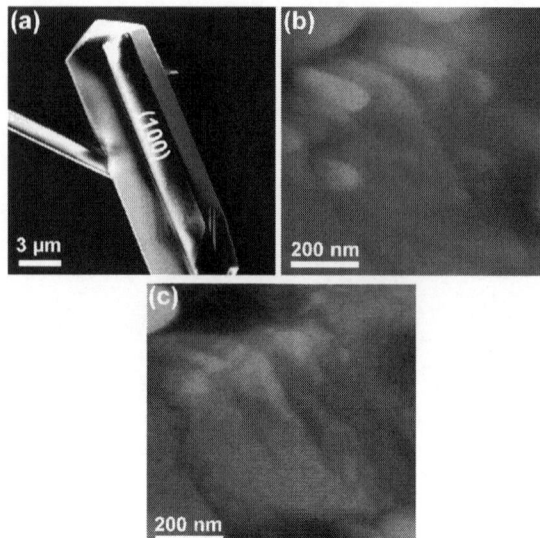

Figure 5.19 Crystal shape and (100) face of HAP. (a) SEM image of the crystal shows a hexagonal prism with six lateral (100) faces. The crystal was synthesized by the homogeneous releasing method.[128] (b) AFM image of HAP (100) face in water. (c) The surface of HAP (100) face in (b) after being exposed to 744 μM AmelC aqueous solution for 30 min. The surface contour is less recognizable in (c) than in (b) due to binding of AmelC to the face. Copyright (2011) Wiley. Used with permission from ref. 33.

$$f_{\text{pull-off}} = \frac{kT}{x_\beta} \ln\left(\frac{r_f}{r_0}\right), \quad r_0 = \frac{kT}{x_\text{a}} \cdot \frac{1}{\tau_D \exp(E_0/kT)}$$

Here, k is the Boltzmann constant, T is absolute temperature, x_β is the distance between the bound state and the transition state, τ_D is the characteristic diffusion time of motion in the system, and E_0 is the activation energy barrier. As described, when r_f increases, $f_{\text{pull-off}}$ increases. Due to this kinetic characteristic of the unbinding process occurring during the AFM force measurements, W is greater than the equilibrium free energy of binding Δ_{GB} by an amount of dissipated heat, which decreases with decreasing r_f. As r_f approaches zero, the rate of bond breaking approaches its equilibrium value and W done on the system should approach Δ_{GB} as shown in Fig. 5.18c.

The solid curve in Fig. 5.18c is a fit to the data according to a theoretical model of Friddle et al.,[33] which treats the forced desorption process as kinetic escape over a potential barrier like that in the model of Evans and Ritchie[53] (see the Supporting Information of ref. 34 for details). The model predicts two major regimes: a non-linear regime at large r_f, characterized primarily by the kinetics of forced desorption, and a linear regime at small r_f where the desorption and adsorption rates are comparable. The linear regime appears as a plateau in the plot of mean W versus log (r_f) and tends asymptotically to the equilibrium free energy difference between the adsorbed and desorbed states. The fit to the data is excellent, and the asymptote gives $\Delta_{GB} = -27.6$ kcal mol^{-1}. This result was used to constrain molecular models of AmelC-(100) HAP binding, thus enabling the models to predict binding free energies for other faces that could not be accessed experimentally. These and other results[22,44,45] demonstrate that AFM force spectroscopy can be successfully used to investigate various protein–materials interactions.

5.6 Conclusions and Future Studies

This chapter has demonstrated that through its capability to image in liquid environments with nanometer-scale spatial resolution and to measure forces of interactions with pN resolution, the AFM is playing a pivotal role in new discoveries and understanding in biological fields such as protein assembly, biomineralization, and biomolecular interactions. By utilizing calibration curves of protein heights versus molecular weights obtained from the AFM height measurements, the molecular weights of unknown globular proteins and degree of oligomerization of known proteins can be determined. This approach provided proof that seed size in HurPrP^C aggregation in vitro that is induced by its conformational change from α-helix–rich into β-sheet–rich structure, is trimeric or tetrameric, as expected for HuPrP^C aggregation in vivo. However, the tapping mode used in that study to image individual proteins on the mica surface using a conventional AFM required several minutes to obtain an image. If one plans to investigate the detailed dynamics of protein aggregation processes on surfaces in real time, the imaging speed of the conventional AFM in tapping mode is likely to prove insufficient. In this case, the high-speed AFMs[129-131] that have been developed recently may provide the solution.

Application of in situ AFM to investigations of biomineralization has proven successful as described here with the example of COM. The influence of biomolecules contained in urine such as citrate, OPN, and THP on the COM growth process was observed by in situ AFM, providing new insights into their roles in the regulation of renal stone formation in vivo. Most AFM studies of solution crystal growth including the COM work described here have been performed using contact mode, which generally provides higher resolution than the tapping mode. Using contact mode, one can generate a high-resolution image at small scan size (for example, 200 nm × 200 nm) even every few seconds. Thus, the conventional AFM with the contact operational mode will continue to be extremely useful in obtaining detailed information on the dynamics of mineral growth processes.

In the last part of this chapter, we introduced an application of AFM force measurement to investigation of biomolecule–biomineral interactions. Quantitatively probing the binding energy of AmelC to the (100) face of HAP crystals demonstrated the unique ability of the AFM to obtain face-specific binding energies at the single molecule level, as compared to bulk measurements that give an average over all faces. Clearly none of the findings described here could have been obtained without using AFM. Moreover, the AFM is being increasingly combined together with other imaging modalities, such as fluorescence microscopy, resulting in instruments with unprecedented capabilities for biological investigations.[132-136] Based on all these considerations, it is evident that this powerful tool will continue to contribute to new discoveries and advancements in our understanding of fundamental mechanisms underlying processes occurring in biological systems.

References

1. Binnig, G., Quate, C. F., and Gerber, C. H. (1986). Atomic force microscope, *Phys. Rev. Lett.*, **56**(9), 930–933.

2. Ohnesorge, F., and Binnig, G. (1993). True atomic resolution by atomic force microscopy through repulsive and attractive forces, *Science*, **260**(5113), 1451–1456.

3. Giessibl, F. J., Hembacher, S., Bielefeldt, H., and Mannhart, J. (2000). Subatomic features on the silicon (111)-(7 × 7) surface observed by atomic force microscopy, *Science*, **289**(5478), 422–425.

4. Plomp, M., Leighton, T. J., Wheeler, K. E., and Malkin, A. J. (2005). The high-resolution architecture and structural dynamics of *Bacillus* spores, *Biophys. J.*, **88**(1), 603–608.

5. Radmacher, M., Fritz, M., and Hansma, P. K. (1995). Imaging soft samples with the atomic force microscope: gelatin in water and propanol, *Biophys. J.*, **69**(1), 264–270.

6. Domke, J., Parak, W. J., George, M., Gaub, H. E., and Radmacher, M. (1999). Mapping the mechanical pulse of single cardiomyocytes with the atomic force microscope, *Eur. Biophys. J.*, **28**(3), 179–186.

7. Lekka, M., Laidler, P., Gil, D., Lekki, J., Stachura, Z., and Hrynkiewicz, A. Z. (1999). Elasticity of normal and cancerous human bladder cells studied by scanning force microscopy, *Eur. Biophys. J.*, **28**(4), 312–316.

8. Raman, A., Trigueros, S., Cartagena, A., Stevenson, A. P. Z., Susilo, M., Nauman, E., and Contera, S. A. (2011). Mapping nanomechanical properties of live cells using multi-harmonic atomic force microscopy, *Nat. Nanotechnol.*, **6(12), 809–814.**

9. Lang, K. M., Hite, D. A., Simmonds, R. W., McDermott, R., Pappas, D. P., and Martinis, J. M. (2004). Conducting atomic force microscopy for nanoscale tunnel barrier characterization, *Rev. Sci. Instrum.*, **75**(8), 2726–2731.

10. Gross, L., Mohn, F., Liljeroth, P., Pepp, J., Giessibl, F. J., Meyer, G. (2009). Measuring the charge state of an adatom with noncontact atomic force microscopy, *Science*, **324**(5933), 1428–1431.

11. Rugar, D., Mamin, H. J., Guethner, P., Lambert, S. E., Stern, J. E., McFadyen, I., and Yogin, T. (1990). Magnetic force microscopy: general principles and application to longitudinal recording media, *J. Appl. Phys.*, **68**(3), 1169–1183.

12. Drake, B., Prater, C. B., Weisenhorn, A. L., Gould, S. A. C., Albrecht, T. R., Quate, C. F., Cannell, D. S., Hansma, H. G., and Hansma, P. K. (1989). Imaging crystals, polymers, and processes in water with the atomic force microscope, *Science*, **243**(4898), 1586–1589.

13. Gratz, A. J., Manne, S., and Hansma, P. K. (1991). Atomic force microscopy of atomic-scale ledges and etch pits formed during dissolution of quartz, *Science*, **251**(4999), 1343–1346.

14. Noy, A., Zepeda, S., Orme, C. A., Yeh, Y., and De Yoreo, J. J. (2003). Entropic barriers in nanoscale adhesion studied by variable temperature chemical force microscopy, *J. Am. Chem. Soc.*, **125**(5), 1356–1362.

15. De Yoreo, J. J., and Dove, P. M. (2004). Shaping crystals with biomolecules, *Science*, **306**(5700), 1301–1302.

16. Radmacher, M., Tillmann R. W., Fritz, M., and Gaub, H. E. (1992). From molecules to cells: Imaging soft samples with the atomic force microscope, *Science*, **257**(5078), 1900–1905.

17. Radmacher, M., Fritz, M., Hansma, H. G., and Hansma, P. K. (1994). Direct observation of enzyme activity with the atomic force microscope, *Science*, **265**(5178), 1577–1579.

18. Henderson, E. (1994). Imaging of living cells by atomic force microscopy, *Prog. Surf. Sci.*, **46**(1), 39–60.

19. Hansma, H. G. and Hoh, J. H. (1994). Biomolecular imaging with the atomic force microscope, *Annu. Rev. Biophys. Biomol. Struct.*, **23**, 115–139.

20. Schneider, S. W., Sritharan, K. C., Geibel, J. P., Oberleithner, H., and Jena, B. P. (1997). Surface dynamics in living acinar cells imaged by atomic force microscopy: Identification of plasma membrane structures involved in exocytosis, *Proc. Natl. Acad. Sci. USA*, **94**(1), 316–321.

21. Cho, S.-J., Jeftinija, K., Glavaski, A., Jeftinija, S., Jena, B. P., and Anderson, L. L. (2002). Structure and dynamics of the fusion pores in live GH-secreting cells revealed using atomic force microscopy, *Endocrinology*, **143**(3), 1144–1148.

22. Merkel, R., Nassoy, P., Leung, A., Ritchie, K., and Evans, E. (1999). Energy landscapes of receptor-ligand bonds explored with dynamic force spectroscopy, *Nature*, **397**(6714), 50–53.

23. Kowalewski, T., and Holtzman, D. M. (1999). In situ atomic force microscopy study of Alzheimer's β-amyloid peptide on different substrates: New insights into mechanism of β-sheet formation, *Proc. Natl. Acad. Sci. USA*, **96**(7), 3688–3693.

24. Zhu, M., Han, S., Zhou, F., Carter, S. A., and Fink, A. L., (2004). Annular oligomeric amyloid intermediates observed by in situ atomic force microscopy, *J. Biol. Chem.*, **279**(23), 24452–24459.

25. Plomp, M., Leighton, T. J., Wheeler, K. E., Hill, H. D., and Malkin, A. J. (2007). In vitro high-resolution structural dynamics of single germinating bacterial spores, *Proc. Natl. Acad. Sci. USA*, **104**(23), 9644–9649.

26. Chung, S., Shin, S.-H., Bertozzi, C. R., and De Yoreo, J. J. (2010). Self-catalyzed growth of S layers via an amorphous to crystalline transition limited by folding kinetics *Proc. Natl. Acad. Sci. USA*, **107**(38), 16536–16541.

27. Cho, K. R. Huang, Y., Yu, S., Yin, S., Plomp, M., Qiu, S. R., Lakshminarayanan, R., Moradian-Oldak, J., Sy, M.-S., and De Yoreo, J. J. (2011). A multistage pathway for human prion protein aggregation in vitro: from multimeric seeds to β-oligomers and non-fibrillar structures, *J. Am. Chem. Soc.*, **133**(22), 8586–8593.

28. Uchihashi, T., Lino, R., Ando, T., and Noji, H. (2011). High-speed atomic force microscopy reveals rotary catalysis of rotorless F_1-ATPase, *Science*, **333**(6043), 755–758.

29. Cross, S. E., Jin, Y.-S., Rao, J., and Gimzewski, J. K. (2007). Nanomechanical analysis of cells from cancer patients, *Nat. Nanotechnol.*, **2** (12), 780–783.

30. Cross, S. E., Jin, Y.-S., Lu, Q.-Y., Rao, J., and Gimzewski, J. K. (2011). Green tea extract selectively targets nanomechanics of live metastatic cancer cells, *Nanotechnology*, **22**(21), 215101(9 pp).

31. Hölscher, H., Gotsmann, B., Allers, W., Schwarz, U. D., Fuchs, H., and Wiesendanger, R. (2001). Measurement of conservative and dissipative tip-sample interaction forces with a dynamic force microscope using the frequency modulation technique, *Phys. Rev. B*, **64**(7), 075402 (6 pp).

32. Dietzel, D., Feldmann, M., Herding, C., Schwarz, U. D., and Schirmeisen, A. (2010). Quantifying pathways and friction of nanoparticles during controlled manipulation by contact-mode atomic force microscopy, *Tribol. Lett.*, **39**(3), 273–281.

33. Friddle, R. W., Battle, K., Trubetskoy, V., Tao, J., Salter, E. A., Moradian-Oldak, J., De Yoreo, J. J., and Wierzbicki, A. (2011). Single-molecule determination of the face-specific adsorption of amelogenin's C-terminus on hydroxyapatite, *Angew. Chem. Int. Ed.*, **50**(33), 7541–7545.

34. Malkin, A. J., Kuznetsov, Y. G., Land, T. A., De Yoreo, J. J., and McPherson, A. (1995). Mechanisms of growth for protein and virus crystals, *Nat. Struct. Biol.*, **2**(11), 956–959.

35. Teng, H. H., Dove, P. M., Orme, C. A., De Yoreo, J. J. (1998). Thermodynamics of calcite growth: baseline for understanding biomineral formation, *Science*, **282**(5389), 724–727.

36. Orme, C. A., Noy, A., Wierzbicki, A., McBride, M. T., Grantham, M., Teng, H. H., Dove, P. M., De Yoreo, J. J. (2001). Formation of chiral morphologies through selective binding of amino acids to calcite surface steps, *Nature*, **411**(6839), 775–779.

37. De Yoreo, J. J., Orme, C. A., and Land, T. A. (2001). Using atomic force microscopy to investigate solution crystal growth, in *Advances in Crystal Growth Research* (ed. Sato, K., Nakajima, K., and Furukawa, Y.), Elsevier Science Amsterdam, pp. 361–380.

38. Gliko, O., Reviakine, I., Vekilov, P. G. (2003) Stable equidistant step trains during crystallization of insulin, *Phys. Rev. Lett.*, **90**(22), 225503 (4 pp.).

39. De Yoreo, J. J., and Vekilov, P. (2003). Principles of crystal nucleation and growth, in *Biomineralization* (ed. Dove, P. M., De Yoreo, J. J., and Weiner, S.), *Mineral Soc. America*, 54, Washington, DC, pp. 57–93.

40. Qiu, S. R., Wierzbicki, A., Orme, C. A., Cody, A. M., Hoyer, J. R., Nancollas, G. H., Zepeda, S., De Yoreo, J. J. (2004). Molecular modulation of calcium oxalate crystallization by osteopontin and citrate, *Proc. Natl. Acad. Sci. USA*, **101**(7), 1811–1815.

41. Weaver, M. L., Qiu, S. R., Hoyer, J. R., Casey, W. H., Nancollas, G. H., and De Yoreo, J. J. (2007). Inhibition of calcium oxalate monohydrate growth by citrate and the effect of the background electrolyte, *J. Cryst. Growth,* **306**(1), 135–145.

42. Weaver, M. L., Qiu, S. R., Hoyer, J. R., Casey, W. H., Nancollas, G. H., and De Yoreo, J. J. (2009). Surface aggregation of urinary proteins and aspartic acid-rich peptides on the faces of calcium oxalate monohydrate investigated by in situ force microscopy, *Calcif. Tissue Int.*, **84**(6), 462–473.

43. Rimer, J. D., An, Z., Zhu, Z., Lee, M. H., Goldfarb, D. S., Wesson, J. A., and Ward, M. D. (2010). Crystal growth inhibitors for the prevention of L-cystine kidney stones through molecular design, *Science*, **330**(6002), 337–341.

44. Sheng, X., Ward, M. D., and Wesson, J. A. (2003). Adhesion between molecules and calcium oxalate crystals: critical interactions in kidney stone formation, *J. Am. Chem. Soc.*, **125** (10), 2854–2855.

45. Sheng, X., Jung, T., Wesson, J. A., and Ward, M. D. (2005). Adhesion at calcium oxalate crystal surfaces and the effect of urinary constituents, *Proc. Natl. Acad. Sci. USA*, **102**(2), 267–272.

46. Digital Instruments, Multimode SPM instruction manual, Version 4.31ce.

47. Plomp, M. (1999). *Crystal Growth Studied on a Micrometer Scale*, PhD thesis, University of Nijmegen, Nijmegen, the Netherlands.

48. Cho, K. R. (2010). *Molecular Scale Materials Assembly Investigated by Atomic Force Microscopy*, PhD thesis, University of California, Los Angeles, Los Angeles, CA, U. S. A.

49. Bezanilla, M., Drake, B., Nudler, E., Kashlev, M., Hansma, P. K., and Hansma, H. G. (1994). Motion and enzymatic degradation of DNA in the atomic force microscope, *Biophys. J.*, **67**(6), 2454–2459.

50. Radmacher, M., Cleveland, J. P., and Hansma, P. K. (1995). Improvement of thermally-induced bending of cantilevers used for atomic force microscopy, *Scanning*, **17**(2), 117–121.

51. Noy, A., Vezenov, D. V., and Lieber, C. M. (1997). Chemical force microscopy, *Annu. Rev. Mater. Sci.*, **27**, 381–421.

52. Hänggi, P., Talkner, P., and Borkovec, M. (1990). Reaction-rate theory: fifty years after Kramers, *Rev. Mod. Phys.*, **62**(2), 251–341.

53. Evans, E., and Ritchie, K. (1997). Dynamic strength of molecular adhesion bonds, *Biophys. J.*, **72**(4), 1541–1555.

54. Evans, E. (1998). Energy landscapes of biomolecular adhesion and receptor anchoring at interfaces explored with dynamic force spectroscopy, *Faraday Discuss.*, **111**, 1–16.

55. Manelyte, L., Urbanke, C., Giron-Monzon, L., Friedhoff, P. (2006). Influence of the N-terminal domain on the aggregation properties of the prion protein Structural and functional analysis of the MutS C-terminal tetramerization domain, *Nucleic. Acids. Res.*, **34**(18), 5270–5279.

56. Maiti, N. R., and Surewicz, W. K. (2001). The role of disulfide bridge in the folding and stability of the recombinant human prion protein, *J. Biol. Chem.*, **276**(4), 2427–2431.

57. Wilkins, D. K., D. K., Grimshaw, S. B., Receveur, V., Dobson, C. M., Jones, J. A., and Smith, L. J. (1999). Hydrodynamic radii of native and denatured proteins measured by pulse field gradient NMR techniques, *Biochemistry*, **38**(50), 16424–16431.

58. Rosenqvist, E., JØssang, T., and Feder, J. (1987). Thermal-properties of human-IgG, *Mol. Immunol.*, **24**(5), 495–501.

59. Petsev, D. N., Thomas, B. R., Yau, S. T., and Vekilov, P. G. (2000). Interactions and aggregation of apoferritin molecules in solution: effects of added electrolytes, *Biophys. J.*, **78**(4), 2060–2069.

60. Kuznetsov, Yu. G., Malkin, A. J., Lucas, R. W., Plomp, M., and McPherson, A. (2001). Imaging of viruses by atomic force microscopy, *J. Gen. Virol.*, **82**, 2025–2034.

61. Feeling-Taylor, A. R., Yau, S.-T., Petsev, D. N., Nagel, R. L., Hirsch, R. E., and Vekilov, P. G. (2004). Crystallization mechanisms of hemoglobin C in the R State, *Biophys. J.*, **87**(4), 2621–2629.

62. Prusiner, S. B., Mckinley, M. P., Bowman, K. A., Bolton, D. C., Bendheim, P. E., Groth, D. F., and Glenner, G. G. (1983). Scrapie prions aggregate to form amyloid-like birefringent rods, *Cell*, **35**(2), 349–358.

63. Pan, K.-M., Baldwin, M., Nguyen, J., Gasset, M., Serban, A., Groth, D., Mehlhorn, I., Huang, Z., Fletterick, R. J., Cohen, F. E. (1993). Conversion of α-helices into β-sheets features in the formation of the scrapie prion proteins, *Proc. Natl. Acad. Sci. USA*, **90**(23), 10962–10966.

64. Prusiner, S. B. (1998). Prions, *Proc. Natl. Acad. Sci. USA*, **95**(23), 13363–13383.

65. Wille, H., Michelitsch, M. D., Guenebaut, V., Supattapone, S., Serban, A., Cohen, F. E., Agard, D. A., Prusiner, S. B. (2002). Structural studies of the scrapie prion protein by electron crystallography, *Proc. Natl. Acad. Sci. USA*, **99**(6), 3563–3568.

66. Caughey, B., Lansbury, P. T., Jr. (2003). Protofibrils, pores, fibrils, and neurodegeneration: separating the responsible protein aggregates from the innocent bystanders, *Annu. Rev. Neurosci.*, **26**, 267–298.

67. Chesebro, B., Trifilo, M., Race, R., Meade-White, K., Teng, C., LaCasse, R., Raymond, L., Favara, C., Baron, G., Priola, S., Caughey, B., Masliah, E., and Oldstone, M. (2005). Anchorless prion protein results in infectious amyloid disease without clinical scrapie, *Science*, **308**(5727), 1435–1439.

68. Silveira, J. R., Raymond, G. J., Hughson, A. G., Race, R. E., Sim, V. L., Hayes, S. F., and Caughey, B. (2005). The most infectious prion protein particles, *Nature*, **437**(7056), 257–261.

69. Brundin, P., Melki, R., and Kopito, R. (2010). Prion-like transmission of protein aggregates in neurodegenerative diseases, *Nat. Rev. Mol. Cell. Biol.*, **11**(4), 301–307.

70. Yu, S., Yin, S., Li, C., Wong, P., Chang, B., Xiao, F., Kang, S.-C., Yan, H., Xiao, G., Tien, P., and Sy, M.-S. (2007). Aggregation of prion protein with insertion mutations is proportional to the number of inserts, *Biochem. J.*, **403**, 343–351.

71. Yin, S., Yu, S., Li, C., Wong, P., Chang, B., Xiao, F., Kang, S.-C., Yan, H., Xiao, G., Grassi, J., Tien, P., and Sy, M.-S. (2006). Prion proteins with insertion mutations have altered N-terminal conformation and increased ligand binding activity and are more susceptible to oxidative attack, *J. Biol. Chem.*, **281**(16), 10698–10705.

72. Swietnicki, W., Morillas, M., Chen, S. G., Gambetti, P., and Surewicz, W. K. (2000). Aggregation and fibrillization of the recombinant human prion protein huPrP90-231, *Biochemistry*, **39**(2), 424–431.

73. Baskakov, I. V., Legname, G., Baldwin, M. A., Prusiner, S. B., and Cohen, F. E. (2002). Pathway complexity of prion protein assembly into amyloid, *J. Biol. Chem.*, **277**(24), 21140–21148.

74. Sokolowski, F., Modler, A. J., Masuch, R., Zirwer, D., Baier, M., Lutsch, S., Moss, D. A., Gast, K., and Naumann, D. (2003). Formation of critical oligomers is a key event during conformational transition of recombinant Syrian hamster prion protein, *J. Biol. Chem.*, **278**(42), 40481–40492.

75. Frankenfield, K. N., Powers, E. T., and Kelly, J. W. (2005). Influence of the N-terminal domain on the aggregation properties of the prion protein, *Protein. Sci.*, **14**(8), 2154–2166.

76. Bocharova, O. V., Breydo, L., Parfenov, A. S., Salnikov, V. V., and Baskakov, I. V. (2005). In vitro conversion of full-length mammalian prion protein produces amyloid form with physical properties of PrP^{Sc}, *J. Mol. Biol.*, **346**(2), 645–659.

77. Yin, S., Pham, N., Yu, S., Li, C., Wong, P., Chang, B., Kang, S.-C., Biasini, E., Tien, P., Harris, D. A., and Sy, M.-S. (2007). Human prion proteins with pathogenic mutations share common conformational changes resulting in enhanced binding to glycosaminoglycans, *Proc. Natl. Acad. Sci. USA*, **104**(18), 7546–7551.

78. Wang. F., Wang, X., and Ma, J. (2011). Conversion of bacterially expressed recombinant prion protein, *Methods*, **53**(3), 208–213.

79. Jarrett, J. T., and Lansbury, P. T., Jr. (1993). Seeding "one-dimensional crystallization" of amyloid: A pathogenic mechanism in Alzheimer's disease and scrapie?, *Cell*, **73**(6), 1055–1058.

80. Nelson, R., Sawaya, M. R., Balbirnie, M., Madsen, A. Ø., Riekel, C., Grothe, R., and Eisenberg, D. (2005). Structure of the cross-β spine of amyloid-like fibrils, *Nature*, **435**(7043), 773–778.

81. Chothia, C., and Janin, J. (1975). Principles of protein-protein recognition, *Nature*, **256**(5520), 705–708.

82. Govaerts, C., Wille, H., Prusiner, S. B., and Cohen, F. E. (2004). Evidence for assembly of prions with left-handed β-helices into trimers, *Proc. Natl. Acad. Sci. USA*, **101**(22), 8342–8347.

83. Aggeli, A., Nyrkova, I. A., Bell, M., Harding, R., Carrick, L., McLeish, T. C. B., Semenov, A. N., and Boden, N. (2001). Hierarchical self-assembly of chiral rod-like molecules as a model for peptide β-sheet tapes, ribbons, fibrils, and fibers, *Proc. Natl. Acad. Sci. USA*, **98**(21), 11857–11862.

84. Grégoire, C., Marco, S., Thimonier, J., Duplan, L., Laurine. E., Chauvin, J.-P., Michel, B., Peyrot, V., and Verdier, J.-M. (2001). Three-dimensional structure of the lithostathine protofibril, a protein involved in Alzheimer's disease, *EMBO J.*, **20**(13), 3313–3321.

85. Dussol, B., Geider, S., Lilova, A., Leonetti, F., Dupuy, P., Daudon, M., Berland, Y., Dagorn, J. C., and Verdier, J. M. (1995). Analysis of the soluble organic matrix of five morphologically different kidney stones: evidence for a specific role of albumin in the constitution of the stone protein matrix, *Urol. Res.*, **23**(1), 45–51.

86. Lieske, J. C., Leonard, R., and Toback, F. G. (1995). Adhesion of calcium oxalate monohydrate crystals to renal epithelial cells is inhibited by specific anions, *Am. J. Physiol.*, **268**(4), F604–F612.

87. Wesson, J. A., Worcester, E. M., Wiessner, J. H., Mandel, N. S., and Kleinman, J. G. (1998). Control of calcium oxalate crystal structure and cell adherence by urinary macromolecules, *Kidney Int.*, **53**(4), 952–957.

88. Kumar, V., Farell, G., and Lieske, J. C. (2003). Whole urinary proteins coat calcium oxalate monohydrate crystals to greatly decrease their adhesion to renal cells, *J. Urol.*, **170**(1), 221–225.

89. Kumar, V., Pena de la Vega, L., Farell, G., and Lieske, J. C. (2005). Urinary macromolecular inhibition of crystal adhesion to renal epithelial cells is impaired in male stone formers, *Kidney Int.*, **68**(4), 1784–1792.

90. Kok, D. J, Papapoulos, S. E., and Bijvoet, O. L. M. (1986). Excessive crystal agglomeration with low citrate excretion in recurrent stone formers, *Lancet,* **1**(8489), 1056–1058.

91. Nicar, M. J., Hill, K., and Pak, C. Y. C. (1987). Inhibition of spontaneous precipitation of calcium oxalate in vitro, *J. Bone Miner. Res.*, **2**(3), 215–220.

92. Hess, B., Zipperle, L., and Jaeger, P. (1993). Citrate and calcium effects on Tamm-Horsfall glycoprotein as a modifier of calcium oxalate crystal aggregation, *Am. J. Physiol.*, **265**(6), F784–F791.

93. Pak, C. Y. C. (1994). Citrate and renal calculi: an update, *Miner. Electrolyte Metab.*, **20**(6), 371–377.

94. Singh, K., DeVouge, M. W., and Mukherjee, B. B. (1990). Physiological properties and differential glycosylation of phosphorylated and nonphosphorylated forms of osteopontin secreted by normal rat kidney cells, *J. Biol. Chem.*, **265**(30), 18696–18701.

95. Worcester, E. M., and Beshensky, A. M. (1995). Osteopontin inhibits nucleation of calcium oxalate crystals, *Ann. N. Y. Acad. Sci.*, **760**, 375–377.

96. Hoyer, J. R., Otvos, L., Jr., and Urge, L. (1995). Osteopontin in urinary stone Formation, *Ann. N. Y. Acad. Sci.*, **760**, 257–265.

97. Hoyer, J. R., Pietrzyk, R. A., Liu, H., and Whitson, P. A. (1999). Effects of microgravity on urinary osteopontin, *J. Am. Soc. Nephrol.*, **10**, S389–S393.

98. Wesson, J. A., Johnson, R. J., Mazzali, M., Beshensky, A. M., Steitz, S., Giachelli, C. M., Liaw, L., Alpers, C. E., Couser, W. G., Kleinman, J. G., and Hughes, J. (2003). Osteopontin is a critical inhibitor of calcium oxalate crystal formation and retention in renal tubules, *J. Am. Soc. Nephrol.*, **14**(1), 139–147.

99. Shiraga, H., Min, W., VanDusen, W. J., Clayman, M. D., Miner, D., Terrell, C. H., Sherbotie, J. R., Foreman, J. W., Przysiecki, C., Neilson, E. G., and Hoyer, J. R. (1992). Inhibition of calcium oxalate crystal growth in vitro by uropontin: another member of the aspartic acid-rich protein superfamily, *Proc. Natl. Acad. Sci. USA*, **89**(1), 426–430.

100. Paulhac, P., Desgrandchamps, F., Dumas, J. P., Teillac, P., Le Duc, A., Colombeau, P. (2002). Role of uropontin in calcium oxalate lithogenesis, *Prog. Urol.*, **12**(1), 114–117.

101. Tamm, I., and Horsfall, F. L., Jr. (1950). Characterization and separation of an inhibitor of viral hemagglutination present in urine, *Proc. Soc. Exp. Biol. Med.*, **74**(1), 108–114.

102. Worcester, E. M., Nakagawa, Y., Wabner, C. L, Kumar, S., and Coe, F. L. (1988). Crystal adsorption and growth slowing by nephrocalcin, albumin, and Tamm-Horsfall protein, *Am. J. Physiol.*, **255**(6), F1197–F1205.

103. Mo, L., Huang, H. Y., Zhu, X. H., Shapiro, E., Hasty, D. L., and Wu, X. R. (2004). Tamm-Horsfall protein is a critical renal defense factor protecting against calcium oxalate crystal formation, *Kidney Int.*, **66**(3), 1159–1166.

104. Deganello, S., and Piro, O. E. (1981). The crystal structure of calcium oxalate monohydrate (whewellite), *N. Jb. Miner. Mh.*, **H**2, 81–88.

105. Cody, A. M., Horner, H. T., and Cody, R. D. (1982). SEM study of the fine surface features of synthetic calcium oxalate monohydrate crystals, *Scanning Electron. Microsc.*, **1**, 185–197.

106. Millan, A. (2001). Crystal growth shape of whewellite polymorphs: influence of structure distortions on crystal shape, *Cryst. Growth Des.*, **1**(3), 245–254.

107. Jung, T., Sheng, X., Choi, C. K., Kim, W. S., Wesson, J. A., and Ward, M. D. (2004). Probing crystallization of calcium oxalate monohydrate and the role of macromolecule additives with in situ atomic force microscopy, *Langmuir*, **20**(20), 8587–8596.

108. De Yoreo, J. J., Qiu, S. R., and Hoyer, J. R., (2006). Molecular modulation of calcium oxalate crystallization, *Am. J. Physiol. Renal. Physiol.*, **291**(6): F1123–F1132.

109. Qiu, S. R., Wierzbicki, A., Salter, E. A., Zepeda, S., Orme, C. A., Hoyer, J. R., Nancollas, G. H., Cody, A. M., and De Yoreo, J. J. (2005). Modulation of calcium oxalate monohydrate crystallization by citrate through selective binding to atomic steps, *J. Am. Chem. Soc.*, **127**(25), 9036–9044.

110. Zuckerman, J. M., and Assimos, D. G. (2009). Hypocitraturia: pathophysiology and medical management, *Rev. Urol.*, **11**(3), 134–144.

111. Shirane, Y., and Kagawa, S. (1993). Scanning electron-microscopic study of the effect of citrate and pyrophosphate on calcium-oxalate crystal morphology, *J. Urol.*, **150**(6), 1980–1983.

112. Wierzbicki, A., Sikes, C. S., Sallis, J. D., Madura, J. D., Stevens, E. D., and Martin, K. L. (1995). Scanning electron-microscopy and molecular modeling of inhibition of calcium-oxalate monohydrate crystal-oxalate by citrate and phosphocitrate, *Calcif. Tissue Int.*, **56**(4), 297–304.

113. Fisher, L. W., Hawkins, G. R., Tuross, N., and Termine, J. D. (1987). Purification and partial characterization of small proteoglycans I and II, bone sialoproteins I and II, and osteonectin from the mineral compartment of developing human bone, *J. Biol. Chem.*, **262**(20), 9702–9708.

114. Prince, C. W., Oosawa, T., Butler, W. T., Tomana, M., Bhown, A. S., Bhown, M., and Schrohenloher, R. E. (1987). Isolation, characterization, and biosynthesis of a phosphorylated glycoprotein from rat bone, *J. Biol. Chem.*, **262**(6), 2900–2907.

115. Singh, K., DeVouge, M. W., and Mukherjee, B. B. (1990). Physiological properties and differential glycosylation of phosphorylated and nonphosphorylated forms of osteopontin secreted by normal rat kidney cells, *J. Biol. Chem.*, **265**(30), 18696–18701.

116. Kasugai, S., Todescan, R., Nagata, T., Yao, K. L., Butler, W. T., and Sodek, J. (1991). Expression of bone matrix proteins associated with mineralized tissue formation by adult rat bone marrow cells in vitro:

inductive effects of dexamethasone on the osteoblastic phenotype, *J. Cell. Physiol.*, **147**(1), 111–120.

117. Hess, B., Nakagawa, Y., and Coe, F. L. (1989). Inhibition of calcium oxalate monohydrate crystal aggregation by urine proteins, *Am. J. Physiol.*, **257**(1), F99–F106.

118. Linde, A., and Lundgren, T. (1995). From serum to the mineral phase. The role of the odontoblast in calcium transport and mineral formation, *Int. J. Dev. Biol.*, **39**(1), 213–222.

119. Weiner, S., Veis, A., Beniash, E., Arad, T., Dillon, J. W., Sabsay, B., and Siddiqui, F. (1999). Peritubular dentin formation: crystal organization and the macromolecular constituents in human teeth. *J. Struct. Biol.*, **126**(1), 27–41.

120. Hunter, G. K., Hauschka, P. V., Poole, A. R., Rosenberg, L. C., and Goldberg, H. A. (1996). Nucleation and inhibition of hydroxyapatite formation by mineralized tissue proteins, *Biochem. J.*, **317**, 59–64.

121. He, G., Dahl, T., Veis, A., and George, A. (2003). Nucleation of apatite crystals in vitro by self-assembled dentin matrix protein 1, *Nat. Mater.*, **2(8), 552–558.**

122. Baht, G. S., O'Young, J., Borovina, A., Chen, H., Tye, C. E., Karttunen, M., Lajoie, G. A., Hunter, G. K., and Goldberg, H. A. (2010). Phosphorylation of Ser[136] is critical for potent bone sialoprotein-mediated nucleation of hydroxyapatite crystals, *Biochem. J.*, **428, 385–395.**

123. Aoba, T., Fukae, M., Tanabe, T., Shimizu, M., and Moreno, E. C. (1987). Selective adsorption of porcine-amelogenins onto hydroxyapatite and their inhibitory activity on hydroxyapatite growth in supersaturated solutions, *Calcif. Tissue Int.*, **41**(5), 281–289.

124. Moradian-Oldak, J., Bouropoulos, N., Wang, L., and Gharakhanian, N. (2002). Analysis of self-assembly and apatite binding properties of amelogenin proteins lacking the hydrophilic C-terminal, *Matrix. Biol.*, **21**(2), 197–205.

125. Shaw, W. J., Campbell, A. A., Paine, M. L., and Snead, M. L. (2004). The COOH terminus of the amelogenin, LRAP, is oriented next to the hydroxyapatite surface, *J. Biol. Chem.*, **279**(39), 40263–40266.

126. Shaw, W. J., and Ferris, K. (2008). Structure, orientation, and dynamics of the C-terminal hexapeptide of LRAP determined using solid-state NMR, *J. Phys. Chem. B*, **112**(51), 16975–16981.

127. Shaw, W. J., and Ferris, K., Tarasevich, B., Larson, J. L. (2008). The structure and orientation of the C-terminus of LRAP, *Biophys. J.*, **94**(8), 3247–3257.

128. Tao, J. H., Jiang, W. G., Pan, H. H., Xu, X. R., and Tang, R. K. (2007). Preparation of large-sized hydroxyapatite single crystals using homogeneous releasing controls, *J. Cryst. Growth*, **308**(1), 151–158.

129. Ando, T., Kodera, N., Takai, E., Maruyama, D., Saito, K., and Toda, A. (2001). A high-speed atomic force microscope for studying biological macromolecules, *Proc. Natl. Acad. Sci. USA*, **98**(22), 12468–12472.

130. Carberry, D. M., Picco, L., Dunton, P. G., and Miles, M. J. (2009). Mapping real time images of high speed AFM using multitouch control, *Nanotechnology*, **20**(43), 434018(5 pp).

131. Casuso, I., Kodera, N., Le Grimellec, C., Ando, T., and Scheuring, S. (2009). Contact-mode high resolution high speed atomic force microscopy movies of the purple membrane, *Biophys. J.* **97**(5), 1354–1361.

132. Chiantia, S., Ries, J., Chwastek, G., Carrer, D., Li, Z., Bittman, R., and Schwille, P. (2008). Role of ceramide in membrane protein organization investigated by combined AFM and FCS, *Biochim. Biophys. Acta.*, **1778**(5), 1356–1364.

133. Oreopoulos, J., and Yip, C. M. (2009). Comninatorial microscopy for the study of protein membrane interactions in supported lipid bilayers: order parameter measurements by combined polarized TIRFM/AFM, *J. Struct. Biol.*, **168**(1), 21–36.

134. Mangold, S., Harneit, K., Rohwerder, T., Claus, G., Sand, W. (2008). Novel combination of atomic force microscopy and epifluorescence microscopy for visualization of leaching bacteria on pyrite, *Appl. Environ. Microbiol.*, **74**(2), 410–415.

135. Frankel, D. J., Pfeiffer, J. R., Surviladze, Z., Johnson, A. E., Oliver, J. M., Wilson, B. S., and Burns, A. R. (2006). Revealing the topography of cellular membrane domains by combined atomic force microscopy/ fluorescene imaging, *Biophys. J.*, **90**(7), 2404–2413.

136. Madl, J., Rhode, S., Stangl, H., Stockinger, H., Hinterdorfer, P., Schütz, G. J., and Kada, G. (2006). A combined optical and atomic force microscope for live cell investigations, *Ultramicroscopy*, **106**(8–9), 645–651.

Chapter 6

High-Resolution Imaging of Amylin Aggregation and Internalization in Pancreatic Cells: Implications in Health and Disease

Saurabh Trikha, Sanghamitra Singh, and Aleksandar M. Jeremic

Department of Biological Sciences,
The George Washington University, Washington, DC 20052

jerema@gwu.edu

Human islet amyloid polypeptide (hIAPP), or amylin, is a pancreatic hormone and the main constituent of amyloid deposits in type-2 *diabetes mellitus* (TTDM). The pathological characteristics of TTDM include progressive beta (β)-cell dysfunction, loss of β-cell mass, and formation of toxic amylin oligomers and fibrils or islet amyloidosis. In this chapter, we describe several routinely and less commonly used methods and approaches such as spectroscopy and high-resolution scanning microscopy to investigate amylin aggregation in vitro and in situ. We also discuss how combination of atomic force microscopy (AFM), confocal microscopy, and fluorescence/CD spectroscopy can be utilized in coherent fashion to examine amylin aggregation, trafficking, and toxicity at nm-resolution and in real

NanoCellBiology: Multimodal Imaging in Biology and Medicine
Edited by Bhanu P. Jena and Douglas J. Taatjes
Copyright © 2014 Pan Stanford Publishing Pte. Ltd.
ISBN 978-981-4411-79-0 (Hardcover), 978-981-4411-80-6 (eBook)
www.panstanford.com

time. These technological advancements provide a novel insight into formation and pro-apoptotic action of toxic amylin oligomers and aggregates, knowledge which is critical for the development of novel therapeutics for the treatment of TTDM.

6.1 Role of Amylin Aggregation in Etiology of Type-2 Diabetes Mellitus

Human islet amyloid polypeptide, or amylin, is 37-aa pancreatic peptide hormone produced and co-secreted with insulin by islet β-cells. Insulin and amylin act in synergistic manner to regulate glucose levels in the blood.[1] In late-onset of type-2 *diabetes mellitus* (TTDM), however, amylin polymerizes to form amyloid deposits or soluble oligomers in islets of *Langerhans*, which are toxic to β-cells.[2-4] Amyloid deposits have been found in over 90% of type-2 diabetic patients, and approximately ~7% in nondiabetic controls. Pathological and clinical studies suggest islet amyloid to be an important risk factor for the development of TTDM.[5-6] The onset of TTDM is characterized by three determining factors: the insufficient ability of pancreatic β-cells to secrete insulin, decreased insulin sensitivity of peripheral tissues, and the deposition of amylin-derived aggregates or amyloid.[7-9] In spite of well-established roles of first two factors in the etiology of diabetes, there is still a critical gap in the knowledge that centers on why amylin oligomerizes and aggregates in the pancreas and how these two processes contribute to the disease, which are the focus of this chapter.

Studies underscore the important role of amylin in the progression of the TTDM as there is a positive relationship between amylin extracellular deposition and the clinical severity of diabetes revealed both in humans and in rodent models.[5,6,10-12] Consistent with the above, a transgenic rat expressing human amylin shows a decreased β-cell mass, an increase in β-cell apoptosis and development of islet aggregates.[11] Interestingly, reports from several laboratories suggest that intermediate sized soluble amylin oligomers rather than mature insoluble fibrils comprise the main cytotoxic species.[4,13,14] However, past and current studies provide compelling evidence for amyloid contribution to the pathogenesis of TTDM.[15-18] Collectively, these studies indicate that amylin aggregates can compromise cell viability alone or together with soluble toxic oligomers, which requires further investigation. Pathological mechanisms may involve amylin oligomers and small aggregates formed intracellularly, extracellularly, or both.[11,13,16,18–20] Current studies performed in primates strongly support the concept that developing islet amyloidosis and β-cell apoptosis are two key determinants of islet of Langerhans dysfunction.[21,22] Studies show that direct contact of amylin aggregates with β-cell membranes is required to elicit apoptosis.[2,3,20] Amylin-evoked membrane destabilization and cation channel formation in cell membranes was proposed as one cytotoxic mechanism.[23–25] Other mechanisms include endoplasmic reticulum stress response,[26] activations of stress-activated kinases,[3,27,28] calcium mobilization,[29] and induction of reactive oxidative stress species (ROS) or radicals.[30,31] An array of rapidly developing spectroscopy and high-resolution microscopy techniques discussed below spurred marked progress in amyloid field in last decade.

6.2 Biochemical Approaches to Analyze Conformational Changes and Kinetics of Amylin Aggregation in Solution

Thioflavin-T (Th-T) fluorescent assay was traditionally used as a method for investigating aggregation properties of amyloid peptides including human amylin.[32] In combination with CD spectroscopy, the Th-T aggregation assay provides simple yet cost-effective approach to investigate kinetics and mechanism of amylin oligomerization and polymerization in solution (often referred to as aggregation), including the conformational changes associated with amylin polymerization.[33–38] In a typical experiment, amylin is added to buffered solution containing 5–10 μM concentrations of freshly prepared Th-T solution in buffer. Increase in Th-T fluorescence over time (Fig. 6.1a), reflecting amylin aggregation and fibril formation, is routinely determined using commercially available spectrofluorimeters or microplate readers with excitation and emission set at 450 and 482 nm, respectively. A typical kinetical profile of amylin aggregation in solution is depicted (Fig. 6.1). The traces of an organic solvent (1–2%) in the reaction buffer such as HFIP that is required to fully dissolve the peptide, accelerates amylin aggregation (shortens lag phase) from couple of hours to several minutes (Fig. 6.1a). The increase in amylin aggregation is initiated by the peptide's conformational transition from soluble (random coil or α-helical) to nonsoluble β-sheet-rich conformation (Fig. 6.1b). In contrast, rat amylin featuring three key Pro substitutions in the amyloidogenic region (Fig. 6.1c) lacks the propensity to adopt β-sheet conformation and hence is nonamyloidogenic (Figs. 6.1a,b).

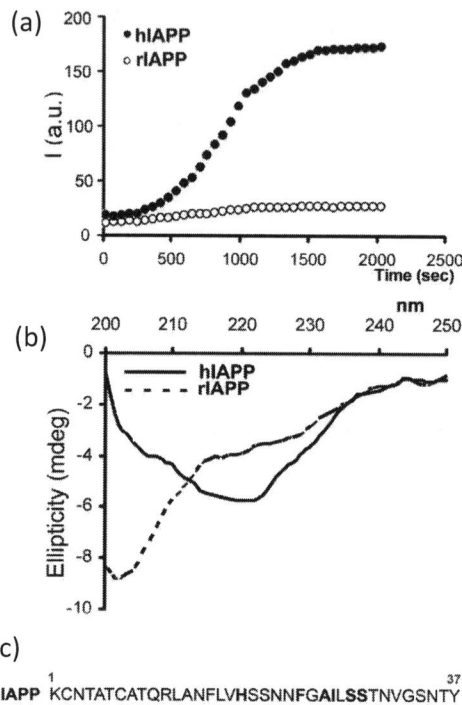

(c)

hIAPP KCNTATCATQRLANFLVHSSNNFGAILSSTNVGSNTY

rIAPP KCNTATCATQRLANFLVRSSNNLGPVLPPTNVGSNTY

Figure 6.1   Aggregation of human amylin coincides with the changes of its secondary structure. (a) Kinetics and extent of aggregation of human amylin (hIAPP) and rat amylin (rIAPP) in PBS as a function of time. Thioflavin-T fluorescent assay reveals fibrilogenesis of 20 μM human amylin in solution (closed circles) and lack of aggregation of non-amyloidogenic rat amylin (20 μM; open circles). (b) Far-UV CD spectra of human and rat amylin taken after 20 min in PBS solution. Note the absorption minimum at ~220 nm for human but not rat amylin, typical for peptides and proteins adopting β-sheet conformation. (c) Full amino acid sequences of human (hIAPP) and rat (rIAPP) amylin are depicted. Species-specific amino-acids in polypeptide chains are bolded for clarity.

Increasing the salt concentrations in the incubation medium (to screen out electrostatic interactions in solution) decreased both the rate and the extent of human amylin aggregation.[37] Thus, amylin aggregation inversely correlates with ionic strength of its solvent, which suggests that electrostatic interactions between amino-acid residues, together with aromatic and hydrophobic interactions, play a major role in self-association or polymerization of human amylin in solution.[37–43]

6.3 High-Resolution Force Imaging of Amylin Aggregation

Although spectroscopic approaches have provided important information regarding molecular forces, kinetics, and conformations driving amylin aggregation in solution,[32–43] they cannot reveal how amylin and other amyloid peptides self-associate into larger complexes, nor can they reveal the molecular architecture of these large supramolecular amyloid assemblies. Without this information, the process of amylin aggregation and amyloid formation in tissues cannot be fully understood. The unique capability of atomic force microscope (AFM) to directly monitor changes in the conformation or aggregation state of macromolecules, and to study dynamic aspects of molecular interactions in their physiological buffer environment has allowed examination of amylin aggregates at ~5 nm lateral and <1 nm vertical resolution, respectively (Figs. 6.2 to 6.6).[37–39,42,44–47] The 3D time-lapse AFM imaging of amylin aggregation has provided new insights into the molecular mechanism of amyloid assembly.[37,38,44,45] In these studies, time-lapse AFM operating either in contact or tapping mode was used to investigate the organization of amylin aggregates on mica (Figs. 6.2 and 6.3) and on planar lipid membranes (Figs. 6.4 to 6.6), two surfaces bearing distinct physicochemical properties. Using high-resolution scanning parameters (512 × 512 lines per image, 1 Hz), the rate and the extent of amylin aggregation and formation of fibrils "the fibril growth" can be quantitatively measured by AFM for several hours with the speed of acquisition of ~5 min/image.[37,38] Amplitude AFM micrographs reveal structural transitions of amylin on mica from small spherical oligomers to extended fibrils on mica over a 30 min time period (Fig. 6.2a). After acquiring micrographs, the size of individual fibrils and oligomers (i.e., radius, length and height), deposited on mica (Fig. 6.2b) or on planar membranes (Fig. 6.5), can be determined using the section analysis tool (Veeco, Santa Barbara, CA). Cross-sectional analysis of amylin

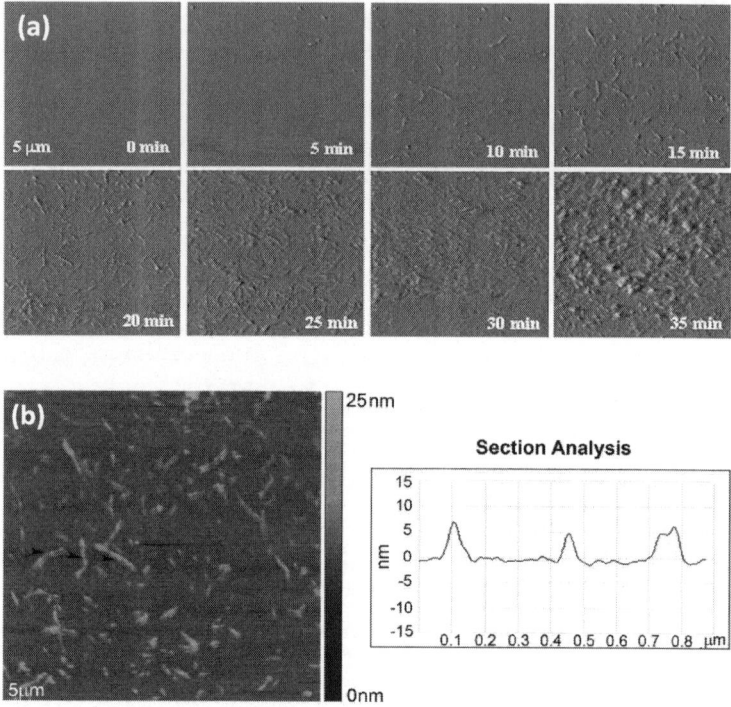

Figure 6.2   Time-lapse AFM imaging of amylin aggregation on solid surfaces are shown. (a) Structural intermediates, oligomers and fibrils, are resolved during amylin aggregation on mica by AFM (tapping mode amplitude images). Note a time-dependent transition of human amylin from small round oligomers (0–10 min) into fibrils during early-mid stage of amylin aggregation (10–25 min). Late–stage of amylin aggregation (25–35 min) is characterized by accumulation of massive peptide deposits on the mica surface. All micrographs are 5 × 5 μm². (b) Height-AFM micrograph depicts fibril formation and their homogeneous distribution across mica surface in PBS buffer 20 min following the addition. Differences in fibril heights are represented using pseudo-color scale (0–25 nm, right bar). Micrograph is 5 × 5 μm² scale. Cross-sectional analysis of three mature fibrils (depicted by arrowheads on the left) reveals similarity in fibrils size (height and width only, right panel).

aggregates revealed that amylin fibrils, varied by length, and consistently measured 90–110 nm in width and 5–6 nm in height, respectively (Fig. 6.2b).[37] In addition to amplitude AFM images (Fig. 6.2a) the height-AFM micrographs (Figs. 6.2b and 6.3a,b) show changes in fibril height during amylin aggregation. Changes in particle height are more visible in the height imaging mode as compared to amplitude images, which are better suited for imaging the fine morphological details of amylin aggregates. Some fibrils were relatively short (less than 200 nm), whereas some fibrils extended over 500 nm in length (Fig. 6.2 and 6.3). Using time-lapse AFM imaging, one may determine the changes in size of individual fibrils over time or amylin fibril growth curve (Fig. 6.3c). To construct fibril growth curves, the average size of oligomers and fibrils needs to be determined for each time point (Figs. 6.2b and 6.3a,b), and plotted (Fig. 6.3c). In the presence of 1–2% HFIP, which accelerates amylin aggregation, massive amyloid-like amylin deposits generally develop after 30 min of incubation (Fig. 6.2a, 30–35 min), thus precluding the monitoring of fibril extension for extended period of time. Nevertheless, AFM resolved amylin structural intermediates prior to amyloid accumulation. Formation of fibrils occurred in two distinct phases: initially through the deposition of small spherical oligomers having a diameter (width) and height of 47 ± 7 nm and 3.4 ± 0.3 nm, respectively (Fig. 6.2a, 0–5 min; Fig. 6.3c). During the next 5 min, oligomers almost doubled in size (diameter: 89 ± 13 nm, height: 5.5 ± 0.4 nm; Figs. 6.2a and 6.3c) followed by oligomers bi-directional extensions into a fibril at average fibrilization rate of 21 nm/min (Fig. 6.2a, 10–25 min; Fig. 6.3c). Growth curves revealed two distinct phases in amylin aggregation: the first phase or oligomer growth is characterized by the large change in oligomers height and width within the first 10 min of amylin aggregation (reaching ~90% of its maximum values, Fig. 6.3c, inset), accounting only for ~20% of maximum fibril length during that period, and second phase or fibrils growth, when fibrils rapidly elongated by doubling their extension rate from 9 nm/min (0–10 min interval, Fig. 6.3c) to 21 nm/min (10–25 min interval, Fig. 6.3c). Taken together these results suggest that fibrils are formed on mica by longitudinal extension of full-grown oligomers rather than by lateral association of protofibrils as suggested by Goldsbury and colleagues.[44] Hence, amylin fibrilization depends on formation of "building block" oligomers or nuclei measuring approximately ~6 nm in height and ~90 nm in diameter that once formed align and elongate into a fibril (Figs. 6.3a–c), a scenario originally proposed by Aebi and co-workers.[45] However, which of these two likely scenarios drives amylin aggregation in the pancreas, a lateral association of protofibrils,[44,48] or longitudinal extension of full-width oligomers observed in our AFM studies[37] and also reported by Green et al.,[45] remains to be determined.

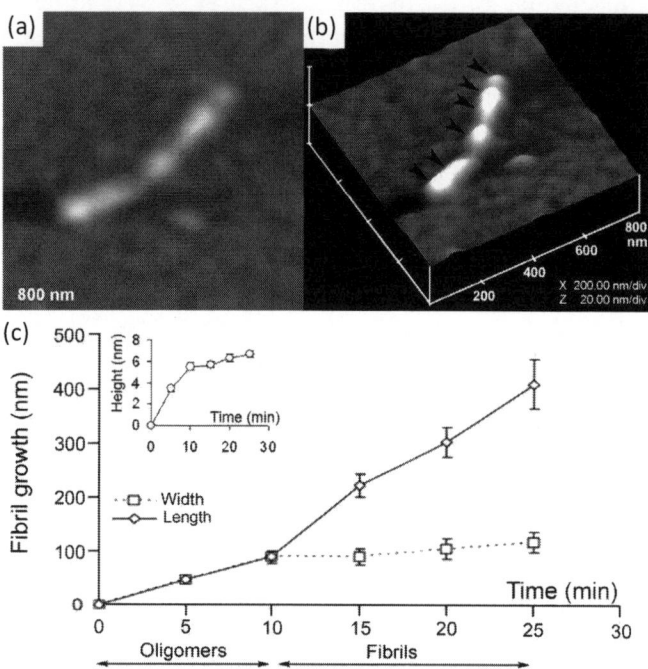

Figure 6.3   Amylin phase transitions during aggregation. The 2D (a) and 3D (b) AFM images of a single full-grown fibril on mica showing arrangement of several amylin oligomers and their bi-directional extension into a fibril (depicted by arrowheads). Micrographs are 800 × 800 nm^2 scale. (c) Fibril growth curves depict two phases of amylin aggregation, an early first phase (0–10 min) characterized by oligomers formation, followed by oligomers extension and fibrils formation (10–25 min, second phase or fibril maturation). Note the significant increase in oligomer heights (Fig. 6.3c, inset) and widths during the first phase of amylin aggregation, and an abrupt increase in fibrils length following formation of full-size oligomers (second phase, Fig. 6.3c). Data represent mean particle size at each time point (mean ± SEM), obtained from three independent time-lapse AFM experiments.

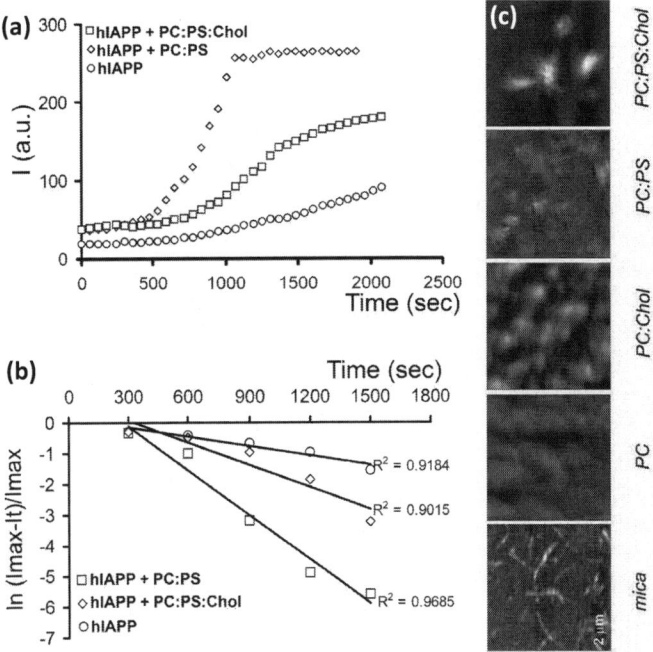

Figure 6.4 Distinct modulatory effects of membrane cholesterol and anionic phospholipids on amylin aggregation are demonstrated. (a) Thioflavin-T fluorescent assay reveals slow aggregation of 5 μM human amylin in solution (circles). Presence of 100 μM phosphatidylcholine:phosphatidylserine (PC: PS) lipid vesicles (diamonds) in incubating solution (PBS, 1% HFIP) potentiates amylin aggregation that was reversed by inclusion of cholesterol in the lipid vesicles (PC:PS:Chol, squares). (b) Kinetics of amylin aggregation is regulated by membrane cholesterol. Presence of negatively charged PC:PS vesicles (squares) in the incubation buffer increased the rate of amylin aggregation (circles) by more than four times. The stimulatory effect of anionic PS on amylin aggregation was decreased by ~30% by inclusion of cholesterol in the vesicle membrane (PC: PS:cholesterol LUV, diamonds). All three curves follow first-order kinetics (R^2 > 0.9, Fig. 6.4b). (c) Cholesterol regulates amylin aggregation and deposition on planar membranes. Note the accumulation of amylin aggregates in the membrane microdomains, and the lack of amylin aggregates in surrounding membrane areas when cholesterol is present (PC:Chol and PC:PS:Chol) Also note the absence of elongated fibrils during amylin aggregation on planar membranes. All micrographs are 2 × 2 μm² scale, and are taken at the same time point (20 min) during amylin aggregation.

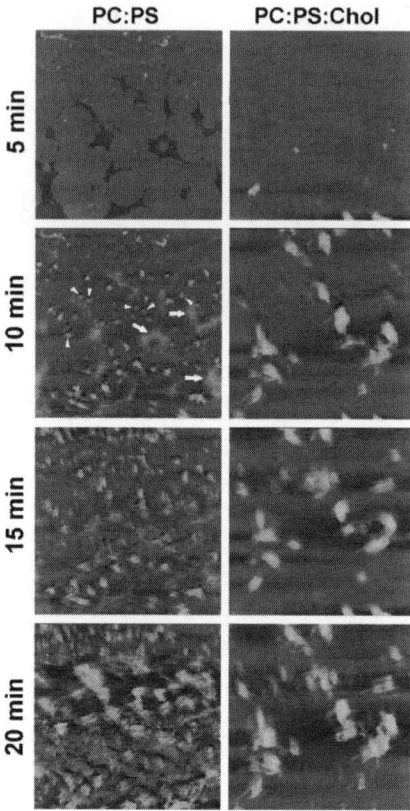

Figure 6.5 Dynamics and organization of amylin aggregates on planar membranes. Amylin (20 μM) at time zero was injected into the imaging chamber and the peptide membrane assembly was monitored in real time by time-lapse AFM. Amylin initially self-assembles into spherical oligomers on negatively charged PC: PS planar membranes, and rarely on the mica surface (5 min, left panel). Mica and membranes above are represented by darker and lighter colors, respectively. At the 10 min time point the bilayer completely fused and covered the mica. Amylin oligomers then associated to form channel-like structures (arrowheads, 10 min left panel). Amylin also accumulated as unstructured diffuse amylin aggregates (arrows, left panel). On anionic membranes that contained cholesterol (PC:PS:Chol) amylin heterogeneously aggregated over the membrane surface (right panel). The presence of cholesterol in the membrane stimulated the formation of ~300–500 nm amylin clusters (10–20 min, right panel). All micrographs are 5 × 5 μm², and are taken at the same time (5 min) intervals.

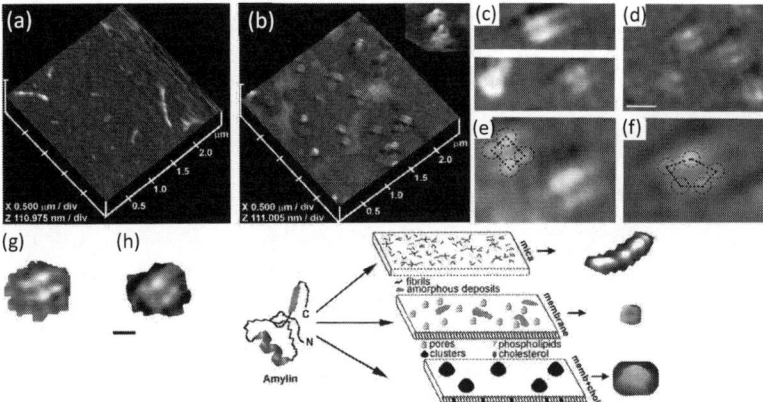

Figure 6.6 AFM analysis of membrane-directed supramolecular amylin assembly. (a) Alignment of individual amylin oligomers into mature elongated fibrils on mica is shown. (b) On planar PC: PS membranes, however, amylin oligomers formed spherical supramolecular structures. Channel-like topology of two amylin supramolecular complexes featuring a central pore is shown (B, inset). (c, d) High-resolution 2D AFM micrographs of several amylin supramolecular complexes on PC:PS membranes are shown. Bar is 100 nm. (e–g) Tetrameric (e) and pentameric (f) subunits of individual amylin supramolecular complexes are outlined for clarity. Bar is 50 nm. Note the absence of elongated amylin fibrils on the planar membranes (b–f). AFM images were taken 10 min after amylin addition to the imaging chamber. (h) Proposed pathway of amylin polymerization and accumulation on distinct surfaces. Collectively, studies indicate that the form and amount of amylin deposits are correlated with the distinct physicochemical properties of the supporting surface. On stiff polar mica surface amylin monomers (left) associate into spherical oligomers that align and elongate overtime to produce mature fibrils. Thus formed fibrils randomly distribute across the underlying mica (top panel, h). On soft planar membranes, amylin self-assembles into globular and highly symmetrical supramolecular structures featuring a central pore (middle panel, h). Unstructured amorphous amylin aggregates are also formed on this surface (middle panel). Incorporation of cholesterol into planar membranes redirects amylin surface deposition by stimulating formation of larger, but fewer amylin clusters (bottom panel, h). Consequently, the membrane surface area free of amylin deposits increases significantly, which diminishes amylin accumulation. Three major polymorphic forms, a fibril, a pore and a single cluster formed during amylin polymerization on different surfaces, are presented top to bottom (AFM micrographs, right panel, h).

6.4 Regulatory Role of Membranes in Amylin Oligomerization and Aggregation

Given that human amylin is a positively charged (cationic) peptide, it can be expected that biological structures bearing strong negative charges such as anionic lipids, the major component of cellular membranes, may modulate amylin aggregation in solution. This idea is supported by several biochemical studies showing accelerated amylin aggregation and conformational change in the presence of anionic lipids such as PS.[33-38] In marked contrast to peptide alone, presence of negatively charged liposomes (PC:PS) in the incubation solution potentiated amylin aggregation by increasing both the extent (Fig. 6.4a) and the rate of amylin aggregation (Fig. 6.4b).[38] Amylin aggregated at a slow rate following first order law with $k = 0.0010$ s^{-1}, whereas presence of PC:PS liposomes accelerated amylin aggregation by ~450% to $k = 0.0045$ s^{-1} (Figs. 6.4a,b). Zwitteronic PC liposomes do not significantly affect the rate or the extent of amylin aggregation in solution.[37] Cholesterol, another essential component of cellular membranes, attenuated stimulatory effect of PS on amylin aggregation in solution by ~30%, $k = 0.0032$ s^{-1} (PC:PS:cholesterol, Figs. 6.4a,b), a decrease comparable to the inhibitory effect of cholesterol on amylin deposition across planar membranes (PC:PS:cholesterol, Fig. 6.4c).[37,38] Applying the calculated rate constants to Arrhenius equation it can be inferred that presence of the negatively charged PC:PS vesicles decreases energy of activation (Ea) of the aggregation process by $\Delta E_a = -3.7$ kJ/mol as compared to amylin alone, which in turn increases the rate of amylin aggregation by more than four times. In contrast, inclusion of cholesterol in PC:PS vesicles reverses their stimulatory effect on amylin aggregation by increasing activation energy by $\Delta E_a = 845$ J/mol. Both thermodynamic values were calculated for amylin aggregation at room temperature (25°C).[37] Interestingly, despite marked difference in their aggregation rates, all three curves exhibit first-order kinetics ($R^2 > 0.9$; Fig. 6.4b). These results suggest that membranes modulate amylin aggregation by changing the activation energy of the amylin phase transition from random coil to β-sheets rather than by changing its mechanistic pathway. In support of this idea, in cholesterol-containing membranes the rate of PS-stimulated amylin conformational transition from random coil to β-sheets was reduced.[37]

6.5 Supramolecular Assembly of Amylin on Neutral, Anionic, and Lipid Raft Membranes

Planar membranes represent a valuable tool for assessing many protein-lipid interactions, which we used in the recent studies to mimic interactions between amylin and β-cell plasma membrane that are likely to occur during amylin aggregation in the pancreas in patients with TTDM.[37,38] We prepared chemically distinct planar membranes and liposomes in order to investigate if and how surface charges, hydrophobicity and other intermolecular forces contribute to amylin aggregation (Figs. 6.4 to 6.6). On stiff and polar surfaces such as mica, AFM revealed that amylin aggregates are homogeneously deposited across the mica surface (Figs. 6.2 and 6.3). We observed a similar uniform surface distribution of amylin aggregates on other polar and charged supports such as neutral PC and anionic PC:PS membranes (Fig. 6.4c). Interestingly, the structure of amylin aggregates varied according to the physicochemical properties of the support used in the study: we consistently observed accumulation of fibrils on stiff and polar surface like mica, a buildup of short thick aggregates on soft zwitteronic/negatively charged PC and PC:PS membranes, whereas short and condensed amyloid structures preferentially deposited in the microdomains of the membranes containing cholesterol (Fig. 6.4c). Furthermore, while amylin oligomers are initially formed on planar membranes (Fig. 6.5, 5 min),[38] they do not elongate into extended fibrils in stark contrast to amylin aggregation on mica (Fig. 6.4c). Thus, different physicochemical properties of underlying surfaces dictated organization and morphological properties of self-assembled amylin aggregates.

Can these spectroscopic and microscopic findings be reconciled? Mica, like most planar membranes, has hydrophilic surface and is overall negatively charged (moderate negative surface charges 0.1 electron per nm^2 have been reported for mica).[49]

Mica also exhibits unusually high attractive van der Waals forces (Nonretarded Hamaker constants of $A = 2 \times 10^{-20}$ J).[50] Thus, while the combination of aromatic, electrostatic and van der Waals forces may drive amylin-mica interactions and amylin aggregation,[37–39,51] combination of surface charges, aromatic interactions and hydrophobic forces may account for amylin–membrane interactions and amylin aggregation on planar membranes (Figs. 6.4 and 6.5).[37,38] In addition to the contribution of these long-range intermolecular forces to amylin–surface interactions and amylin deposition, other surface properties such as membrane mobility and fluidity (i.e., membrane liquid ordered/disordered domains) and/or membrane elasticity may possibly contribute to amylin aggregation, as explained in the next paragraph. It is worth noting here that the structure of amylin aggregates and their deposition patterns on planar membranes obtained in these AFM studies[37,38] resemble the amorphous amyloid deposits that occur in diabetes,[52] indicating that reconstituted approach is of pathophysiological relevance. Taken together, these findings postulate an important and possibly regulatory role of pancreatic cell plasma membrane constituents, anionic lipids and cholesterol, in the development of pancreatic amyloidosis and TTDM, as explained below.

6.6 Distinct Roles of Anionic Lipids and Cholesterol in Supramolecular Assembly of Amylin on Planar Membranes

AFM, operating in real-time and under physiological conditions in buffer, has recently revealed new qualitative and quantitative information of amylin aggregation on planar membranes bearing distinct physical properties. In these studies, neutral and negatively charged planar membranes that contained or lacked cholesterol were used. AFM studies were followed by single particle analysis of amylin aggregates. Amylin was injected into perfusion chamber containing planar membranes, and dynamics of amylin aggregation were monitored by time-lapse AFM operated in the tapping mode.[38] AFM revealed transitions of small 25–35 nm diameter spherical oligomers (formed during the first 5 min) into larger 90–130 nm supramolecular complexes on anionic phosphatidylcholine:phosphatidylserine (PC:PS) membranes (Fig. 6.5, 10 min, arrowheads). Amylin oligomers formed during the first 5 min (Fig. 6.5, left panel) and were morphologically similar to oligomers initially assembled on mica (Fig. 6.2). In marked contrast to amylin fibrilization on mica, amylin oligomers did not align and elongate into fibrils on anionic membranes but rather assembled into channel- or pore-like structures (Fig. 6.5, 10 min; arrowheads).[38] Interestingly, AFM revealed that amylin oligomers preferentially deposited on planar lipidic membranes and much less frequently (<3% particles) on mica surface (Fig. 6.5, 5 min).[38] This result indicates that amylin interacts with membranes early during the oligomeric stage of amylin aggregation. Image analysis revealed a characteristic fourfold rotational symmetry for amylin supramolecular complexes featuring a central pore (Fig. 6.6b, inset). The majority of self-assembled amylin complexes on planar PC:PS membranes exhibited tetrameric and at times pentameric globular organization (Figs. 6.6c–g). Tetrameric amylin complexes accounted for ~95% of all supramolecular amylin structures assembled on the membranes. Less than 5% of amylin complexes were pentamers (7 out of 163 particles examined).[38]

Hence, after nucleation, amylin oligomers directly assembled into symmetrical channel-like structures on pre-formed planar membranes (Fig. 6.6b).[38] This may be relevant for the pathology of diabetes, as amylin and other amyloid proteins interact with cellular membranes and are cytotoxic when assembled into oligomers.[4,53] Interestingly, the sizes of amylin globular particles assembled on planar membranes (Figs. 6.6c–g) were in the same range (20–40 nm) as the soluble intermediate-sized cytotoxic amylin particles reported earlier.[4] As expected from the peptide's amphiphilic nature, cytotoxic amylin oligomers readily formed ion-permeable channels in bilayers and in cell membranes.[23,24] Besides forming channel-like structures, amylin also accumulates on membranes as unstructured amorphous amylin aggregates (Fig. 6.5, left panel; arrows), resembling in size (300–500 nm) and morphology the amyloid deposits often associated with TTDM.[7,52] Incorporation of cholesterol into anionic membranes (PC:PS:cholesterol) redirected the surface distribution of amylin aggregates (Fig. 6.5, right panel). While oligomers were observed again during the first 5 min of aggregation on cholesterol containing membranes, amylin further aggregated and concentrated in discrete areas measuring 300–500 nm in diameter (Fig. 6.5, 10–20 min; PC:PS:Chol).[38]

To further understand how phospholipids and cholesterol modulate amylin surface deposition and to learn more about the regulatory mechanisms driving aggregation, we also measured amylin accumulation on planar membranes using single particle analysis in these studies.[38] Over time there was a significant increase in the height of amylin aggregates due to the large clustering effect of cholesterol (Fig. 6.5, right panel). This was accompanied by an overall decrease in amylin deposition across the planar membranes.[38] As amylin aggregated and accumulated in some membrane areas, other regions of the membrane were virtually depleted of amylin aggregates.[38] Consequently, amylin capacity to form an extensive network of amyloid aggregates on the membrane was diminished in membranes that contained cholesterol (PC:PS: Chol and PC:Chol; Figs. 6.4c and 6.5). Amylin aggregation, as for most amyloid proteins, is nucleation-dependent.[54,55] Consistent with the "nucleation" hypothesis, amylin seeding was diminished in the presence of cholesterol: a sevenfold decrease in the number of amylin particles was observed on planar anionic membranes that contained cholesterol (Figs. 6.4 and 6.5).[38] This phenomenon may also account for the observed prolonged lag phase (nucleation phase) during amylin polymerization in solution elicited by liposomes of the same composition.[37] As the number of amylin particles diminished in the presence of cholesterol, the size and average volume increased over time (Fig. 6.5).[38] The mean volume V_m of amylin particles was significantly larger on lipidic membranes than on mica (Fig. 6.4). A further, much larger increase in particle size was detected on membranes containing cholesterol (PC:PS: Chol, Fig. 6.5, right panel). However, the total volume V_t of amylin aggregates on the cholesterol-containing membranes significantly decreased when compared to cholesterol-lacking membranes due a large decrease in the amylin seeding capacity.[38] Thus, in the presence of cholesterol, amylin aggregated and accumulated on planar membranes as submicron-sized protein clusters, which served as template for ongoing amylin binding and aggregation. Comparisons of AFM micrographs and amylin growth curves on surfaces bearing different physicochemical properties revealed that amylin monomers polymerize via two distinct mechanisms: on stiff and polar mica, amylin formed fibrils by longitudinal bi-directional extensions of full-grown spherical oligomers or nuclei measuring ~6 nm in height and ~90 nm in diameter,[37,45] and on soft negatively charged PC:PS-planar membranes, amylin formed pore-like supramolecular structures that self-assembled from ~25–35 nm diameter globular subunits or oligomers (Fig. 6.6h).[38] AFM revealed another important feature of amylin aggregates on planar membranes. Amorphous deposits and channel-like structures that were formed during the early (5–10 min) stage of amylin aggregation did not convert into new structures, although the total amount and size of amorphous aggregates increased over time (Figs. 6.5 and 6.6).[38] These findings indicate the important contribution and long-lasting effect of lipids and cholesterol in the regulation of amylin aggregation.

Collectively, these results demonstrate the intrinsic ability of cholesterol to regulate amylin aggregation and deposition on membranes irrespective of the chemical composition or charge on the membrane. Consistent with this idea, both amylin and β-amyloid self-assembled into nonfibrilogenic, channel-like structures on neutral and anionic membranes (Figs. 6.4c, 6.5, and 6.6),[56,57] indicating that physical rather than specific chemical properties of the membrane determine its susceptibility to distinct amyloid forms. For example, β-amyloid oligomers but not fibrils were found to be enriched in neuronal membrane lipid-rafts,[58] a discrete segments of membrane with phospholipids in a liquid-ordered phase.[59] Cholesterol is a well-known regulator of membrane fluidity. It is found in lipid rafts and many endomembranes, where it establishes a sorting platform for a scaffold of various protein-lipid complexes important for cell signaling, endocytosis and other essential physiological processes.[59] It is, therefore, reasonable to propose that cholesterol, by modulating membrane fluidity and/or membrane curvature,[60,61] also regulates amylin–lipid interactions and amylin aggregation observed in our studies.[37,38] This conclusion, however, does not exclude involvement of certain, in particular anionic, lipids and surfaces charges in mediating amylin–membrane interactions and amylin aggregation, as reported previously.[34,37,38,54]

6.7 High-Resolution Microscopy of Amylin Aggregation and Internalization in Cells

The above-described reconstituted approach composed of amylin, phospholipids, and cholesterol allowed examination and modeling of amylin–membrane interactions at nanometer-resolution (Fig. 6.6h), which may be relevant for amylin–cell interactions and amylin aggregation in situ. Although the formation of extracellular and intracellular amylin oligomers and fibrils were documented in the past mainly by EM microscopy,[62,63] whether and how human amylin interacts with cellular membranes remained largely unexplored. A recent study provided evidence that human amylin self-assembles into toxic oligomers on plasma membranes of rat and human pancreatic cells,[64] a process of which is remarkably similar to amylin aggregation on artificial planar membranes. Analogous to synthetic membranes,[37,38] amylin oligomers clustered into submicron domains on PM that consequently limited the membrane area occupied by amylin aggregates (Fig. 6.7b).[64] Confocal microscopy analysis of amylin aggregates on PM of cultured rat β-cell line RIN-m5F and human pancreatic islets demonstrated that cholesterol was chiefly responsible for amylin oligomer clustering and heterogeneous accumulation on their cell PM (Fig. 6.7b).[64] By contrast, clustering and accumulation of amylin monomers on PM were little or not modulated by PM cholesterol (Fig. 6.7a).[64] Image and single particle analysis revealed that the mean particle area of PM-bound amylin oligomers (Fig. 6.7b, right boxed panel) in cholesterol-depleted cells (BCD/Lov) decreased significantly as compared to control cells.[64] Moreover, the number of amylin oligomer clusters or puncta on the PM of cholesterol-depleted cells increased by threefold relative to control cells. Consequently, amylin oligomer cell surface coverage increased by about twofold in cells with reduced PM cholesterol content as compared to control cells.[64] This inhibitory effect of BCD/Lov on clustering of amylin oligomers on the PM was also observed at earlier time points, 30 min and 3 h. These results demonstrated that the seeding (nucleation) capacity of amylin oligomers and their ability to form a dense network of amyloid aggregates on the PM were augmented in cells with impaired cholesterol homeostasis. Clustering of amylin oligomers on the PM was fully restored following replenishment of PM cholesterol (Fig. 6.7b), indicating that amylin oligomer deposition on the PM is modulated by cholesterol and is reversible.[64]

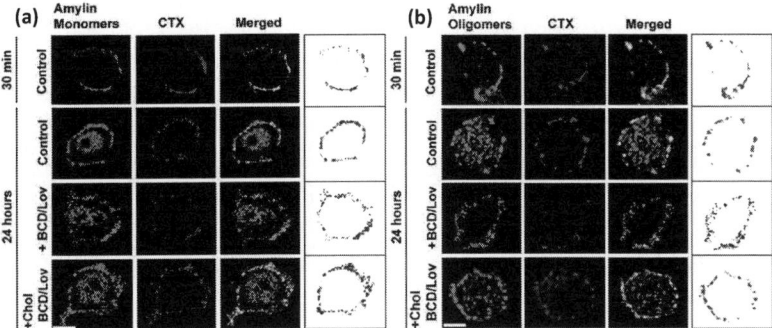

Figure 6.7 Confocal microscopy analysis of amylin monomer, oligomer and CTX distribution on the cell PM. (a) Confocal microscopy demonstrates internalization of amylin monomers both in controls (hA) and cholesterol-depleted cells (hA + BCD/Lov). No appreciable change in PM-binding pattern of amylin monomers (right boxes) was noticed upon cholesterol-depletion. (b) In contrast to monomers, binding and clustering of amylin oligomers into microdomains on the cell PM requires cholesterol. Characteristic binding profiles of amylin oligomers on the cell PM for each treatment (within boxes, right panel) are rendered in gray tones for easier particle comparisons, which are presented side by side with the original fluorescence images (left panels). Note the time-dependent increase in the number of internalized amylin oligomers (control, 30 min vs. 24 h, left panel, B), which prevents accumulation of amylin oligomers on the cell PM (control, 30 min vs. 24 h, right box, B). Bar, 5 μm.

These confocal microscopy studies further revealed that PM cholesterol is required for uptake and clearance of toxic oligomers from the PM, which protects cells from oligomer-induced apoptosis.[64] Although amylin oligomer clusters are detected on the cell surface as early as 10 min following amylin addition to the cultures, the number and size of these clusters do not increase significantly over time in cells with intact PM cholesterol due to uptake of amylin monomers and oligomers by cells.[64] These findings indicate the existence of a cholesterol-sensitive clearance mechanism in rat insulinoma and human islet cells that prevents extracellular accumulation of toxic amylin oligomers, impairment of which may contribute to the β-cell loss and islet dysfunction.[4,13,14,63,64] In support of this idea, studies show an impaired clearance but not production of brain β-amyloid in patients with Alzheimer's disease.[65] Confocal microscopy studies further reveal that amylin oligomers and lipid raft marker, cholera toxin (CTX), bind to the same microdomains on the cell PM, possibly lipid rafts, prior to their internalization.[64] Cholesterol-rich lipid rafts appear to be a major port of entry for amylin oligomers since oligomer internalization is associated with a marked reduction (~50%) in colocalization between remaining amylin oligomers at the cell PM and the lipid raft marker CTX. This conclusion is further supported by the inhibitory action of the well-known lipid raft disrupting agents, BCD and Lov, on amylin oligomer internalization. In agreement with the colocalization data, depletion of PM cholesterol in cells by BCD/Lov decreased the size of amylin oligomer clusters, which become more numerous and occupy more of the PM. Conversely, cholesterol supplementation reduced amylin oligomer seeding and accumulation on the PM as revealed by a marked approximately threefold drop in the number of amylin oligomers on the PM, and by a comparable twofold (approximately) decrease in the amylin oligomer surface coverage as compared to cholesterol-depleted membranes.[64] Importantly, the extent of amylin oligomer accumulation and amylin-mediated cell death are inversely proportional to the magnitude of reduction of cellular, and in particular PM, cholesterol levels. Thus, PM cholesterol is required for uptake and clearance of toxic oligomers from the PM, and protects cells from oligomer-induced apoptosis.[64] The causality between cholesterol-mediated amylin oligomer turnover and toxicity are yet to be determined in the future studies, the knowledge of which will help us to decipher why and how amyloid is formed in islets of Langerhans and will clarify its role in TTDM.

Acknowledgment

This work was supported by the George Washington University UFCC grant and the ICR Basic Science Islet Distribution Program (to A. J.).

References

1. Wagoner, P. K., Chen, C., Worley, J. F., Dukes, I. D., and Oxford, G. S. (1993). Amylin modulates beta-cell glucose sensing via effects on stimulus-secretion coupling. *Proc Natl Acad Sci USA*, **90**, 9145–9149.

2. Lorenzo, A., Razzaboni, B., Weir, G. C., and Yankner, B. A. (1994). Pancreatic islet cell toxicity of amylin associated with type-2 diabetes mellitus. *Nature,* **368**, 756–760.

3. Zhang, S., Liu, J., Dragunow, M., and Cooper, G. J. (2003). Fibrillogenic amylin evokes islet beta-cell apoptosis through linked activation of a caspase cascade and JNK1. *J Biol Chem,* **278**, 52810–52519.

4. Janson, J., Ashley, R. H., Harrison, D., McIntyre, S., and Butler, P. C. (1999). The mechanism of islet amyloid polypeptide toxicity is membrane disruption by intermediate-sized toxic amyloid particles. *Diabetes,* **48**, 491–498.

5. Narita, R., Toshimori, H., Nakazato, M., Kuribayashi, T., Toshimori, T., Kawabata, K., Takahashi, K., and Masukura, S. (1992). Islet amyloid polypeptide (IAPP) and pancreatic islet amyloid deposition in diabetic and non-diabetic patients. *Diabetes Res Clin Pract,* **15**, 3–14.

6. Zhao, H. L., Lai, F. M., Tong, P. C., Zhong, D. R., Yang, D., Tomlinson, B., and Chan, J. C. (2003). Prevalence and clinicopathological characteristics of islet amyloid in Chinese patients with type 2 diabetes. *Diabetes,* **52**, 2759–2766.

7. Clark, A., and Nilsson, M. R. (2004). Islet amyloid: a complication of islet dysfunction or an aetiological factor in Type 2 diabetes? *Diabetologia,* **47**, 157–169.

8. Hull, R. L., Westermark, G. T., Westermark, P., and Kahn, S. E. (2004). Islet amyloid: a critical entity in the pathogenesis of type 2 diabetes. *J Clin Endocrinol Metab,* **89**, 3629–3643.

9. Haataja, L., Gurlo, T., Huang, C. J., and Butler, P. C. (2008). Islet amyloid in type 2 diabetes, and the toxic oligomer hypothesis. *Endocr Rev,* **29**, 303–316.

10. de Koning, E. J., Bodkin, N. L., Hansen, B. C., and Clark, A. (1993). Diabetes mellitus in Macaca mulatta monkeys is characterised by islet amyloidosis and reduction in beta-cell population. *Diabetologia,* **36**, 378–384.

11. Janson, J., Soeller, W. C., Roche, P. C., Nelson, R. T., Torchia, A. J., Kreutter, D. K., and Butler, P. C. (1996). Spontaneous diabetes mellitus in transgenic mice expressing human islet amyloid polypeptide. *Proc Natl Acad Sci USA,* **93**, 7283–7288.

12. Hoppener, J. W., Oosterwijk, C., Nieuwenhuis, M. G., Posthuma, G., Thijssen, J. H., Vroom, T. M., Ahren, B., and Lips, C. J. (1999). Extensive islet amyloid formation is induced by development of Type II diabetes mellitus and contributes to its progression: pathogenesis of diabetes in a mouse model. *Diabetologia,* **42**, 427–434.

13. Butler, A. E., Jang, J., Gurlo, T., Carty, M. D., Soeller, W. C., and Butler, P. C. (2004). Diabetes due to a progressive defect in beta-cell mass in rats transgenic for human islet amyloid polypeptide (HIP Rat): a new model for type 2 diabetes. *Diabetes, 53*, 1509–1516.

14. Konarkowska, B., Aitken, J. F., Kistler, J., Zhang, S., and Cooper, G. J. (2006). The aggregation potential of human amylin determines its cytotoxicity towards islet beta-cells. *FEBS J, 273*, 3614–3624.

15. Bennett, R. G., Hamel, F. G., and Duckworth, W. C. (2003). An insulin-degrading enzyme inhibitor decreases amylin degradation, increases amylin-induced cytotoxicity, and increases amyloid formation in insulinoma cell cultures. *Diabetes, 52*, 2315–2320.

16. Zraika, S., Hull, R. L., Udayasankar, J., Clark, A., Utzschneider, K. M., Tong, J., Gerchman, F., and Kahn, S. E. (2007). Identification of the amyloid-degrading enzyme neprilysin in mouse islets and potential role in islet amyloidogenesis. *Diabetes, 56*, 304–310.

17. Engel, M. F., Khemtemourian, L., Kleijer, C. C., Meeldijk, H. J., Jacobs, J., Verkleij, A. J., de Kruijff, B., Killian, J. A., and Hoppener, J. W. (2008). Membrane damage by human islet amyloid polypeptide through fibril growth at the membrane. *Proc Natl Acad Sci USA, 105*, 6033–6038.

18. Khemtemourian, L., Killian, J. A., Hoppener, J. W., and Engel, M. F. (2008). Recent insights in islet amyloid polypeptide-induced membrane disruption and its role in beta-cell death in type 2 diabetes mellitus. *Exp Diabetes Res, 2008*, 421287.

19. Potter, K. J., Scrocchi, L. A., Warnock, G. L., Ao, Z., Younker, M. A., Rosenberg, L., Lipsett, M., Verchere, C. B., and Fraser, P. E. (2009). Amyloid inhibitors enhance survival of cultured human islets. *Biochim Biophys Acta, 1790*, 566–574.

20. Ritzel, R. A., Meier, J. J., Lin, C. Y., Veldhuis, J. D., and Butler, P. C. (2007). Human islet amyloid polypeptide oligomers disrupt cell coupling, induce apoptosis, and impair insulin secretion in isolated human islets. *Diabetes, 56*, 65–71.

21. Nakamura, S., Okabayashi, S., Ageyama, N., Koie, H., Sankai, T., Ono, F., Fujimoto, K., and Terao, K. (2008). Transthyretin amyloidosis and two other aging-related amyloidoses in an aged vervet monkey. *Vet Pathol, 45*, 67–72.

22. Guardado-Mendoza, R., Davalli, A. M., Chavez, A. O., Hubbard, G. B., Dick, E. J., Majluf-Cruz, A., Tene-Perez, C. E., Goldschmidt, L., Hart, J., Perego, C., Comuzzie, A. G., Tejero, M. E., Finzi, G., Placidi, C., La Rosa, S., Capella, C., Halff, G., Gastaldelli, A., DeFronzo, R. A., and Folli, F. (2009). Pancreatic islet amyloidosis, beta-cell apoptosis, and alpha-cell

proliferation are determinants of islet remodeling in type-2 diabetic baboons. *Proc Natl Acad Sci USA,* **106**, 13992–13997.

23. Mirzabekov, T. A., Lin, M. C., and Kagan, B. L. (1996). Pore formation by the cytotoxic islet amyloid peptide amylin. *J Biol Chem,* **271**, 1988–1992.

24. Brender, J. R., Hartman, K., Reid, K. R., Kennedy, R. T., and Ramamoorthy, A. (2008). A single mutation in the nonamyloidogenic region of islet amyloid polypeptide greatly reduces toxicity. *Biochemistry,* **47**, 12680–12688.

25. Casas, S., Novials, A., Reimann, F., Gomis, R., and Gribble, F. M. (2008). Calcium elevation in mouse pancreatic beta cells evoked by extracellular human islet amyloid polypeptide involves activation of the mechanosensitive ion channel TRPV4. *Diabetologia,* **51**, 2252–2262.

26. Huang, C. J., Lin, C. Y., Haataja, L., Gurlo, T., Butler, A. E., Rizza, R. A., and Butler, P. C. (2007). High expression rates of human islet amyloid polypeptide induce endoplasmic reticulum stress mediated beta-cell apoptosis, a characteristic of humans with type 2 but not type 1 diabetes. *Diabetes,* **56**, 2016–2027.

27. Zhang, S., Liu, J., MacGibbon, G., Dragunow, M., and Cooper, G. J. (2002). Increased expression and activation of c-Jun contributes to human amylin-induced apoptosis in pancreatic islet beta-cells. *J Mol Biol,* **324**, 271–285.

28. Subramanian, S. L., Hull, R. L., Zraika, S., Aston-Mourney, K., Udayasankar, J., and Kahn, S. E. (2012). cJUN N-terminal kinase (JNK) activation mediates islet amyloid-induced beta cell apoptosis in cultured human islet amyloid polypeptide transgenic mouse islets. *Diabetologia,* **55**, 166–174.

29. Huang, C. J., Gurlo, T., Haataja, L., Costes, S., Daval, M., Ryazantsev, S., Wu, X., Butler, A. E., and Butler, P. C. (2010). Calcium-activated calpain-2 is a mediator of beta cell dysfunction and apoptosis in type 2 diabetes. *J Biol Chem,* **285**, 339–348.

30. Janciauskiene, S., and Ahren, B. (2000). Fibrillar islet amyloid polypeptide differentially affects oxidative mechanisms and lipoprotein uptake in correlation with cytotoxicity in two insulin-producing cell lines. *Biochem Biophys Res Commun,* **267**, 619–625.

31. Zraika, S., Hull, R. L., Udayasankar, J., Aston-Mourney, K., Subramanian, S. L., Kisilevsky, R., Szarek, W. A., and Kahn, S. E. (2009). Oxidative stress is induced by islet amyloid formation and time-dependently mediates amyloid-induced beta cell apoptosis. *Diabetologia,* **52**, 626–635.

32. Munishkina, L. A., and Fink, A. L. (2007). Fluorescence as a method to reveal structures and membrane-interactions of amyloidogenic proteins. *Biochim Biophys Acta,* **1768**, 1862–1885.

33. Knight, J. D., and Miranker, A. D. (2004). Phospholipid catalysis of diabetic amyloid assembly. *J Mol Biol,* **341**, 1175–1187.

34. Jayasinghe, S. A., and Langen, R. (2005). Lipid membranes modulate the structure of islet amyloid polypeptide. *Biochemistry,* **44**, 12113–12119.

35. Knight, J. D., Hebda, J. A., and Miranker, A. D. (2006). Conserved and cooperative assembly of membrane-bound alpha-helical states of islet amyloid polypeptide. *Biochemistry,* **45**, 9496–9508.

36. Jayasinghe, S. A., and Langen, R. (2007). Membrane interaction of islet amyloid polypeptide. *Biochim Biophys Acta,* **1768**, 2002–2009.

37. Cho, W. J., Jena, B. P., and Jeremic, A. M. (2008). Nano-scale imaging and dynamics of amylin–membrane interactions and its implication in type II diabetes mellitus. *Methods Cell Biol,* **90**, 267–286.

38. Cho, W. J., Trikha, S., and Jeremic, A. M. (2009). Cholesterol regulates assembly of human islet amyloid polypeptide on model membranes. *J Mol Biol,* **393**, 765–775.

39. Marek, P., Abedini, A., Song, B., Kanungo, M., Johnson, M. E., Gupta, R., Zaman, W., Wong, S. S., and Raleigh, D. P. (2007). Aromatic interactions are not required for amyloid fibril formation by islet amyloid polypeptide but do influence the rate of fibril formation and fibril morphology. *Biochemistry,* **46**, 3255–3261.

40. Nanga, R. P., Brender, J. R., Xu, J., Veglia, G., and Ramamoorthy, A. (2008). Structures of rat and human islet amyloid polypeptide IAPP(1–19) in micelles by NMR spectroscopy. *Biochemistry,* **47**, 12689–12697.

41. Keller, A., Fritzsche, M., Yu, Y. P., Liu, Q., Li, Y. M., Dong, M., and Besenbacher, F. (2011). Influence of hydrophobicity on the surface-catalyzed assembly of the islet amyloid polypeptide. *ACS Nano,* **5**, 2770–2778.

42. Yanagi, K., Ashizaki, M., Yagi, H., Sakurai, K., Lee, Y. H., and Goto, Y. (2011). Hexafluoroisopropanol induces amyloid fibrils of islet amyloid polypeptide by enhancing both hydrophobic and electrostatic interactions. *J Biol Chem,* **286**, 23959–23966.

43. Doran, T. M., Kamens, A. J., Byrnes, N. K., and Nilsson, B. L. (2012). Role of amino acid hydrophobicity, aromaticity, and molecular volume on IAPP(20–29) amyloid self-assembly. *Proteins,* **80**, 1053–1065.

44. Goldsbury, C., Kistler, J., Aebi, U., Arvinte, T., and Cooper, G. J. (1999). Watching amyloid fibrils grow by time-lapse atomic force microscopy. *J Mol Biol,* **285**, 33–39.

45. Green, J. D., Goldsbury, C., Kistler, J., Cooper, G. J., and Aebi, U. (2004). Human amylin oligomer growth and fibril elongation define two distinct phases in amyloid formation. *J Biol Chem,* **279**, 12206–12212.

46. Jha, S., Sellin, D., Seidel, R., and Winter, R. (2009). Amyloidogenic propensities and conformational properties of ProIAPP and IAPP in the presence of lipid bilayer membranes. *J Mol Biol,* **389**, 907–920.

47. Weise, K., Radovan, D., Gohlke, A., Opitz, N., and Winter, R. Interaction of hIAPP with model raft membranes and pancreatic beta-cells: cytotoxicity of hIAPP oligomers. *Chembiochem,* **11**, 1280–1290.

48. Goldsbury, C. S., Cooper, G. J., Goldie, K. N., Muller, S. A., Saafi, E. L., Gruijters, W. T., Misur, M. P., Engel, A., Aebi, U., and Kistler, J. (1997). Polymorphic fibrillar assembly of human amylin. *J Struct Biol,* **119**, 17–27.

49. Pashley, R. M. (1981). DLVO and hydration forces between mica surfaces in Li$^+$, Na$^+$, K$^+$, and Cs$^+$ electrolyte solutions: a correlation of double-layer and hydration forces with surface cation exchange properties. *J Coll Interf Sci,* **2**, 531–546.

50. Israelachvili, J. (1992). *Intermolecular and Surface Forces*. Academic Press, London, UK.

51. Porat, Y., Mazor, Y., Efrat, S., and Gazit, E. (2004). Inhibition of islet amyloid polypeptide fibril formation: a potential role for heteroaromatic interactions. *Biochemistry,* **43**, 14454–14462.

52. Jaikaran, E. T., and Clark, A. (2001). Islet amyloid and type 2 diabetes: from molecular misfolding to islet pathophysiology. *Biochim Biophys Acta,* **1537**, 179–203.

53. Kayed, R., Head, E., Thompson, J. L., McIntire, T. M., Milton, S. C., Cotman, C. W., and Glabe, C. G. (2003). Common structure of soluble amyloid oligomers implies common mechanism of pathogenesis. *Science,* **300**, 486–489.

54. Padrick, S. B., and Miranker, A. D. (2002). Islet amyloid: phase partitioning and secondary nucleation are central to the mechanism of fibrillogenesis. *Biochemistry,* **41**, 4694–4703.

55. Khemtemourian, L., Killian, J. A., Hoppener, J. W., and Engel, M. F. (2008). Recent insights in islet amyloid polypeptide-induced membrane disruption and its role in beta-cell death in type 2 diabetes mellitus. *Exp Diabetes Res,* **2008**, 421287.

56. Lin, H., Bhatia, R., and Lal, R. (2001). Amyloid beta protein forms ion channels: implications for Alzheimer's disease pathophysiology. *FASEB J,* **15**, 2433–2444.

57. Quist, A., Doudevski, I., Lin, H., Azimova, R., Ng, D., Frangione, B., Kagan, B., Ghiso, J., and Lal, R. (2005). Amyloid ion channels: a common structural link for protein-misfolding disease. *Proc Natl Acad Sci USA,* **102**, 10427–10432.

58. Simons, K., and Toomre, D. (2000). Lipid rafts and signal transduction. *Nat Rev Mol Cell Biol,* **1**, 31–39.

59. Schneider, A., Schulz-Schaeffer, W., Hartmann, T., Schulz, J. B., and Simons, M. (2006). Cholesterol depletion reduces aggregation of amyloid-beta peptide in hippocampal neurons. *Neurobiol Dis,* **23**, 573–577.

60. Chen, Z., and Rand, R. P. (1997). The influence of cholesterol on phospholipid membrane curvature and bending elasticity. *Biophys J,* **73**, 267–276.

61. Smith, P. E., Brender, J. R., and Ramamoorthy, A. (2009). Induction of negative curvature as a mechanism of cell toxicity by amyloidogenic peptides: the case of islet amyloid polypeptide. *J Am Chem Soc,* **131**, 4470–4478.

62. Clark, A., Cooper, G. J., Lewis, C. E., Morris, J. F., Willis, A. C., Reid, K. B., and Turner, R. C. (1987). Islet amyloid formed from diabetes-associated peptide may be pathogenic in type-2 diabetes. *Lancet,* **2**, 231–234.

63. Gurlo, T., Ryazantsev, S., Huang, C. J., Yeh, M. W., Reber, H. A., Hines, O. J., O'Brien, T. D., Glabe, C. G., and Butler, P. C. (2010). Evidence for proteotoxicity in beta cells in type 2 diabetes: toxic islet amyloid polypeptide oligomers form intracellularly in the secretory pathway. *Am J Pathol,* **176**, 861–869.

64. Trikha, S., and Jeremic, A. M. (2011). Clustering and internalization of toxic amylin oligomers in pancreatic cells require plasma membrane cholesterol. *J Biol Chem,* **286**, 36086–36097.

65. Mawuenyega, K. G., Sigurdson, W., Ovod, V., Munsell, L., Kasten, T., Morris, J. C., Yarasheski, K. E., and Bateman, R. J., Decreased clearance of CNS beta-amyloid in Alzheimer's disease. *Science,* **330**, 1774.

Chapter 7

Repair of Nanodefects in a Two-Dimensional Crystal Anticoagulant Shield in the Antiphospholipid Syndrome: Novel Molecular Strategies Assessed by Atomic Force Microscopy

Douglas J. Taatjes,[a] Anthony S. Quinn,[a] Xiao-Xuan Wu,[b]
Han-Mou Tsai,[c] and Jacob H. Rand[b]

[a]*Department of Pathology, and Microscopy Imaging Center,*
College of Medicine, University of Vermont, Burlington, VT, USA
[b]*Department of Pathology, Montefiore Medical Center,*
Albert Einstein College of Medicine, Bronx, NY, USA
[c]*iMAH Hematology Associates, New Hyde Park, NY, USA*

douglas.taatjes@uvm.edu

The development of the atomic force microscope (AFM) helped usher in the era of "nanotechnology" with its ability to measure (1) atomic interactions between molecules and (2) image components at the molecular level. In this chapter, we describe the use of the AFM to investigate novel treatment strategies for the autoimmune thrombotic condition known as the antiphospholipid (aPL) syndrome (APS). This coagulation-based disorder has been hypothesized to be caused by disruption of an anticoagulant two-dimensional

NanoCellBiology: Multimodal Imaging in Biology and Medicine
Edited by Bhanu P. Jena and Douglas J. Taatjes
Copyright © 2014 Pan Stanford Publishing Pte. Ltd.
ISBN 978-981-4411-79-0 (Hardcover), 978-981-4411-80-6 (eBook)
www.panstanford.com

crystalline lattice of annexin V (AnxA5) present on the surface of placental trophoblasts and damaged endothelial cells. We present evidence supporting the premise that antibody-induced defects to this AnxA5 crystal shield may be "patched" through treatment with the synthetic antimalarial drug hydroxychloroquine, or with a phospholipid-binding phage-display heptapeptide. Critical to the outcome of the experiments was the ability to use high-resolution atomic force microscopy as a "nanoinstrument" to image reaction components in a non-fixed fluid environment.

7.1   Introduction

The atomic force microscope (AFM), a member of the family of scanning probe microscopes[1] and introduced in 1986,[2] has had a profound impact on molecular and cell biological research. The possibility, afforded by AFM, to image biological molecular interactions at a resolution of tens of nanometers, has created a realm of new high-resolution imaging capabilities that was previously unattainable. Protein–protein, protein–lipid, and protein–DNA interactions can now be readily assessed in a hydrated environment with AFM. Temporal studies, such as the action of enzymes are also feasible, within the limitations imposed by the scan rate of the instrument; recently, these temporal limitations have been largely overcome by the introduction of high-speed AFM scanners.[3,4] AFM has also established a niche in translational basic/clinical research. For instance, we have employed AFM as a key instrument for the investigation of a mechanism of a human thrombotic disease, the antiphospholipid (aPL) syndrome (APS).[5–8] APS is an enigmatic autoimmune disorder in which patients present with thrombosis and/or recurrent pregnancy losses in tandem with laboratory evidence for the presence of autoantibodies in the blood. Although several mechanisms have been proposed to explain the clinical manifestations of the syndrome,[9] a common denominator is that these antibodies have been shown to recognize proteins (the most important of which is β_2-glycoprotein I (β_2GPI)) that bind to anionic phospholipids. We have hypothesized that in the presence of the autoantibodies and β_2GPI, annexin A5 (AnxA5), which forms an anticoagulant protein crystal shield on the surface of placental trophoblasts, is disrupted, resulting in defects within the shield that expose phospholipid moieties that accelerate coagulation-based enzymatic reactions.[10] Indeed, our AFM investigations, using a simulacrum, corroborated this hypothesis by demonstrating aPL antibody-mediated disruption of a 2D AnxA5 crystal lattice formed on a planar lipid layer.[5] This result led us to consider potential therapeutic strategies that might "patch" the anticoagulant AnxA5 lattice on cell membranes damaged by aPL autoantibodies. In this chapter, we present our efforts to date in this regard, emphasizing results obtained with AFM nanotechnology. For a more thorough treatment of AFM techniques in general, and with respect to investigations of APS specifically, please refer to our recently published reviews.[8,11]

7.2 Materials and Methods

Our AFM investigations of APS have relied upon development of an in vitro simulacrum for this syndrome.[8,11] Briefly, a planar lipid layer, consisting of 30% phosphatidylserine (PS)/70% phosphatidyl choline (PC) is formed on a freshly cleaved muscovite mica platform. Upon incubation with AnxA5, 2D crystals are formed over this planar lipid layer as AnxA5 binds to the polar heads of PS moieties (the details of the crystallization process are described in Reviakine et al., 2000[12]). Various components can be subsequently added and ensuing molecular interactions are monitored via temporal AFM imaging techniques in a hydrated environment. To summarize the components involved in these AFM experiments:

(1) Phosphatidylserine/phosphatidylcholine (PS/PC)
  - Planar lipid layers formed on freshly cleaved mica discs
(2) Annexin V (AnxA5)
  - High affinity for phosphatidyl serine polar heads (anionic)
  - Forms anticoagulant shield over membrane
(3) Antiphospholipid (aPL) antibodies
  - Recognize phospholipid-binding cofactor proteins (primarily β_2GPI)
  - aPL mAbs are generated from APS patient peripheral blood mononuclear cells

(4) β_2GPI cofactor

- Also known as apoplipoprotein H
- Phospholipid-binding protein
- Primary antigenic target for aPL antibodies
- Presence required for APS effect of aPL antibodies

Imaging reported in this chapter was performed with a Veeco BioScope I AFM, mounted on an Olympus IX70 inverted microscope. Samples were imaged in fluid using "tapping mode" conditions with a fluid cell to minimize probe–sample interactions, while providing high-resolution images.

7.3   Results

7.3.1   AFM Assessment of the Modulation of APS by Pharmacological and Molecular Intervention

The APS disorder is currently treated with anticoagulant drugs (for a recent review, see Tripodi et al., 2011[13]), a treatment that is effective, but is associated with the inherent risks of hemorrhagic complications and does not target earlier steps in the mechanism of this disease process. These limitations led us to consider the question of whether putative candidate therapies that target specific aPL-mediated disease processes might be detected through AFM imaging. We present our recent AFM efforts directed at molecular interventions to repair the defects in the AnxA5 2D crystal anticoagulant shield initiated by exposure to β_2GPI and aPL monoclonal antibodies.

7.3.1.1   Hydroxychloroquine-mediated increase of AnxA5 binding

We considered hydroxychloroquine (HCQ), a synthetic antimalarial drug, to be a particularly attractive candidate molecule for this repair mechanism because retrospective clinical studies have indicated that there is a reduction in the prevalence of thrombosis among patients who take the drug for other clinical indications[14,15]; prospective clinical trials have not yet been done, but are currently being planned by the international collaborative network APS ACTION (http://www.apsaction.org).

HCQ, a 4-aminoquinoline molecule, is a synthetic amphiphilic antimalarial drug that has come to be widely used for nonsteroidal immunosuppressive treatment of systemic lupus erythematosus (SLE) and rheumatoid arthritis.[16-21] Importantly, there has been significant experience with its use for SLE during pregnancy, a critical parameter for its proposed use to treat APS. Moreover, HCQ has been demonstrated to reduce thrombosis in an animal model of APS.[22] Since the recognition of β_2GPI by anti-aPL antibodies plays a central role in the aPL syndrome, we investigated whether the reported reduction of thrombosis might be attributable to a HCQ-mediated impact on the formation of β_2GPI-aPL antibody complexes on planar lipid layers.

HCQ was prepared in HEPES-buffered saline (HBS) to yield final concentrations of 1 mg/mL or 1 µg/mL. In these experiments, β_2GPI was added to a planar lipid layer (PS/PC), and incubated overnight at 4°C. The remainder of the experiment including the varying orders of addition of the proteins and the drug was performed the following day, as previously described.[6] Qualitatively, the addition of HCQ to pre-formed β_2GPI-anti-aPL antibody complexes resulted in an attenuation of complex height.[6] Moreover, quantitative analysis of 40 randomly selected β_2GPI-anti-aPL antibody complexes revealed that the complex height was reduced from a mean value of 17.7 (±5.8) nm to 9.3 (±2.4) nm 30 min following the addition of HCQ.[6] In a further set of experiments, β_2GPI cofactor was added to a planar lipid layer at room temperature, followed approximately 45 min later by the addition of aPL antibody, resulting in the appearance of "toroid-like" complexes (Fig. 7.1). This was followed approximately 30 min later by the addition of HCQ, once again resulting in dissolution of the complexes (Fig. 7.1). These AFM imaging results showing the effect of HCQ on β_2GPI–anti-aPL antibody complexes were corroborated by other experimental data obtained via: (1) ellipsometry[6] (see Fig. 2 in the reference); (2) the binding of aPL antibodies to THP-1 monocytes[6] (see Fig. 3 in the reference); and (3) numerous clinical assays used for APS.[6]

Encouraged by these results, we next sought to determine whether the AnxA5 anticoagulant shield disruption promoted by exposure to β_2GPI–anti-aPL antibody complexes might be treated/reversed by addition of HCQ. For these experiments, AnxA5 2D crystals were formed on planar lipid layers, followed by addition of β_2GPI and anti-aPL antibodies in the presence of HCQ. Remarkably, we observed that HCQ disrupted these immune complexes[7] and that a second layer of AnxA5 crystals formed around and over the β_2GPI–anti-aPL antibody complexes, in essence producing a "patch" of the damaged planar lipid layer induced by binding of the β_2GPI cofactor–antibody complex.[7] Other experiments were then performed in which the order of addition of the reagents was altered. For instance, in the experiment depicted in Fig. 7.2, β_2GPI was added to a planar lipid layer and incubated overnight at 4°C, followed by the addition of anti-aPL antibodies. After the

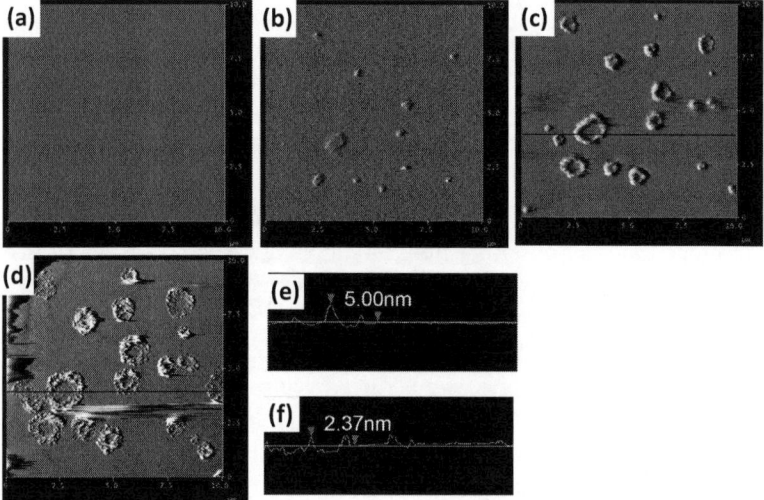

Figure 7.1 The panel of AFM (a–d) depicts the interactions of plasma protein cofactor β_2 glycoprotein I (β_2GPI) with PS/PC (30% phosphatidylserine and 70% phosphatidylcholine) planar lipid layer and monoclonal antiphospholipid antibody IS4, and the subsequent effects of addition of the drug hydroxychloroquine (HCQ). (a) lipid membrane alone. (b) Interaction of β_2GPI with lipid membrane, presenting oligomer conglomerates. (c) Toroid formations of aPL antibody–β_2GPI complexes. (d) The post-addition of HCQ reduces and removes aPL antibody–β_2GPI complexes from the lipid bilayer, with voids resulting. (e–f) Off-line cross-sectional measurements from corresponding height mode images of indicated toroid structures (black line across image in "d") before (e) and after (f) administration of HCQ. This sectional analysis reveals an approximate >50% reduction in height of the complexes.

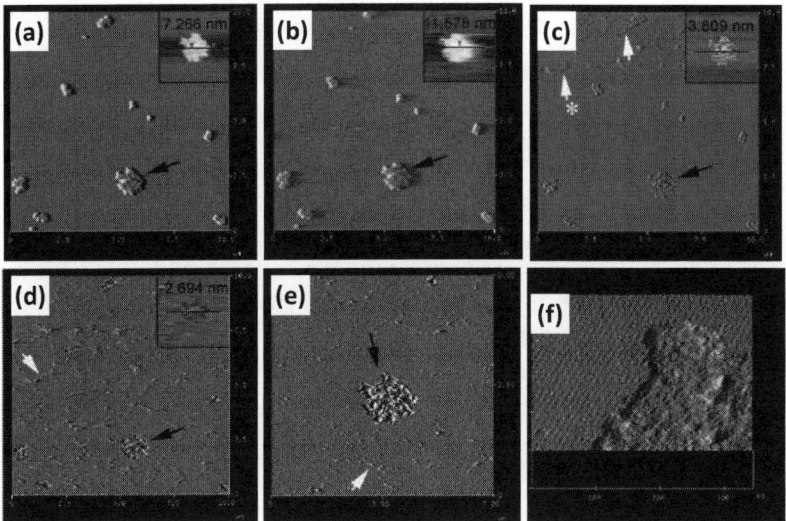

Figure 7.2 AFM images showing AnxA5 2D crystal formation in the presence of HCQ-dissociated aPL IgG–β_2GPI complexes. (a) Binding of β_2GPI to phospholipid bilayer resulted in formation of distinct structures. (b) Binding of aPL IgG mAb, to β_2GPI prebound to phospholipid results in enlarged structures of increased height as indicated by insets displaying cross-sectional measurements of a specific complex (black arrow). (c, d, e, f) The effect of addition of HCQ on the aPL mAb and β_2GPI complexes, followed by the addition of AnxA5; concurrently, AnxA5 forms with distinctive grain boundaries (white arrows). White arrow with asterisk indicates advancing edge of a coalescing AnxA5 island. After addition of HCQ, the indicated immune complex (black arrows) along with others, was significantly eroded. Further disintegration of complex erosion and reduction is followed by height analysis as displayed in panel insets. (f) Zoomed image of area of interest (edge of dissolving complex) indicated by black arrow in image e, displaying transitional formation of the AnxA5 2D crystalline lattice around and within the reduced immune complex. (a–d) 10 μm square images were electronically zoomed from original 30 μm × 30 μm scans. (a–d) insets: 3 μm square height images were zero lattened and show quantitative analysis of the effects of antibody on bound cofactor (antigen) and HCQ on immune complex on indicated structure (black arrow). (e) Original 5 μm × 5 μm scan size. (f) A three-dimensional 350 nm × 350 nm image electronically zoomed from an original 500 nm × 500 nm scan. (a–e) Images were minimally processed to remove scan lines. (f) 2D fast Fourier transform was applied.

formation of β_2GPI–anti-aPL antibody complexes, HCQ was added (which began dissolution of the complexes as shown in Fig. 7.2c), followed shortly thereafter by addition of AnxA5. Two-dimensional crystals of AnxA5 formed monolayers over the planar lipid surface (seen at the top of 7.2c), with some components of the dissolving β_2GPI–anti-aPL antibody complexes pushed to the peripheries of the growing AnxA5 islands (Fig. 7.2d). A high-resolution image of the formed AnxA5 crystal is shown in Fig. 7.2f, with an indication that the crystal has also formed within a partially dissolved β_2GPI–anti-aPL antibody complex. These results are highly compelling and have prompted us to continue investigation of the molecular mechanisms responsible for this "patching" phenomenon. To this end, our current efforts are directed toward elucidating the mechanisms responsible for the "patching" effect HCQ manifests toward the damaged AnxA5 crystal formed on artificial lipid bilayers. Specifically, we are investigating the molecular interactions between HCQ and the aPL immune complexes, utilizing chemical modifications of the HCQ molecule, other 4-aminoquinoline analog molecules, and with recombinant forms of the proteins.

The data described above confirm our hypothesis that patient-derived aPL antibodies disrupt the AnxA5 crystal shield and also demonstrate that exploration of this mechanism for thrombosis can yield innovative and practical treatment approaches that may have the possible benefit, with a drug that is already available and FDA-approved for other uses, of reducing thrombosis without the bleeding risks of long term anticoagulation.

7.3.1.2 Reversal of AnxA5 resistance with "patch treatment" for AnxA5 defects

In addition to exploring the use of agents that promote the binding of AnxA5 to the phospholipid bilayer (such as HCQ described above), we also investigated whether an alternative form of "patch treatment" might be used to cover gaps in the AnxA5 crystals that leave PS exposed. We sought to identify novel PS ligands through phage display technology[23] to screen a Ph.D.-7 random heptapeptide library (New England Biolabs), containing ~1.8 × 10^9 sequences, utilizing the following novel panning strategy:

(1) adding the phage to PS-coated polystyrene beads; (2) adding AnxA5 to elute PS-bound phage; (3) subtracting PC-bound phage by absorption with a PC bilayer; (4) panning with PS/PC bilayer; (5) adding AnxA5 to elute PS-bound phage from PS/PC; and finally (6) a second round of subtraction of PC-bound phage.

The results from this panning procedure yielded the following consensus heptapeptide sequence: his-trp-X-X-X-X-arg (hwxxxxr), characterized by the presence of cationic amino acids at its ends and hydrophobic amino acids in the center. Initially, the capacity of these consensus phages to bind to PS/PC bilayers was assayed with ellipsometry. The results demonstrated that the consensus phages significantly bound to the PS/PC bilayer, whereas the non-consensus control peptides displayed little or no binding (not shown).

Next, we performed high-resolution imaging of the binding of the consensus sequence phages to planar PS/PC layers by AFM. Using our previously developed (and described above) simulacrum for these AFM studies, PS/PC planar lipid layers were formed on a freshly cleaved mica surface, 14×10^6 consensus phage particles were then added, followed finally by the addition of AnxA5. Consensus phage (hwkwplr) particles bound directly to the PS/PC layer (Fig. 7.3), carpeting the entire surface of the bilayer. Interestingly, upon addition of AnxA5, 2D AnxA5 crystals rapidly formed on the PS/PC planar lipid layer, forcing the previously bound phage particles to the borders of the incompletely formed individual 2D crystal assemblages (Fig. 7.3). We had earlier demonstrated by temporal AFM imaging that the formation of 2D AnxA5 crystals on planar PS/PC lipid layers progressed from individual nucleation sites.[5] Moreover, we also showed that defects existed at the border zones (furrows) between adjacent forming crystals where they were not precisely aligned.[5] Thus, at higher magnification (Fig. 7.3f), individual phage particles expressing the consensus hwkwplr peptide can be observed binding to the crystal defects in these border zones. In a similar set of experiments, addition of control phage particles lacking the consensus PS-binding sequence showed no change on the planar lipid layer (Fig. 7.4). Addition of AnxA5 resulted in the formation of 2D AnxA5 crystals on the planar lipid layer as normally observed (Fig. 7.4).

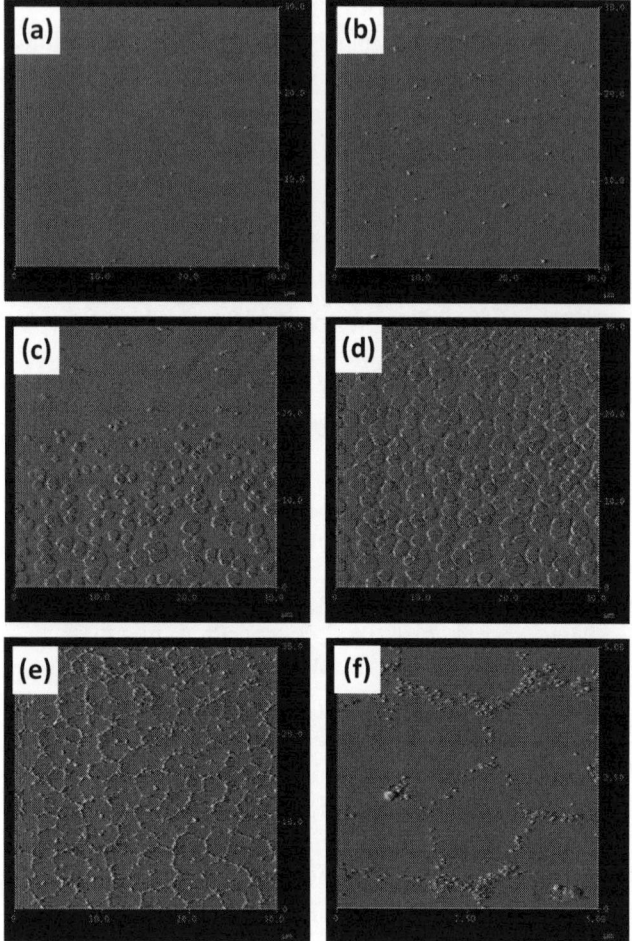

Figure 7.3   Temporal imaging displaying phage binding to planar PS/PC lipid layer and its displacement by post-addition of AnxA5. (a) Lipid surface only. (b) Phage particles with consensus PS-specific sequence binding to lipid surface. (c) Initial dynamic formation of AnxA5 crystal coalescing from nucleation sites. (d) Continued AnxA5 coalescence displaces and relocates lightly bound phage. (e) Loosely bound phage structures are relocated to AnxA5 furrow sites between adjacent coalesced AnxA5 crystalline patches. (f) High-magnification (5 μm × 5 μm) scan of completed AnxA5 crystal with ridge lines of consensus phage particles. (a–f) Amplitude images; (a–e) 30 μm × 30 μm scans.

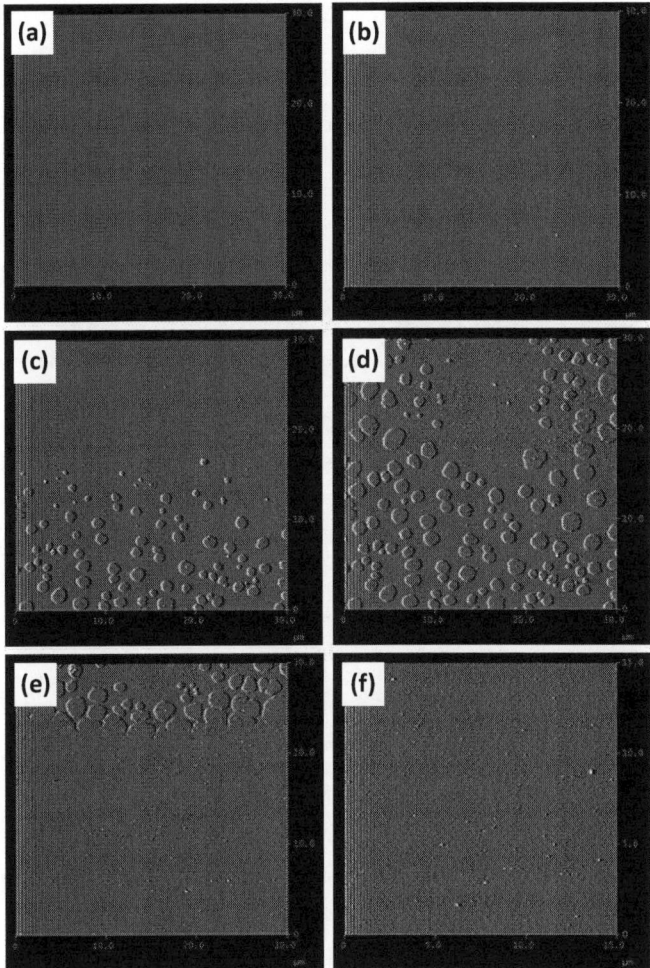

Figure 7.4 Temporal sequence showing absence of binding of control phage particles, lacking consensus sequence, to planar PS/PC lipid membrane and complete formation of AnxA5 crystal. (a) Lipid surface only. (b) Lipid surface following addition of control phage particles; no binding is apparent. (c) Initial dynamic formation of AnxA5 2D crystal coalescing from nucleation sites. (d) Continued AnxA5 coalescence with time. (e) Coalescing AnxA5 2D crystal nearing completion. (f) 15 μm × 15 μm scanned image of AnxA5 2D crystal on PS/PC lipid layer displaying a homogenous surface devoid of ridge lines. (a–e) Amplitude images; 30 μm × 30 μm scans.

These observations led us next to the question as to whether this consensus heptapeptide sequence, which binds to PS residues and marks sites of defective AnxA5 crystallization, might also be able to patch aPL antibody-induced defects of AnxA5 crystallization. In essence, could the consensus peptide reverse "AnxA5 resistance" and "repair" the anticoagulant shield? Addition of the sequence—hwkwplr—(phage concentration 5×10^9/mL) reversed a functional assay that reflects disruption of AnxA5 crystallization—an assay for the aPL-mediated resistance to AnxA5 anticoagulant activity.[24–28] As a control reaction, addition of an irrelevant heptapeptide (tktptvl) at the same concentration of 5×10^9/mL resulted in no such reversal effect. Moreover, from a clinical standpoint, plasmas from three patients with APS showed significant resistance to AnxA5 anticoagulant activity at baseline and with addition of the control sequence; however, addition of phage particles displaying the consensus hwkwplr sequence restored anticoagulant activity to the same level as the control plasmas. These results are impressive "proof of concept" for the idea that aPL mediated defects in the AnxA5 crystal can be "repaired" and can also reverse the thrombogenic procoagulant effects of aPL antibodies. Additionally, these experiments demonstrate the feasibility of the new concept of using "patching agents" to fill in gaps in the crystalline AnxA5 arrays and offer a non-anticoagulant approach for the treatment of APS. Such an approach could offer the benefits of a treatment that is without the risks of hemorrhage that are inherent to all of the current anticoagulant therapies for this disorder.

7.4 Conclusions and Outlook

APS has been an enigmatic and relatively intractable condition, treated to date only with anticoagulant regimens that are associated with potentially serious long-term risks. It would therefore be desirable to provide alternative treatment options that target specific steps in the disease mechanism(s). If APS does indeed represent antibody-mediated unmasking of an anticoagulant shield of AnxA5 on cell membranes, then "patching" strategies may represent ideal targets for therapeutic intervention. The two novel "patching" methods described in this chapter offer initial strategies to address this clinical conundrum. We anticipate that future studies, which will continue to include high-resolution atomic force microscopy will successfully elucidate the molecular mechanisms underpinning these novel "patching" approaches to treat this thrombotic disorder.

Acknowledgment

The work described in this chapter was supported in part by a grant from the National Institutes of Health/National Heart, Lung & Blood Institute (R01 HL061331).

References

1. Binnig, G., Rohrer, H., Gerber, C., and Weibel, E. (1982). Surface studies by scanning tunneling microscopy. *Phys. Rev. Lett.*, **49**, 57–61.

2. Binnig, G., Quate, C. F., and Gerber, C. (1986). Atomic force microscope. *Phys. Rev. Lett.*, **56**, 930–933.

3. Katan, A. J., and Dekker, C. (2011). High-speed AFM reveals the dynamics of single biomolecules at the nanometer scale. *Cell*, **23**, 979–982.

4. Miyagi, A., Ando, T., and Lyubchenko, Y. L. (2011). Dynamics of nucleosomes assessed with time-lapse high-speed atomic force microscopy. *Biochemistry*, **50**, 7901–7908.

5. Rand, J. H., Wu, X.-X., Quinn, A. S., Chen, P. P., McCrae, K. R., Bovill, E. G., and Taatjes, D. J. (2003). Human monoclonal antiphospholipid antibodies disrupt the annexin A5 anticoagulant shield on phospholipid bilayers. Evidence from atomic force microscopy and functional assay. *Am. J. Pathol.*, **163**, 1193–1200.

6. Rand, J. H., Wu, X.-X., Quinn, A. S., Chen, P. P., Hathcock, J. J., and Taatjes, D. J. (2008a). Hydroxychloroquinine directly reduces the binding of antiphospholipid antibody–β_2-glycoprotein I complexes to phospholipid bilayers. *Blood*, **112**, 1687–1695.

7. Rand, J. H., Wu, X.-X., Quinn, A. S., Ashton, A. W., Chen, P. P., Hathcock, J. J., Andree, H. A. M., and Taatjes, D. J. (2010a). Hydroxychloroquine protects the annexin A5 anticoagulant shield from disruption by antiphospholipid antibodies: evidence for a novel effect for an old antimalarial drug. *Blood*, **115**, 2292–2299.

8. Quinn, A. S., Wu, X.-X., Rand, J. H., and Taatjes, D. J. (2012a). Insights into the pathophysiology of the antiphospholipid syndrome provided by atomic force microscopy. *Micron*, **43**, 851–862.

9. Rand, J. H. (2002). Molecular pathogenesis of the antiphospholipid syndrome. *Circ. Res.*, **90**, 29–37.

10. Rand, J. H., Wu, X.-X., Quinn, A. S., and Taatjes, D. J. (2010b). The annexin A5-mediated pathogenic mechanism in the antiphospholipid syndrome: role in pregnancy losses and thrombosis. *Lupus*, **19**, 460–469.

11. Quinn, A. S., Rand, J. H., Wu, X.-X., and Taatjes, D. J. (2012b). Viewing dynamic interactions of proteins and a model lipid membrane with atomic force microscopy. In *Cell Imaging Techniques: Methods and Protocols* (D. J. Taatjes and J. Roth, eds.), *Methods in Molecular Biology*, **Vol. 931**, Springer Science + Business Media, pp. 259–293.

12. Reviakine, I., Bergsma-Schutter, W., Mazeres-Dubut, C., Govorukhina, N., and Brisson A. (2000). Surface topography of the p3 and p6 annexin V crystal forms determined by atomic force microscopy. *J. Struct. Biol.*, **131**, 234–239.

13. Tripodi, A., de Groot, P. G., and Pengo, V. (2011). Antiphospholipid syndrome: laboratory detection, mechanisms of action and treatment. *J. Intern. Med.*, **270**, 110–122.

14. Petri, M. (1996). Thrombosis and systemic lupus erythematosus: the Hopkins Lupus Cohort perspective. *Scand. J. Rheumatol.*, **25**, 191–193.

15. Erkan, D., Yazici, Y., Peterson, M. G., Sammaritano, L., and Lockshin, M. D. (2002). A cross-sectional study of clinical thrombotic risk factors and preventive treatments in antiphospholipid syndrome. *Rheumatology (Oxford)*, **41**, 924–929.

16. Rothfield, N. (1988). Efficacy of antimalarials in systemic lupus erythematosus. *Am. J. Med.*, **85**, 53–56.

17. Nayak, V., and Esdaile, J. M. (1996). The efficacy of antimalarials in systemic lupus erythematosus. *Lupus*, **5**(Suppl 1), S23–S27.

18. Tsakonas, E., Joseph, L., Esdaile, J. M., Choquette, D., Senécal, J. L., Cividino, A., Danoff, D., Osterland, C. K., Yeadon, C., and Smith, C. D. (1998). A long-term study of hydroxychloroquine withdrawal on exacerbations in systemic lupus erythematosus. The Canadian Hydroxychloroquine Study Group. *Lupus*, **7**, 80–85.

19. Molad, Y., Gorshtein, A., Wysenbeek, A. J., Guedj, D., Majadla, R., Weinberger, A., and Amit-Vazina, M. (2002). Protective effect of hydroxychloroquine in systemic lupus erythematosus. Prospective long-term study of an Israeli cohort. *Lupus*, **11**, 356–361.

20. Fessler, B. J., Alarcón, G. S., McGwin, G. Jr, Roseman, J., Bastian, H. M., Friedman, A. W., Baethge, B. A., Vilá, L., and Reveille, J. D. (2005). LUMINA Study Group. Systemic lupus erythematosus in three ethnic groups: XVI. Association of hydroxychloroquine use with reduced risk of damage accrual. *Arthritis Rheum.*, **52**, 1473–1480.

21. Clowse, M. E., Magder, L., Witter, F., and Petri, M. (2006). Hydroxy-chloroquine in lupus pregnancy. *Arthritis Rheum.*, **54**, 3640–3647.

22. Edwards, M. H., Pierangeli, S., Liu, X., Barker, J. H., Anderson, G., and Harris, E. N. (1997). Hydroxychloroquine reverses thrombogenic properties of antiphospholipid antibodies in mice. *Circulation,* **96**, 4380–4384.

23. O'Neil, K. T., Hoess, R. H., Jackson, S. A., Ramachandran, N. S., Mousa, S. A., and DeGrado, W. F. (1992). Identification of novel peptide antagonists for GPIIb/IIIa from a conformationally constrained phage peptide library. *Proteins*, **14**, 509–515.

24. Rand J. H., Wu X.-X., Lapinski R., van Heerde W. L., Reutelingsperger C. P., Chen P. P., and Ortel, T. L. (2004). Detection of antibody-mediated reduction of annexin A5 anticoagulant activity in plasmas of patients with the antiphospholipid syndrome. *Blood*, **104**, 2783–2790.

25. Rand, J. H., Wu, X.-X., Quinn, A. S., and Taatjes, D. J. (2008b). Resistance to annexin A5 anticoagulant activity: a thrombogenic mechanism for the antiphospholipid syndrome. *Lupus*, **17**, 922–930.

26. de Laat, B., Wu, X.-X., van Lummel, M., Derksen, R. H., de Groot, P. G., and Rand, J. H. (2007). Correlation between antiphospholipid antibodies that recognize domain I of β_2-glycoprotein I and a reduction in the anticoagulant activity of annexin A5. *Blood*, **109**, 1490–1494.

27. Hunt, B. J., Wu, X.-X., de Laat, B., Arslan, A. A., Stuart-Smith, S., and Rand. J. H. (2011). Resistance to annexin A5 anticoagulant activity in women with histories for obstetric antiphospholipid syndrome. *Am. J. Obstet. Gynecol.*, **205**, 485, e17–e23.

28. Wahezi, D. M., Ilowite, N. T., Rajpathak, S., and Rand, J. H. (2012). Prevalence of annexin A5 resistance in children and adolescents with rheumatic diseases. *J. Rheumatol.*, **39**, 382–388.

Chapter 8

A Novel Approach to Study Molecular Evolution: Detection of Ancestral Conformation Hidden in Present-Day Proteins Using Antibody as Nanostructure Probes

Jian-Ping Jin

Department of Physiology, Wayne State University School of Medicine,
Detroit, Michigan 48201, USA

jjin@med.wayne.edu

Physical remains of ancestor nucleotides and proteins are largely unavailable; thus, sequence comparison between homologous genes in present-day organisms forms the basis for the current knowledge of molecular evolution. However, variation in protein three-dimensional structure is the foundation for functional diversity and pathological mutations. To understand the evolutionary relationship of three-dimensional structures in related proteins would have major impacts on our understanding of the diversity as well as conservation of biological processes. A protein may contain ancestor conformations that have been allosterically suppressed or modified by evolutionarily added structures. Using polyclonal and monoclonal antibodies as three-dimensional nanostructure

NanoCellBiology: Multimodal Imaging in Biology and Medicine
Edited by Bhanu P. Jena and Douglas J. Taatjes
Copyright © 2014 Pan Stanford Publishing Pte. Ltd.
ISBN 978-981-4411-79-0 (Hardcover), 978-981-4411-80-6 (eBook)
www.panstanford.com

probes to detect such conformation in proteins after removing the suppressor/modifier structure, we have demonstrated the feasibility to detect evolutionarily suppressed ancestor-like conformational states in proteins. In addition to identifying structural modifications that were critical to the emerging of diverged proteins, investigating protein evolution using this novel approach will help to understand the origin as well as functional potential of existing protein structures.

8.1 Evolution Is data: Can We Get Fossil-Like Information for the Ancestor of a Present-Day Protein?

The study of biological evolution depends crucially on fossil evidence. However, material remains of ancestor nucleotides and proteins are largely unavailable, thus, the evolutionary relationship between proteins is primarily evaluated by sequence comparisons between homologous genes in present-day species. Although this approach has provided information for how evolution has worked,[1] it has two intrinsic limitations: One is that the sequences of homologous genes in present-day species have all gone through the same length of evolutionary time and the counterpart in lower organisms do not exactly equal to the ancestor of a protein in higher organisms. The other is that the functional evolution of a protein is determined by changes in three-dimensional structures; thus, the primary structural relationship indicated in sequence comparisons provides only limited information.

To further investigate how the three-dimensional structure or conformation of a protein has evolved would significantly strengthen our understanding of evolutionary relationships indicated by using nucleotide and amino acid sequence comparisons. Protein three-dimensional structures may contain also allosteric information that reflects the evolutionary past. The growing but still small database of high-resolution three-dimensional structures, especially allosteric states, determined by X-ray crystallography and nuclear magnetic resonance spectroscopy is currently insufficient for the study of molecular evolution and a higher throughput approach is needed.

8.2 Antibodies as Three-Dimensional Nanostructure Probes

The binding affinity of an antibody to an antigen depends on three-dimensional structure fits between the antibody variable region (the paratope) and a specific epitope on the antigen protein. The degree of cross-reactivity of an antibody to homologous proteins reflects their structural similarity and diversity.[2] This "immunological distance" has been used to evaluate the phylogeny of homologous proteins.[3–5] While the cross-reactivity of polyclonal antisera reflects the overall structural similarity of related proteins, monoclonal antibodies (mAbs)[6] detect similarities and differences in specific epitopic structures to provide information for their evolutionary conservation and variation. An important extension of the application of antibody based analysis of protein structures is that in addition to detecting static structures, antibodies against allosteric epitopes provide sensitive probes for measuring conformational changes in a protein.[7–9] This approach enables us to explore a novel detection of ancestor structures hidden in present-day proteins.

8.3 Enzyme-Linked Immunosorbant Assay for Protein Epitope Analysis

Protein epitopic structures can be studied by microtiter plate enzyme-linked immunosorbant assay (ELISA) that effectively detects conformational differences and changes under native conditions.[7,10] ELISA epitope analysis can sensitively and quantitatively compare the relative binding affinity of a specific antibody probe to the corresponding epitopes in evolutionarily related proteins or protein isoforms. The assay can be readily carried as standard antibody affinity titrations, involving non-covalently immobilization of the proteins to be compared in 96-well microtiter plates, block any remaining free plastic surface with nonionic detergent, incubation with serial dilutions of the specific antibody probes, incubation with enzyme-conjugated second antibody, and substrate reaction. The enzymatic color development is monitored at a series of time points using a microplate reader. The values in the linear course of the color development are used to plot titration curves for comparing the relative binding affinity of the antibodies to the protein variants to be studied. Antibody dilutions at 50% maximum binding are calculated from the titration curves to quantify the relative binding affinity.[11]

The sensitive and rapid throughput ELISA epitope analysis[7,10] can readily detect similarities, differences, and especially changes in protein conformation. This approach is powerful not only in detecting protein allosteric changes during a switch between two functional states[10] but also in evaluating three-dimensional structure changes in geological time scale to identify fossils-like evidence for the study of molecular evolution. Since many mAbs as well as polyclonal antibodies can detect protein conformational changes[7-10,12-14] application of this approach can be explored on many proteins of interest using existing antibodies.

8.4 Exampling Studies of Troponin Subunits and Isoforms

Gene duplication and divergence produce genetic variation and molecular diversity that fuel evolution.[15] Troponin I (TnI) and troponin T (TnT) are two subunits of the troponin complex that regulates the contraction of striated (skeletal and cardiac) muscles.[16] The two proteins function together but play distinct roles in the troponin complex. Genes encoding TnI and TnT are closely linked in the vertebrate genome,[17] suggesting their origin from the duplication of a single ancestral gene followed by neofunctionalization. As linked pairs of genes, TnI and TnT have each evolved into three isoforms with specific expressions in cardiac, slow and fast skeletal muscles,[16] implying a series of duplication events from an ancestral TnI–TnT gene pair followed by subfunctionization and neofunctionalization.

Distinct TnI and TnT isoforms are found in the cardiac and skeletal muscles of hagfish,[11] an elementary vertebrate that shared with true vertebrates a common ancestor lived some half a billion years ago.[18] Although TnI and TnT are significantly diverged proteins, the close physical linkage of TnI and TnT genes and their shared feature of a variable N-terminal region[16] suggest the origination of TnI and TnT from duplication of an ancestor gene. Supporting this hypothesis, phylogenetic analysis of amino acid sequences showed a distant homology between the TnI and TnT gene families.[11] The early emergence and long existence of TnI–TnT gene family encoding genetically and functionally close-linked allosteric protein isoforms provides an excellent system to experimentally demonstrate the evolution of protein three-dimensional structures. This approach is strengthened by the availability of our collection of anti-TnT and anti-TnI antibodies.[11]

8.5 Unidirectional Immunological Cross-Reactivity of Polyclonal Anti-TnI and Anti-TnT Antibodies

While using antibodies to compare epitope structures of TnT and TnI, our first intriguing observation was that the cross-reactivity of an anti-TnI polyclonal antiserum RATnI raised by chicken fast TnI immunization cross-reacted with mouse fast TnT but not cardiac TnT or slow TnT. RATnI did not show detectable cross-reaction to chicken fast TnT,[11] excluding the possibility of TnT contamination in the chicken fast TnI immunogen used in generating the RATnI antiserum. Therefore, the cross-reactivity of RATnI to TnT indicated the presence of TnI-like epitope structures in TnT inherited from a common ancestor. This result supports the notion that TnI and TnT genes were evolved from duplication, other than splitting, of an ancestor gene followed by neofunctionalization.

In contrast, a polyclonal antiserum RATnT raised by fast TnT immunization did not show detectable cross-reactivity to TnI.[11] This unidirectional immunological cross-reactivity between TnI and TnT indicates that the dominant immunogenic epitopes in the TnT immunogen are not present in TnI, which correspond to evolutionarily diverged structures that were absent in the common ancestor of TnI and TnT.

Troponin I is the inhibitory subunit of troponin complex whose primary function is an inhibitor of actin–myosin interactions.[19] Consistent with this conserved function, the presence of TnI-like epitope structure in present day TnI and TnT further suggested a hypothesis that the ancestor of TnI and TnT was likely a TnI-like inhibitory protein.

The data that RATnI discriminately cross-reacted to fast TnT than cardiac or slow TnT[11] indicated a shorter immunological distance to fast TnI from fast TnT than that from cardiac or slow TnT, suggesting a further hypothesis that the original TnI–TnT gene pair evolved from duplication of the ancestral troponin gene was a fast TnI–TnT-like pair and the present-day fast TnI and fast TnT genes had less neofunctionalization from the original gene pair as compared to that of the other two TnI–TnT gene pairs.

8.6 Detection of Evolutionarily Suppressed TnI-Like Epitope Structures in TnT

Beyond static immunological distance, we further demonstrated the feasibility of experimentally detecting ancestor-like three-dimensional structure states that are suppressed in the evolutionarily diverged present-day protein. The N-terminal variable region of TnT is an evolutionarily additive structure[20] that conveys conformational and functional modulations.[7,13,14] Therefore, we made N-terminal truncated TnT to examine the effect of removing this evolutionary addition on the three-dimensional structure of the conserved regions of TnT. The results showed whereas no detectable cross-reaction between polyclonal antiserum RATnI with intact cardiac TnT, deletion of the N-terminal region of cardiac TnT resulted in significant binding of RATnI, indicating an unsuppression of TnI-like (ancestor-like) epitope structure or molecular conformation.

The C-terminal region of TnI is evolutionarily conserved among isoforms and across species.[21,22] Using mAb TnI-1 against a highly conserved C-terminal epitope of TnI as probe, we detected a significant cross-reaction of TnI-1 to N-terminal truncated, but not intact, cardiac TnT.[11] This study further supports the presence of evolutionarily suppressed ancestor-like (TnI-like three-dimensional structure) inherited in present-day TnT, which became detectable after removing an evolutionary determinant. These experiments also demonstrated that after the three-dimensional structure of TnT evolved to become distinct from that of TnI as the lack of cross-reactivity of RATnI and TnI-1 with intact cardiac TnT, removal of the evolutionarily additive and conformationally modulating N-terminal segment resulted in the resuming of an ancestor-like conformation detectable with the anti-TnI antibodies.

8.7 Detection of Evolutionarily Suppressed Fast TnT-Like Epitope Structure in Cardiac TnT

Sequence analysis of the muscle fiber type isoforms of TnI and TnT demonstrated that the divergence of the three muscle type-specific isoforms is greater than the divergence of each isoform in different vertebrate species.[11,23] Although a (fast skeletal [slow skeletal, cardiac]) evolutionary relationship was indicated by a sequence-based study of other muscle proteins,[24] this order of divergence was unclear from the sequence analysis of the muscle type-specific TnI and TnT isoforms.[11]

Immunological distance between fast TnT and cardiac or slow TnT indicated that polyclonal antiserum RATnT raised with fast TnT immunization cross-reacted more strongly to cardiac TnT than that to slow TnT.[11] A mAb T12 raised with fast TnT immunization cross-reacted with cardiac TnT but not slow TnT.[11] The data that fast TnT is related more closely to cardiac TnT than slow TnT suggest that the fast TnI-fast TnT and slow TnI-cardiac TnT gene pairs shared a common ancestor (a fast TnI-fast TnT-like gene pair as suggested above).

This evolutionary lineage was further demonstrated by detecting suppressed fast TnT-like epitope structure in cardiac TnT after removing the evolutionarily additive N-terminal segment. The affinity of mAb T12 to N-terminal truncated cardiac TnT was significantly increased in comparison to that to intact cardiac TnT, whereas N-terminal truncation of slow TnT did not produce such effect.[11] It is worth noting that TnT is an elongated protein in which the N-terminal segment is an extended structure.[25,26] Therefore, removal of the N-terminal domain enhances antibody affinity to the distant T12 epitope structure by modulating molecular conformation other than unmasking a static structure.

8.8 Detection of Evolutionarily Suppressed Cardiac TnT-Like Epitope Structure in Slow TnT

As whether the cardiac TnI-slow TnT gene pair had evolved from duplication of the original fast TnI-fast TnT-like gene pair or from a slow TnI-cardiac TnT-like gene pair, the fact that mAb T12 raised by fast TnT immunization cross-reacted to cardiac TnT but not slow TnT supports the later hypothesis. Along this evolutionary lineage, mAb 1G9 raised with cardiac TnT immunization exhibited a weak cross-reaction to slow TnT but did not react to fast TnT, indicating a presence of cardiac TnT-like epitope in slow but not fast TnT.[11]

Serial N-terminal truncations of slow TnT produced progressive increases in the affinity of mAb 1G9 for an epitope in the C-terminal region when the N-terminal and middle regions were removed. Such conformational effects on mAb 1G9 affinity indicated that the cardiac TnT-like 1G9 epitope structure had significantly diverged in slow TnT, but the evolutionarily suppressed three-dimensional structure/molecular conformtation can restore when the modulating effect of the N-terminal segment was removed. The restoration of cardiac TnT-like three-dimensional structure in slow TnT was further demonstrated by the even higher affinity of mAb 1G9 to isolated C-terminal segment of slow TnT, indicating its nature as a "default" modular structure.[11]

In contrast, removing the N-terminal and middle regions of fast TnT did not produce detectable binding of mAb 1G9. The absence of 1G9 epitope-like structure in fast TnT suggest its emergence after the divergence of the ancestors of slow TnI-cardiac TnT and fast TnI-fast TnT gene pairs.

8.9 Implications

This chapter describes a novel approach to experimentally determine the evolutionary lineage of proteins through the analysis of folded structures. By using specific antibody epitope probes to dissect protein structure and conformational states, we showed an example of the evolutionary pedigree in the TnI–TnT gene family. In addition to demonstrating with three-dimensional structure/ conformation data that TnI and TnT arose from a TnI-like ancestor protein, we detected ancestral relationships from the fast TnI-fast TnT-like gene pair to the slow TnI-cardiac TnT gene pair and from the slow TnI-cardiac TnT-like ancestor gene pair to the cardiac TnI-slow TnT gene pair.

Beyond static state immunological distance, the effect of removing an evolutionarily additive structure from TnT on the detection of evolutionarily suppressed TnI-like epitopic structures inherited from the TnI-like ancestor is consistent with the role of the N-terminal variable region of TnT as a conformational modulator.[7,13,14] In contrast to the commonly seen destructive effects in which deleting a portion of a protein destroys the integrity of antibody epitopes, the dissection of TnT proteins based on their structural and functional domains demonstrated three-dimensional structures with ancestor-like conformational states. This restoring effect indicated that ancestor-like structures/conformations may be retained in a suppressed state in present-day proteins and be exhibited when the evolutionary modulating determinant(s) is removed. The detection of TnI-like epitopes in TnT after removing the evolutionarily added N-terminal modulator reflected structural changes reflecting a reversal of structural evolution to an ancestor-like state.

In addition to providing experimental evidence for the evolution of TnI, TnT and their isoforms, the detection of ancestor-like epitopic structures in present-day proteins by removing evolutionarily additive structures has a twofold significance. First, it suggests a hypothesis that while genetic variations produce gradual changes in the amino acid sequence of proteins to provide the raw material for natural selection and molecular evolution, functional divergence and fitness value are mainly based on three-dimensional structure changes that could be a global conformational change produced by critical sequence changes in a modulator domain. This model may also explain quantum functional changes and leaps in an evolutionary lineage. Our recent data[11] suggested a critical role of "evolutionary determinants" that are new structures capable of tuning the global conformation and function during protein evolution.

Second, these data suggested that the evolution of protein three-dimensional structure is under a selection in which a trait would be fixed and maintained depending on the final functional outcome. Therefore, present-day proteins could retain potentials of conferring ancestor-like three-dimensional structures that are suppressed by conformational modulations occurred during evolution. When the stabilizing effect of the conformational modulator is removed, the suppressed ancestor-like three-dimensional structures will be exhibited. This hypothesis may explain the molecular basis of atavism, in which a mutation that causes the loss of evolutionary modulation in a protein would result in the expression of an ancestral trait in contrast to a radical new trait.

8.10 Applications

Although the post genome era databases allow extensive sequence analyses for investigating evolutionary relationship of proteins or their functional domains, protein three-dimensional structure analysis has a unique value. Ancestral information may be retained in a present-day protein as an ancestor-like alternative conformation that might be more common for one protein than another derived from the same ancestor. The novel approach to experimentally detecting evolutionarily suppressed ancestor-like alternative conformations in present-day proteins makes it feasible to dig "fossil-like" molecular information and study the origin of present-day protein structures. While the protein three-dimensional structure database is rapidly growing, the precise correlation between sequence and folded structure remains to be established. At the present, an effective and practical method to sensitively, quantitatively and readily compare protein three-dimensional structure differences is extremely valuable for the study of protein evolution. The data presented in this chapter have demonstrated that ancestor-like conformational states can be detected with antibodies against epitopes that represent the ancestor-like alternative conformation. In addition, knowing that an ancestor-like alternative conformation is retained in a protein structure will guide the search for the same information "hidden" in the amino acid sequences. The novel approach will also allow investigations for the functional significance of ancestor-like conformational states inherited in present-day proteins as well as contribute to a better understanding of protein structure-function relationships.

8.11 Summary and Perspectives

This chapter presents an effective experimental approach to demonstrate that *Evolution is Data*. The feasibility to detect suppressed history-telling structural states in proteins by removing conformational modulator segments added during evolution revealed a novel mode of protein evolution (summarized in Fig. 8.1). Investigation of the evolution of protein three-dimensional structures opens a door to the study of molecular paleontology despite the absence of fossil materials. In addition to identifying structural modifications that were critical to the emerging of present-day proteins, to understand this novel mode of evolution and the role of conformational modulator sites that determined evolutionary divergence of a protein and the consequence of their manipulation will help to predict the structural and functional potentials of the protein and guide the design of structural modifications and the search for effective drug targets. Mutations in such evolutionarily determining structure of a protein might have high potency to modify phenotypes and cause diseases.

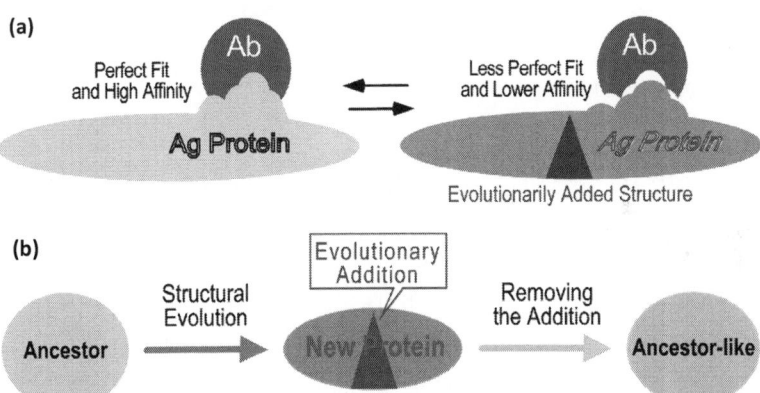

Figure 8.1 (a) Antigen–antibody binding affinity depends on structural fit between the antigenic epitope and the antibody paratope. A change in the conformation of an antigen protein may alter the affinity of an antibody. This feature provides a sensitive, quantitative and highly effective approach to detect and study the evolutionary relationship between related proteins. (b) A model is illustrated to show that the evolution of a protein three-dimensional structure could result from addition and/or alteration of a critical structure that modulates the global conformation and function of the protein. Ancestor-like three-dimensional structure/conformation suppressed in the new protein may restore after removing the effect of the modulatory structure.

Acknowledgment

This research was supported by grants from the National Institutes of Health AR048816, HL078773, HL086720 and HL098945 to J-PJ.

References

1. Benner, SA, Caraco, MD, Thomson, JM, and Gaucher, EA (2002). Planetary biology–paleontological, geological, and molecular histories of life. *Science,* **296**, 864–868.

2. Miller, EJ, and Cohen, AB (1991). Use of antibodies in the study of protein structure and function in lung diseases. *Am. J. Physiol.,* **260**, L1–L12.

3. Li, WH, and Tanimura, M (1987). The molecular clock runs more slowly in man than in apes and monkeys. *Nature,* **326**, 93–96.

4. Kaminogawa, S, Shimoda, M, Kurisaki, J, and Yamauchi, K (1989). Application of a monoclonal antibody to a comparative study of alpha-lactalbumins from various species. *J. Dairy Sci.,* **72**, 1124–1129.

5. Prager, EM (1993). The sequence-immunology correlation revisited: data for cetacean myoglobins and mammalian lysozymes. *J. Mol. Evol.,* **37**, 408–416.

6. Köhler, G, and Milstein, C (1975). Continuous cultures of fused cells secreting antibody of predefined specificity. *Nature,* **256**, 495–497.

7. Wang, J, and Jin, JP (1998). Conformational modulation of troponin T by configuration of the NH_2-terminal variable region and functional effects. *Biochemistry,* **37**, 14519–14528.

8. Jin, JP, Chen, A, Ogut, O, and Huang, QQ (2000). Conformational modulation of slow skeletal muscle troponin T by an NH(2)-terminal metal-binding extension. *Am J. Physiol. Cell Physiol.,* **279**, 1067–1077.

9. Jin, JP, Walsh, MP, Sutherland, C, and Chen, W (2000). A role for serine-175 in modulating the molecular conformation of calponin. *Biochem. J.,* **350**, 579–588.

10. Jin, JP, Chong, SM, and Hossain, MM (2007). Microtiter plate monoclonal antibody epitope analysis of Ca^{2+}- and Mg^{2+}-induced conformational changes in troponin C. *Arch. Biochem. Biophys.,* **466**, 1–7.

11. Chong, SM, and Jin, JP. (2009). To investigate protein evolution by detecting suppressed epitope structures. *J. Mol. Evol.,* **68**, 448–460.

12. Ogut, O, and Jin, JP (1996). Expression, zinc-affinity purification, and characterization of a novel metal-binding cluster in troponin T: metal-stabilized alpha-helical structure and effects of the NH_2-terminal variable region on the conformation of intact troponin T and its association with tropomyosin. *Biochemistry,* **35**, 16581–16590.

13. Jin, JP, and Root, DD (2000). Modulation of troponin T molecular conformation and flexibility by metal ion binding to the NH_2-terminal variable region. *Biochemistry,* **39**, 11702–11713.

14. Biesiadecki, BJ, Chong, SM, Nosek, TM, and Jin, JP (2007). Troponin T core structure and the regulatory NH_2-terminal variable region. *Biochemistry,* **46**, 1368–1379.

15. Raes, J, and Van de Peer, Y (2003). Gene duplication, the evolution of novel gene functions, and detecting functional divergence of duplicates in silico. *Appl. Bioinformatics.,* **2**, 91–101.

16. Jin, JP, Zhang, Z, and Bautista, JA (2008). Isoform diversity, regulation and functional adaptations of troponin and calponin. *Crit. Rev. Eukar. Gene Expr.,* **18**, 93–124.

17. Huang, QQ, and Jin, JP (1999). Preserved close linkage between the genes encoding troponin I and troponin T, reflecting an evolution of adapter proteins coupling the $Ca^{(2+)}$ signaling of contractility. *J. Mol. Evol.,* **49**, 780–788.

18. Fock, U, and Hinssen, H (2002). Nebulin is a thin filament protein of the cardiac muscle of the agnathans. *J. Muscle Res. Cell Motil.,* **23**, 205–213.

19. Perry, SV (1999). Troponin I: inhibitor or facilitator. *Mol. Cell. Biochem.,* **190**, 9–32.

20. Jin, JP, and Samanez, R (2001). Evolution of a metal-binding cluster in the NH(2)-terminal variable region of avian fast skeletal muscle troponin T: functional divergence on the basis of tolerance to structural drifting. *J. Mol. Evol.,* **52**, 103–116.

21. Jin, JP, Yang, F, Yu, Z, Ruse, C, Bond, M, and Chen, A (2001). The highly conserved COOH terminus of troponin I forms a Ca^{2+}-modulated allosteric domain in the troponin complex. *Biochemistry,* **40**, 2623–2631.

22. Akhter, S, Zhang, Z, Jin, JP (2012). The heart-specific NH_2-terminal extension regulates the molecular conformation and function of cardiac troponin I. *Am. J. Physiol. Heart Circ. Physiol.,* **302**, H923–H933.

23. Jin, JP, Chen, A, and Huang, QQ (1998). Three alternatively spliced mouse slow skeletal muscle troponin T isoforms: conserved primary structure and regulated expression during postnatal development. *Gene,* **214**, 121–129.

24. Oota, S, and Saitou N (1999). Phylogenetic relationship of muscle tissues deduced from superimposition of gene trees. *Mol Biol Evol.,* **16**, 856–867.

25. Cabral-Lilly, D, Tobacman, LS, Mehegan, JP, and Cohen, C (1997). Molecular polarity in tropomyosin-troponin T co-crystals. *Biophys. J.,* **73**, 1763–1770.

26. Wendt, T, Guenebaut, V, and Leonard, KR (1997). Structure of the Lethocerus troponin-tropomyosin complex as determined by electron microscopy. *J. Struct. Biol.,* **118**, 1–8.

Chapter 9

mRNA Nanomachines and Stress Reprogramming Following Brain Ischemia

Donald J. DeGracia, Jill T. Jamison, Manupreet Chawla, Monique K. Lewis, Michelle Smith, and Jeffery J. Szymanski

Department of Physiology, Wayne State University School of Medicine, Detroit, MI, 48201, USA

ddegraci@med.wayne.edu

List of Abbreviations

2VO/HT, bilateral carotid artery occlusion with hypovolemic hypotension
40S, small ribosomal subunit
60S, large ribosomal subunit
AA, arachidonic acid
ARE, adenine and uridine rich element
ATF6, activating transcription factor 6
CA/R, cardiac arrest and resuscitation
CA1, region Cornu Ammonis 1 of hippocampus
CA3, region Cornu Ammonis 3 of hippocampus
c_D, proportionality constant of D
COXIV, cytochrome C oxidase subunit IV

NanoCellBiology: Multimodal Imaging in Biology and Medicine
Edited by Bhanu P. Jena and Douglas J. Taatjes
Copyright © 2014 Pan Stanford Publishing Pte. Ltd.
ISBN 978-981-4411-79-0 (Hardcover), 978-981-4411-80-6 (eBook)
www.panstanford.com

c_S, proportionality constant of S

D, total damage

DND, delayed neuronal death

eIF2 αP, eIF2 holoprotein with alpha subunit phosphorylated

eIF2, eukaryotic translation initiation factor 2

eIF2B, eukaryotic translation initiation factor 2B

eIF2α, alpha subunit of eIF2

eIF2αP, phosphorylated eIF2α

eIF4F, eukaryotic translation initiation factor 4F

eIF4G, subunit G of eIF4F

ER, endoplasmic reticulum

erp29, ER protein 29 kDa

ET1, endothelin-1

FISH, fluorescent in situ hybridization

GM130, Golgi marker 130

Hdj1, co-chaperone for HSP70

hsp70/HSP70, inducible heat shock protein 70 kDa mRNA/protein, respectively

I, amount of the injury

I/R, ischemia and reperfusion

IF, immunofluorescence histochemistry

IHC, immunohistochemistry

IP, ischemic preconditioning

IRE1, inositol requiring 1

MCA, middle cerebral artery

MCAO, middle cerebral artery occlusion

mRBPs, mRNA-binding proteins

mRNA, messenger RNA

mRNP, mRNA ribonucleoprotein complexes

n_D, Hill coefficient for D

Neurofilament H/M, neurofilament heavy/medium weight

NIC, non-ischemic controls

n_S, Hill coefficient for S

P bodies, processing bodies

PA, protein aggregates

PABP, poly(A)-binding protein

PDI, protein disulfide isomerase

PERK, PKR-like endoplasmic reticulum eIF2α kinase

pMCAO, permanent MCAO

poly(A), poly-adenylated mRNA
RNS, reactive nitrogen species
ROS, reactive oxygen species
S, total induced stress responses,
S6, small ribosomal subunit protein 6
SG, stress granules
TA, translation arrest
TGN38, trans-Golgi network marker 38
TIA-1, T cell internal antigen 1
tPA, tissue plasminogen activator,
TTP, tristetraprolin
UPC, ubi-protein clusters
UPR, unfolded protein response
UTR, untranslated regions
Θ_D, threshold amount of D inhibiting S by 50%
Θ_S, threshold amount of S inhibiting D by 50%
λ_D, injury constant of D
λ_S, injury constant of S

Cardiac arrest and stroke are two leading causes of morbidity and mortality worldwide. Both conditions induce brain ischemia, a decrease or cessation of brain blood flow, resulting in neuronal death. Depending on the extent and duration of ischemia, cell death may be immediate (necrosis) or delayed. Delayed neuronal death (DND) after ischemia offers opportunity for intervention to prevent cell death; so-called neuroprotection. It is therefore of significant clinical interest to understand DND as a prerequisite to successful neuroprotection. All post-ischemic neurons show a translation arrest (TA), or inhibition of proteins synthesis. Translation arrest is transient in surviving neurons but persists in neurons destined to die by DND. Translation arrest is a marker of the expression of post-ischemic stress responses, the intrinsic genetic programs possessed by neurons to protect themselves and ameliorate the damage wrought by injuries such as ischemia. Transient TA is a marker of successful and persistent TA indicates unsuccessful execution of neuronal stress responses. We trace the evolution of studies of post-ischemic TA, with focus on our recent work on mRNA regulation as a contributor to post-ischemic TA, where ribosomes are physically separated from mRNA in post-ischemic neurons in the form of mRNA granules. The phenomenology of mRNA granules in the context of the molecular biology of mRNA regulation is described. Persistent TA can be linked to a defect in mRNA regulation that precludes the translation of intrinsic neuroprotective genes. The empirical results are abstracted to a nonlinear model of cellular injury that accounts for how injury magnitude determines whether a cell recovers or dies. Though driven by clinical necessity to study brain ischemia, we discover an underlying systems biology that illustrates a complex, nonlinear, networked nanotechnology underlying the binary fate decision of survival or death after cell injury.

9.1 Introduction

Molecular biology has continued to rapidly expand on many fronts over the past 20 years. One of these fronts is of particular importance for the fundamental understanding of general cell biology: the handling and regulation of messenger RNA (mRNA) molecules by cells. Recent advances can broadly be classified into two areas: (1) the discovery of microRNAs and their regulation and (2) the discovery of new, usually transient and labile, subcellular structures involved in mRNA function, a general area that we will term "ribonomics."[1] Some well-studied recent examples of ribonomic structures include stress granules and processing bodies,[2] and other ribonomic structures have been identified. MicroRNAs are widely known because they have engendered the technology of RNA silencing, but from a basic cell biology perspective the microRNA systems are examples of ribonomic systems,[3] and thus are a subset of the more general discovery that cells have evolved a diverse set of structures for regulating mRNA molecules.

In this chapter, we will provide a cursory overview of ribonomic regulation of mRNA as a prelude to discussing our studies of mRNA regulation in brain neurons following brain ischemia and reperfusion (I/R) injury. By 1990, it had been conclusively shown that there is a strict correlation between protein synthesis rates in post-ischemic neurons and their eventual survival or death.[4] In our studies of the translation system of post-ischemic neurons, we discovered that ribonomics holds the key to understanding this correlation between translation and neuronal survival following

I/R. In this chapter, we summarize our main findings on ribonomic studies of brain I/R. As this work is currently under way, the majority of the discussion will focus on the phenomenology of ribonomic structures in post-ischemic neurons.

The phenomenology lends itself an obvious functional interpretation that the ribonomic regulation plays a crucial role in the genetic reprogramming of post-ischemic neurons to a stress response phenotype. The successful expression of this stress response phenotype is the fulcrum between life and death for the post-ischemic neuron. Thus, the ribonomic phenomena sit in a complex web of cellular damage, cellular signaling pathways, and transcriptional and translational regulation. In short, it is part of a complex network of molecular changes induced in post-ischemic neurons, where the global behavior of the network determines cellular outcome in a deterministic fashion.

These results are of interest in a nanotechnology context by illustrating that individual cell behavior, even the decision to live or die, results from the holistic action of all of the cell's internal components, which are interlinked in a complex, nonlinear influence network. While it is possible to excise specific components of the cell and have them function in isolation from the rest of the network, if we wish to technologically emulate the global behaviors of cells, individual subcellular components must be seen as nodes in a complex network of interactions, the global dynamics of which generates cell behavior.

We shall proceed by first giving a general and cursory overview of the networked of structures underlying mRNA regulation. We then discuss our studies of ribonomic structures in the context of the overall understanding of brain I/R injury. Next we interpret the ribonomic results as an additional marker of the genetic reprogramming of the post-ischemic neuron. This allows us to abstract the activity of post-ischemic neurons to a general mathematical model of cell outcome following injury. This model is predicated on the assumption that it is the network as a whole, and not any specific nodes, or nodal motifs, which determines outcome. Such a model provides a layer of abstraction between specific biological instantiations and possible nano-technological implementations any system that might behave with similar dynamics.

9.2 The Complexity of mRNA Regulation

The regulation of mRNA is an active research area, and the molecular biology is complex. Many recent reviews summarize different subareas in much more detail than we shall here and are cited as we proceed. Our goal in this section is to provide a cursory working knowledge of the area.

9.2.1 mRNA Binding Proteins and mRNAs

Cell genomes code a large number of mRNA-binding proteins (mRBPs). For example, it is estimated that ~500[5] or 2.5% of the ~20,000 proteins[6] expressed by C Elegans are mRBPs. In the human genome of ~21,000 genes,[7] 800 of them, ~3.8%, have been identified to be "professional" mRBPs (e.g., with well-defined mRNA-binding domains).[8] This latter number is a lower bound that does not account for alternative spliced variants and "nonprofesional" mRBPs (those that may bind mRNA, but that is not their primary function). There are less than a dozen known mRNA-binding domain motifs present is various numbers and in the mRBPs.[9] The distribution of the mRNA-binding domains generates not only the diversity of mRBPs, but allows them to function pleiotropically in different complexes in a cell context–dependent fashion,[5] that is, mRBPs are generally multifunctional proteins.

The mRBPs bind to cis-acting sequences in mRNA molecules. Such sequences are found in the 3′ and 5′ untranslated regions (UTR) of most mRNAs. An example of a 5′ UTR sequence is the terminal oligopyrimidine tract (5′ TOP) found in ribosomal protein and some translation factor mRNAs.[10] A common 3′ UTR cis-acting sequence is the adenine and uridine rich element (ARE) sequence,[11] responsible, for example, for the rapid degradation of *cfos* mRNA.[12] It is estimated that ~8% of all mRNA contain ARE sequences.[13] The poly(A) tail, and even the coding region of the mRNA, also bind mRBPs.[14] Thus, a second layer of diversity exists in the types and numbers of cis-acting sequences in a given mRNA molecule.

Therefore, the relationship between mRBPs and mRNA is many-to-many: a given mRBP can bind mRNAs coding different proteins, and a single mRNA species can potentially bind many different mRBPs.

To make sense of this complexity, Jack Keene has developed a combinatorial model of mRBPs that draws analogy to how transcription factors act in a combinatorial fashion at gene promoters.[15] In a cell context–dependent fashion, combinations of mRBPs bind mRNA molecules and regulate their function in the cell. The functions mRNAs undergo include transcription, pre-mRNA maturation, nucleocytoplasmic transport, and in the cytosol, translation, silencing (also called "storage"), transport, and degradation.

9.2.2 mRNA Ribonucleoprotein Complexes

The mRBPs and mRNA molecules form mRNA ribonucleoprotein (mRNP) complexes that mediate mRNA functions (Fig. 9.1).[16] The best known mRNP is the **ribosome**, an mRNP that mediates translation. However, at every stage of its life cycle, mRNA is bound

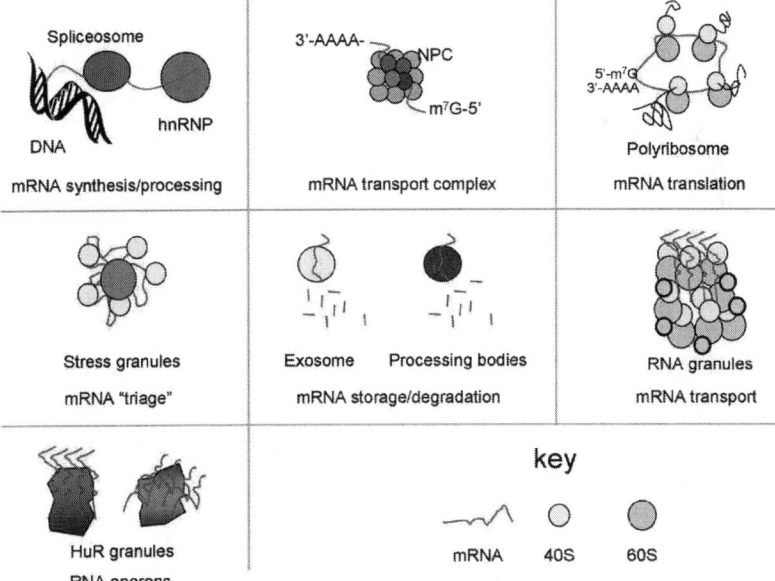

Figure 9.1   Ribonomic structures. Simplified drawings illustrating nine of the known mRNP complexes. Spliceosomes and heavy-nuclear RNPs function in the nucleus. The remainder are cytoplasmic structures. Red lines, yellow and orange circles depict mRNA, 40S and 60S subunits, respectively.

in mRNPs. In the nucleus, the **hnRNP complexes** mediate transcription and pre-mRNA maturation.[17] Specific **mRNA transport complexes** shuttle mRNA from the nucleus to the cytoplasm.[18] Several mRNPs exist in the cytoplasm along with the ribosome. **Processing bodies** (P bodies) are involved in mRNA storage and degradation.[19] **Exosomes**[20] and **RNA-induced silencing complexes** (RISC)[21] mediate mRNA degradation. **Stress granules** are "triage" centers, which route mRNA to different mRNPs in a cell context–dependent manner.[22] Different cell types show specific **silencing complexes**. For example, mRNA is packaged in silencing complexes with mRNPs in oocytes, where the localization of the silencing complexes is critical to subsequent development following fertilization.[23] In neurons, **RNA granules** are mRNPs that contain inactive ribosomes and mRNAs and function to transport the latter to synapses and spines in dendrites far removed from the neuronal cell body.[24]

A particularly relevant mRNP is the **HuR granule**, so called because it contains the mRPB HuR (also called HuA or HNel1), or one of the HuR analogs, HuB, HuC or HuD.[25,26] All of the Hu proteins bind the ARE sequence found in the 3′ UTR of ARE-containing mRNAs.[27] The combinatorial model of Keene is particularly focused on Hu protein–containing mRNPs. For the past two decades, Keene's lab, and those of his students, and other independent laboratories, have build a large body of evidence that HuR granules function as, to use a term introduced by Keene, "RNA operons."[28] Whereas the combinatorial model of mRBPs describes the structural aspect, the term "RNA operon" indicates the functional aspect of mRBP-mRNA interactions. We elaborate these ideas because they factor in substantially to our work on brain ischemia.

A bacterial operon encodes a single contiguous stretch of DNA coding functionally related proteins (see [29] for a contemporary view of operons). The operon is regulated (turned on or off), transcribed and translated as a single unit, thereby allowing the coordination of the coded proteins. In fact, an operon can be thought of as formally analogous to calling a subroutine in a program. A subroutine is a discreet grouping of code that is executed as a unit on an "as needed" basis. An RNA operon has a similar formal structure, and Keene suggests RNA operons are the functional analogs in metazoan cells to the operons in bacteria.

Keene has shown, for example, that the mRNAs coding proteins involved in transcriptional regulation and cell cycle/proliferation share common cis-acting elements, and bind the same mRBP PUM1.[30] This leads to coordinated handling of the mRNAs coding these proteins such that they tend to be translated simultaneously, and also degraded simultaneously. Thereby, because of the common cis-acting sequences, and hence common handling by PUM1, the mRNAs coding proteins regulating cell cycling and proliferation act as a unit, a subroutine. This is one of several empirically demonstrated examples of an RNA operon in eukaryotic systems.[30]

Therefore, the Hu protein–containing mRNP complexes, the HuR granules, are of particular significance because they can function as relatively discreet "subroutines" mediating the coordination of complex programs of cell activity. We show below that post-ischemic neurons appear to be using HuR granules to express stress response programs that clearly represent the cell's attempt to resist ischemia-induced molecular damage.

9.2.3 The Ribonomic Network and the Central Dogma

We previously termed the various mRNA regulatory structures the "ribonomic network,"[1] the set of mRNPs that mediate mRNA functions in a cell context–dependent fashion. The ribonomic network introduces new layers of complexity into cell biology. The classical view of the "central dogma" of molecular biology was

DNA $\rightarrow$ mRNA $\rightarrow$ protein.

While serving to drive research in the early days of molecular biology, this *linear* notion of the core process of cell biology, the expression of the genetic information, is obsolete in light of today's understanding. Upstream of transcriptional activation, it is now well-understood that many cellular signaling pathways operate in parallel and converge to trans-activation of DNA in the nucleus. It is also now well understood that many signaling pathways converge to the ribosomes and control the rate of translation,[31] as well as which mRNAs gain access to the ribosome.[32] When we factor in the ribonomic network, many new layers of regulation get added to our picture of how DNA information converts to cell behavior via protein expression. Indeed, it is the imposition of the ribonomic network between DNA and protein expression that explains the general lack of correlation between mRNA and protein levels in a cell.[33] From a biomedical standpoint, many previously intractable disease states are being revealed as dysfunctions of ribonomic structures and/or regulation.[1,34,35] We now turn to brain I/R as one such example where ribonomics is helping refine our understanding of the nature of the disease process at the cellular level.

9.3 Brain Ischemia

9.3.1 What Is Ischemia?

The term "ischemia" means a pathological reduction, or even complete cessation, of blood flow to an organ. Since the term implies a spectrum of blood flow decrements, four terms are used in conjunction with the term ischemia. **Complete ischemia** is 100% reduction in blood flow, which is zero blood flow. **Incomplete ischemia** is a nonzero blood flow, ranging perhaps from ~50% to >0%. **Global ischemia** means blood flow is reduced to an entire organ. **Focal ischemia** means blood flow is reduced to a subvolume of the organ. When normal blood flow resumes following a period of ischemia, it is called **reperfusion**.

When ischemia occurs to an organ, the delivery of the essential metabolic nutrients O_2 and glucose slows significantly or halts. This in turn slows the rate the cell can perform oxidative metabolism in the mitochondria, the main system used to produce ATP in a metazoan cell. Decreased ATP slows all ATP-dependent processes in the cell. The effect of ischemia on a tissue is dependent on a given tissue's intrinsic ability to synthesize ATP independent of blood flow.

For example, skeletal muscle tissue is relatively resistant to ischemia because it has a robust system for regenerating ATP (the phospho-creatine system), and it stores glycogen, a polymeric form of glucose storage.[36] Thus muscle can breakdown stored glycogen and use it as a carbon source for generating ATP. At the opposite end of the spectrum is brain. Brain has no system for recycling ATP, it has little to no glycogen storage, and under normal conditions, it utilizes ATP at a very high rate compared to other organs (indeed brain consumes 20% of blood flow, but is only 2% of total body weight).[37] Thus, skeletal muscle can withstand ischemia for 4–6 h before irreversible injury sets in.[38] On the other hand, if ischemia is greater than ~10 min in brain, neuronal cell death is inevitable.[39] These two tissues lay at the extremes and all other organ systems fall in between. Heart, kidney, and GI tract are closer to brain in that they are relatively more vulnerable to ischemia.[40]

9.3.2   Clinical Manifestations of Brain Ischemia

Clinically, brain ischemia occurs under two circumstances: stroke and cardiac arrest. Stroke is a form of focal brain ischemia caused by an obstruction in the brain arterial tree (~85% are embolisms and ~15% are clots). The spatial pattern of blood flow reduction is complex in stroke, being ~100% (zero blood flow) at the site of obstruction, and increasing in a (very) roughly spherical direction away from the obstruction. At some distance away from the obstruction, blood flow is normal. Cardiac arrest is the state where the heart ceases to pump blood. The result is ischemia of the entire body and all organs undergo global ischemia during cardiac arrest. Under these circumstances, the graded response of the organs to ischemia comes into play, and brain is the most vulnerable. The blood volume at any instant in the brain contains enough O_2 and glucose to support only ~4 min of normal ATP usage in the brain.[41] By 5 min global ischemia, ATP is 50% its normal level, and ATP levels become zero between 6 and 7 min.[42]

The pathophysiology of brain damage following stroke and cardiac arrest expectedly has some overlap, but there are significant differences as well, and these follow from the difference in spatial distribution and degree of the ischemia. Clinically, stroke patients may not see a physician for up to 24 h following the stroke, allowing the stroke-induced brain damage substantial time to develop. The common clinical finding for stroke is a **core** of necrotic brain tissue around the site of the occlusion. Surrounding the core is an area of reduced blood flow, the **penumbra**. The penumbra is generally viable shortly after the stroke (on the order of a few days), but over time, (weeks) the penumbral tissue can die, thereby expanding the core area with a concomitant increase in dead brain tissue.[43]

Unlike stroke, cardiac arrest victims tend to be hospitalized within minutes following cardiac arrest. Typical EMS time-to-hospital is on the order of 10–20 min.[44] However, unlike stroke, patient survival following cardiac arrest is abysmal. Only ~20%

are successfully resuscitated, and of these, only a meager 5% escape substantial brain damage.[45] A patient must be successfully resuscitated within <5 min to escape brain damage. Even if the resuscitation is successful, but takes >10 min, the patient will invariably suffer brain damage, the severity of which is proportional to cardiac arrest duration. However, unlike stroke, a core is not produced. Instead, there is a graded loss of specific neuron populations as a function of global ischemia duration. At 10 min global ischemia, only one neuron population, the CA1 sector of the hippocampus, will die. The neurons do not die during the ischemia, but die ~3 days following the resuscitation. This phenomenon, (DND), is the hallmark of global brain ischemia,[39] but it also occurs following stroke with the DND of the penumbra surrounding the core.[46]

9.3.3 The Holy Grail of Neuroprotection

Thus, for both stroke and cardiac arrest, the goal of biomedical brain ischemia research is to slow, and preferably, halt the DND component of the neuronal death. This is not an unrealistic goal. The patterns of neuronal death following both focal and global ischemia are highly reproducible in both human patients and in experimental animals. The main idea driving biomedical research into brain ischemia therapies is that the ischemia must induce some specific form of damage in the neurons, where this damage evolves over time, resulting in the DND observed both clinically and experimentally. Then, knowing this damage mechanism, in principle a therapy can be devised to halt it. Such a strategy is called "**neuroprotection.**"

This idea, while sensible on the surface, has however, been a spectacular failure. Considering only stroke, there have been over 100 clinical trials of drugs that have been shown to halt neuronal death in experimental animals, and *every single one* of the clinical trials has failed.[47] Thus, at present, there are no effective clinical pharmacologics to halt brain damage following stroke or cardiac arrest. The only FDA approved stroke treatment is tissue plasminogen activator (tPA), which is a drug that dissolves blood clots. Tissue plasminogen activator does not stop neuronal death, it simply relieves the ischemia by dissolving the obstruction.[48] Tissue plasminogen activator is only applicable to stroke patients where the stroke has been caused by a blood clot and must be applied within 4 h of the stroke; these conditions limit tPA administration to only a small minority of all stroke patients (~5%).[49] In addition to tPA, **therapeutic hypothermia** is emerging as an effective treatment to halt brain damage following cardiac arrest.[50] Unlike tPA, hypothermia is effective at halting neuronal death; however, its mechanism of action is not fully understood; likely it involves pleiotropic effects best understood by the network formalism described below.

A discussion of ischemic therapy merits consideration of the phenomena of **ischemic preconditioning** (IP). IP is the phenomena whereby previous exposure of an organ to a sublethal dose of ischemia can subsequently, within a very specific time frame, completely protect the organ from damage to a lethal dose of ischemia.[51] This is perhaps the most robust form of ischemic neuroprotection known at present. However, there are presently no practical ways to take advantage of IP in the setting of stroke or cardiac arrest. Nonetheless, it serves as an important experimental tool to study the intrinsic ability of neurons to survive ischemia. IP is not a panacea. It is effective only for a limited range of lethal ischemia doses. Above a threshold dose of lethal ischemia, IP is ineffective.[52] But an optimized dose of sublethal ischemia can halt cell death over a range of lethal ischemia doses by essentially 100%. Thus the study of IP is important for guiding the development of neuroprotective strategies.

In spite of the clinical failure, the search for a neuroprotectant for ischemic brain injury has lead to a very detailed understanding of the molecular changes neurons undergo in response to ischemia. In the next section we briefly survey these changes.

9.4   Molecular Understanding of Brain Ischemia

The molecular changes wrought in neurons and other brain cell types by ischemia are complex and multifactorial. Similar to the above discussion on mRNA regulation, we here provide a cursory working knowledge and cite more detailed reviews or original articles as appropriate. Molecular changes are characterized as to whether they happen during ischemia or the subsequent reperfusion period.

9.4.1 Neuronal Cell Biology of the Ischemic Period

Complete ischemia leads to a very rapid depletion of ATP in the brain neurons. Approximately 50% of all ATP utilization in neurons is to maintain ion gradients,[42] so there is rapid dissipation of neuronal ion gradients following the precipitous loss of ATP. Calcium ion (Ca^{2+}) has the largest transmembrane gradient across neurons of ~10,000:1 (extra to intracellular ratios). Thus, there is a large and unregulated influx of Ca^{2+} into the neurons.[53] The rapid rise in Ca^{2+} has pleiotropic negative effects. It induces excitotoxicity, a positive feedback loop of Ca^{2+}-mediated uncontrolled neuro-transmitter release, depolarization, and further transmitter release, until transmitter vesicles are depleted.[54] Vascular cells produce copious nitric oxide (NO) as a protective vasodilatory response, but neurons produce abnormally high levels of NO in response to Ca^{2+} that can act as a free radical and damage cell components.[55] In general, Ca^{2+} activates catabolic enzymes,[56] including proteases (calpains),[57] lipases (phospholipase A2)[58] and nucleases.[59] Ca^{2+}-dependent signaling pathways are also activated including PKC isoforms and Ca^{2+}-calmodulin-dependent enzymes such as CAMKII; both undergo membrane translocation since this is an ATP-independent process.[60] However, since there is no ATP, kinases, even though activated, cannot phosphorylate substrates. At some point, the pro-catabolic state of ischemia must end and reperfusion resume.

9.4.2 Neuronal Cell Damage in the Reperfusion Period

When reperfusion occurs, O_2 and glucose enter neurons whose cell biology has been negatively impacted by ischemia. The breakdown of lipids during ischemia releases free fatty acids, of which arachidonic acid (AA) is a quantitatively significant product. In combination with O_2 and metal ions (Fe^{2+}, Zn^{2+}), AA can serve as a substrate for lipid oxidation.[58] AA can also be enzymatically transformed to eicosanoids that serve as pro-inflammatory mediators.[61] In general, there is a significant burst of non-physiological reactive oxygen species (ROS) with reintroduction of O_2.[62] There is further excess production of NO in neurons, which can serve as a reactive nitrogen species (RNS) and a substrate for nitrosylation of neuronal proteins.[56] Decoupling of ATP production and electron transport causes mitochondria to be a significant ROS source,[63] mitochondrial dysfunction can also induce pro-apoptotic pathways via a number of discreet mechanisms.[64] The ROS and RNS serve to further damage neuronal intracellular structures.

Dysfunctional lipid metabolism during ischemia further affects cell signaling pathways during reperfusion. Build up of inositol can act at endoplasmic reticulum (ER) Ca^{2+} release channels to decrease ER Ca^{2+}.[65]This in turn leads to ER stress and activation of the ER-stress response the unfolded protein response (UPR).[66] PKC and CAMKII pathways, which had been activated during ischemia, undergo degradation during reperfusion, so that the global intracellular signaling patterns found in neurons are altered during reperfusion.[67,68] Many other molecular alterations occur in reperfused neurons,[42,56] but detailing these goes beyond the scope of this Chapter.

An important breakthrough in understanding I/R damage came in the late 1990s, when it was observed that masses of ubiquinated protein, called ubi-protein clusters (UPC), accumulate in all reperfused neurons.[69,70] The formation of UPCs is reversible in neurons that survive the insult. However, in neurons that will die, the UPCs undergo a transformation into protein aggregates (PAs), which continue to accumulate in neurons fated to die. Protein aggregates represent a convergence of different upstream damage mechanisms to protein damage. Many of the above mechanisms damage proteins: proteolysis, abnormal signaling patterns, ROS and RNS; changes in pH and intracellular osmolality during ischemia can also contribute to protein damage.[71]

9.4.3 Intracellular Stress Responses in Reperfused Neurons

The neurons are not passive spectators to the damage wrought by I/R. A number of intrinsic, and often highly evolutionary conserved, intracellular stress responses are activated in post-ischemic neurons. Examples of intracellular stress responses expressed in post-ischemic neurons include the heat shock response,[72] the unfolded protein response,[66] the anti-oxidant response,[56] DNA damage responses,[73] and anti-apoptotic pathways.[74]

Stress responses are accompanied by, and indeed, linked to, major changes in gene expression. Gene expression changes occur in waves in the post-ischemic neuron[75] that can be broadly broken into three phases: (1) immediate early genes (<1 h), (2) stress response genes (<8 h), and (3) cellular repair genes (<24 h to days). Thus, the general response of the neuron is to first halt or eliminate molecular damage, and then repair itself. Through these processes the neuron seeks to *reset* itself to its pre-injury state.

Instead of focusing on details of individual stress responses, we outline a general algorithm capturing their function. Specific stress responses may show variants of this general algorithm and emphasize some steps over others, but the general algorithm provides a framework to make sense out of what is otherwise very complex molecular biology. The stress response algorithm is as follows (Fig. 9.2).

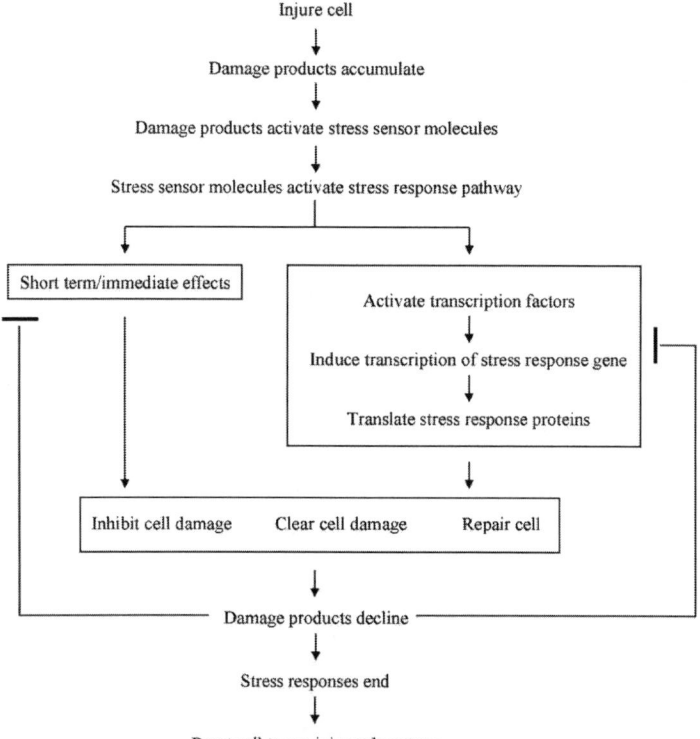

Figure 9.2 An algorithm capturing the salient behavior of intracellular stress responses. An uninjured cell is in a state-state condition of homeostasis. Injury is a perturbation to cell homeostasis. This algorithm shows how the cell "rests" itself following injury to regain homeostasis.

Injury induces damage products that in turn activate sensors of these damage products. The sensors are the most proximal molecules in the stress response pathways. Binding of the damage product activates the sensor, which in turn initiates a cascade of events. The subsequent events can be broken into two broad classes: short-term effects and longer-term transcriptional effects. Short-term effects tend to include activation of enzymes via signal transduction cascades, where such enzymes play some role in inhibiting damage. Transcription events are the direct link to injury-induced changes in gene expression. Stress response pathways generally induce activation of transcription factors that will upregulate stress response genes, leading to new, and transient, mRNA species coding the stress response proteins. The mRNAs are translated at the ribosome, often by special mechanisms that lead to exclusive translation of stress response mRNAs. Translation of stress response proteins is the distal step, producing the functionally active agents (proteins) that will inhibit and clear cell damage and contribute to cell repair. These processes decrease the damage products, thus decreasing the stimuli for the stress response pathways, and via negative feedback, will shut off the stress response pathways.

Because ischemic injury induces many forms of damage, several such stress response pathways run in parallel in post-ischemic neurons. Expressing the molecular biology in algorithmic form makes clear that neurons that successfully execute the stress response pathways will survive the insult. This view also offers the idea that cells do not die exclusively from the accumulation of damage. In addition, dysfunction in stress responses can prevent the neurons from coping with the damage. Thus, cell death can instead be seen to be caused by the interplay between damage and stress response, and is not exclusively caused by accumulation of damage. We return to this point near the end of the chapter, when we present our mathematical model of cell injury.

9.5 Translational Control in Post-Ischemic Neurons and Injured Cells

An essential aspect of post-ischemic molecular biology was omitted from the above narrative. That is the fact that protein synthesis is inhibited in post-ischemic neurons, a condition known as **translation arrest (TA)**. For the first several hours into the reperfusion period translation rates are essentially zero. Translation rates return to pre-injury values in neurons that recover from the ischemia after ~24 h of reperfusion (following 10 min global ischemia).[4,76] In neurons fated to die, protein synthesis rates never recover normal values, and stay persistently decreased all the way until the neurons disintegrate.[4] We can see in Fig. 9.2 that translation is an essential step in expressing stress responses. Translation is a critical bottleneck point where changes in gene expression transform into functional changes in cell biology mediated by the newly translated stress proteins. Thus, the strict correlation between persistent TA and cell death of post-ischemic neurons can readily be linked to a dysfunction in stress responses via the algorithm in Fig. 9.2. In this section, we set the background on post-ischemic TA and then proceed to describe our experimental studies in following sections.

Post-ischemic TA was discovered in 1971, in the context of studying ATP-dependent processes in reperfused neurons.[77] With the onset of reperfusion, ATP levels rapidly recover, and energy charge is essentially normal within 10 min of reperfusion. Ion gradients are re-established within a similar time frame.[42] Many other ATP-dependent biochemical processes resume to pre-injury levels within the first 30 min of reperfusion.[42] Thus, in 1971, it presented a mystery as to why translation did not recover. However, at that time, the understanding of the complex molecular biology of ribosomes was only at a nascent stage and so the molecular understanding did not exist to fully appreciate the observation.

Fast-forwarding to our present understanding, over the past 20–30 years, it has become clear that a large number of injuries, not just ischemia, lead to TA in cells. Stimuli that lead to TA include: heat stress, osmotic stress, heavy metal poisoning (AsO_4^-, Hg^{2+}), nutrient deprivation, mechanical trauma, ethanol exposure, and a whole host of other poisonous substances.[1,22,78] It is now appreciated that the transient TA in injured cells is a marker of the genetic reprogramming to a stress response phenotype depicted in Fig. 9.2. The ostensibly teleological "reasoning" behind the transient TA response is that injured cells do not need to translate their normal complement of proteins. The cell has been injured, and to go on with "business as usual" for the cell would be detrimental. Instead, the cell needs to focus its resources on repairing itself.

Thus, an integral part of the stress response algorithm that is not shown in Fig. 9.2 is that there is a rapid cessation of translation upon cell injury. This diverts the significant energy expenditure due to translation to more pressing matters of cell survival. Aside from energy expenditure, there is a second teleological reason for the transient TA. As the stress-induced mRNAs accumulate following injury, they are allowed exclusive access to the ribosomes. When any set of mRNA gains exclusive access to the ribosomes over all other mRNAs, this is known as **selective translation**. There is thus selective translation of the stress-induced mRNAs so the translated proteins can play their roles in cell recovery. If the stress-induced mRNAs had to compete with the normal compliment of cell mRNAs, this would clearly decrease the efficiency of the stress responses.

In this context, we can recognize that the transient TA in reperfused neurons is related to the selective translation of stress response mRNAs. However, even with this expanded understanding, it does not explain why there is a persistent TA in neurons fated to die.

9.6 Persistent TA in Vulnerable Neurons

Cells fated to die following ischemia are called "vulnerable neurons" and it is these that show persistent TA during reperfusion. It is widely recognized that understanding the mechanism of this persistent TA would be of potential therapeutic benefit, thus the phenomenon has been well studied. Here we briefly summarize the various understandings that have emerged since the original description of post-ischemic TA. There have been five major waves of ideas guiding work in this area: (1) phenomenological observations, (2) ribosome regulation, (3) TA and stress responses, (4) ribosomal sequestration, and (5) mRNA regulatory systems.

9.6.1 Phenomenology of Post-Ischemic TA

In the seminal observation, electron microscopy was used to show that polyribosomes (polysomes) dissociate in the early reperfusion period. Figure 9.3 shows data from our lab illustrating polysome dissociation in reperfused neurons. Even in 1971, it was

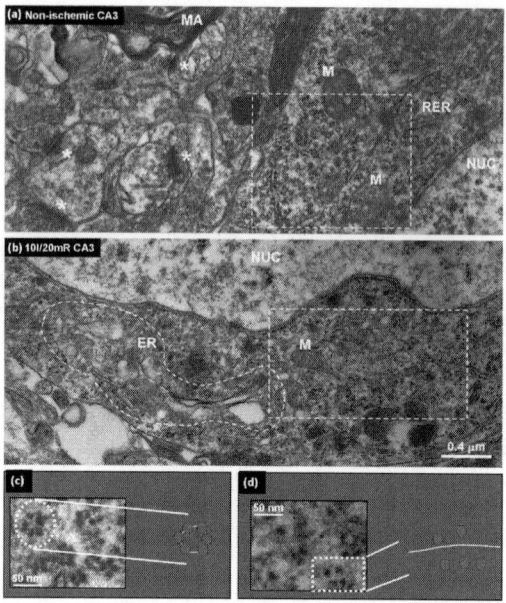

Figure 9.3 Classical electron microscope (EM) appearances of (a) non-ischemic CA3 pyramidal neurons. A number of neuron specific structures are seen to the left: a cross-section through a myelinated axon (MA), and several synaptic contacts (post-synaptic densities marked by asterisks, *); these are outside the neuronal cytoplasm proper. The cell nucleus (NUC) is seen in the lower right. Mitochondria (M) and rough endoplasmic reticulum (RER) are labeled. Inside dashed box are visible many "rosette" (flower-shaped) structures which are the characteristic appearance of polysomes (e.g., when many ribosomes are translating an mRNA molecule in a roughly circularized structure) in EM micrographs. (b) CA3 pyramidal neuron after 10 min ischemia plus 20 min reperfusion. Nucleus fills upper part of panel. Now, in the dashed box, the rosette structures are lost indicating dissociated polysomes; ribosomal subunits appear as randomly distributed spots (~20 nm in diameter) throughout the cytoplasm. A second characteristic of reperfused neurons is distended and vesicularized endoplasmic reticulum (inside area surrounded by the curved dashed line); here the ER has lost its characteristic "flat pancake" structure and appears rounder and more distended. (c) Blow up of "rosettes," taken from dashed box in a, and schematic of how the appearance links to circularized polysomes. (d) Blow up of dissociated polysomes, taken from dashed box in b, with associated schematic. Scale bars in b is 0.4 microns and applies to both a and b. Scale bars in c and d are 50 nm. EM magnification for all images is 15,000×.

appreciated that polysome dissociation is linked to the inhibition of translation initiation. Again, however, the molecular understanding simply was not in place to explain this observation. Thus, through the 1970s and 1980s, studies of TA and brain ischemia established unequivocally the correlation between persistent TA and neuronal vulnerability, but little was advanced in terms of molecular understanding.

9.6.2   Ribosome Regulation and Post-Ischemic TA

By the early 1990s, the molecular biology of ribosomes had advanced sufficiently that these could be studied in reperfused neurons. Hu and Wieloch were the first to show that the activity of the translation initiation factor eIF2B was decreased in reperfused brain homogenates.[79] eIF2B is a guanine nucleotide exchange factor that recycles the initiation factor eIF2.[80] eIF2 in turn delivers the first amino acid, a methionine, to initiating ribosomes, and this is one of two key rate limiting, and highly regulated steps in translation initiation.[80] Decreased eIF2B activity was well known to induced polysome dissociation and TA. The next advance was the observation by Burda and colleagues of the phosphorylation of the alpha subunit of the initiation factor eIF2 [eIF2(αP)].[81] eIF2(αP) is a competitive inhibitor of eIF2B.[80] We confirmed Burda et al.'s observation and added the additional observation that the initiation factor eIF4G underwent degradation as a function of ischemia duration.[82] eIF4G is a subunit of the eIF4F complex, whose function is to deliver the mRNA to initiation ribosomes.[80] The activity of eIF4F is also highly regulated and is the other main rate-limiting step of translation initiation.[80] Thus, these initial molecular studies suggested that regulation of ribosomes played a critical role in post-ischemia TA.

However, as different laboratories, including our own, continued to investigate ribosome regulators, it soon became apparent that these changes were not unique to vulnerable neurons, but occurred in all post-ischemic neurons. We had developed the first phospho-specific eIF2α antisera[83] and, using it for immunohisto-chemical studies, showed that eIF2(αP) occurs in all post-ischemic neurons (Fig. 9.4). Importantly, eIF2α dephosphorylated by ~6 h reperfusion, again, in all post-ischemic neurons.[84] In addition, neuronal vulnerability occurs in animal models of brain ischemia that

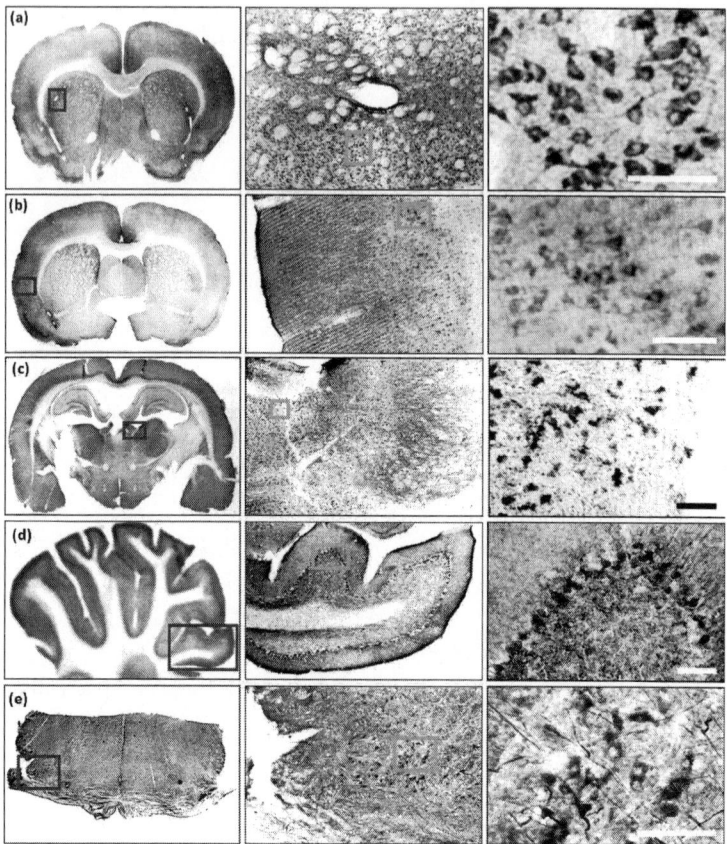

Figure 9.4   Phosphorylation of alpha subunit of eIF2 in rat brain subjected to 10 min CA/R ischemia and 60 min reperfusion, as detected by IHC. Left column shows whole brain slice for orientation. Middle column shows blow up of the boxes in first column. Third column shows cell staining, and blows-ups taken from the boxes in middle column. Brain regions shown are: (a) striatum, third column shows medium spiny neurons, (b) lateral cerebral cortex at level of striatum, third column shows layer V, (c) thalamus, third column shows relay neurons (d) cerebellum, third column shows Purkinje cell layers, and (e) brainstem at the level of the pyramids, third column shows motor nucleus. Scale bars in third column are each 100 microns and apply to the respective panel only. As can be seen, all post-ischemic neurons phosphorylated eIF2α at early reperfusion.

do not lead to appreciable degradation of eIF4G.[76] Thus, by the early 2000 s, the consensus emerged that the well-studied mechanisms of ribosome regulation did not distinguish vulnerable from resistant post-ischemic neurons and some other mechanism must be responsible for the persistent TA of the vulnerable neurons.[85]

9.6.3 Post-Ischemic TA and Stress Responses

The link between TA and stress responses was outlined above. Here we only briefly mention the historical aspects of how this idea emerged.

The first link between these was the observation that inducible hsp70 mRNA is massively transcribed in post-ischemic neurons.[86] However, it is only translated in surviving neurons.[86] In vulnerable neurons, the levels of hsp70 mRNA continue to increase but HSP70 protein is not translated.[86,87] The initial interpretation of these observations was that TA interfered with the successful execution of the heat shock response.[86] However, based on the current knowledge outlined above, we can now state that selective translation of hsp70 mRNA does not occur in vulnerable neurons. This is a seemingly subtle but significant difference in outlook we elaborate further below.

The next link emerged from the observation that depletion of ER Ca^{2+} lead to TA.[88] This led Wulf Paschen to suggest that ER stress was responsible for post-ischemic TA.[89] We were able to confirm this suggestion by showing I/R-induced activation of the ER-associated eIF2α kinase PERK,[90] a now well-studied member of the UPR. However, the unusual circumstance exists in post-ischemic neurons that PERK appears to be activated independently of the other members of the UPR: IRE1 and ATF6.[84] There is still unresolved controversy about some of the molecular details of the UPR in post-ischemic neurons.[87] But now, at least two studies, one from our lab[87] and one from the Hu lab,[91] have compared the magnitude of various stress responses in post-ischemic neurons. By far the largest is the heat shock response, and the UPR, in terms of mRNA transcriptional changes is a magnitude of only a few percent compared to the heat shock response. Thus, the question of whether the UPR is fully expressed in post-ischemic neurons is overshadowed by the fact that, even if it is, it is only a relatively small part of the total stress response, even in neurons that survive the ischemia.

Thus, the link between TA and stress response has not, to this point, been able to reveal why there is persistent TA in vulnerable neurons. Nevertheless, the idea has been crucial for revealing that post-ischemic TA is, in general, an integral part of the post-ischemic stress response, similar to TA in other cell injury systems. This insight has provided the context for understanding the function and role of TA after ischemia. As we elaborate ahead, this line of thinking may yet explain persistent TA in vulnerable neurons, when it is coupled to our modern understanding of mRNA regulation outlined in Section 9.2.

9.6.4   Sequestration of Ribosomes

With the discovery of UPCs and PAs, the Hu lab developed evidence that ribosomes are among the many proteins found in a denatured state in Pas.[92,93] Protein aggregates are specific to vulnerable neurons and not resistant neurons. The sequestration of obviously nonfunctional ribosomal subunits in PAs was termed "co-translational aggregation," and proposed as the cause of persistent TA in vulnerable neurons.[93] This was the first set of observations and the first model that successfully distinguished persistent from transient TA. However, this model, on its own, could not fully account for persistent TA in post-ischemic neurons. Quantitative studies revealed that, at most, ~20% of total ribosomes were present in the PA fraction,[93] leaving the remaining ~80% of ribosomes available for translation. Some co-translational chaperones, such as Hdj1, are more quantitatively sequestered, approaching ~75%,[93] but these are not rate-limiting translational components, and the remaining un-sequestered amounts could still participate in translation, although at lower rates.

Thus, while the model of co-translational aggregation has not fully solved the problem of persistent TA, it has been a major step in that direction. It clearly accounts for some percentage of the persistent TA. Further, the idea that ribosomes could be sequestered into non-functional complex has inspired us to consider this idea in the broader context of mRNA regulation.

To this point in our story, all of the focus has been on ribosome molecular biology. At least in the brain ischemia field, there has been a significant blind spot to the other main part of translation: mRNA. Polysome dissociation implies upstream events causing the dissociation. The ribosome regulatory pathways, as described above, have been well studied in reperfused neurons, but cannot account for the persistent TA of vulnerable neurons. Since 2005 we have performed studies that have shifted focus from the ribosomes to the mRNA. We now describe the main findings of these studies. As our work has progressed, we have come to realize there is a simple insight at the core of our current work: polysome dissociation produced free ribosomal subunits, but it also produces free mRNA. The question then becomes: what happens to the mRNA?

9.7 Stress Granules and Brain Ischemia and Reperfusion

9.7.1 The Fuzzy Border between Sequestration and Regulation

By the early 2000s, it was clear that phosphorylation of the alpha subunit of eIF2 was responsible for polysome dissociation within minutes of the onset of reperfusion. However, determining that PERK was the kinase responsible for this phosphorylation did not ultimately contribute to understanding the difference between transient and persistent TA in vulnerable neurons. Around this time, Hu and colleagues presented the evidence for the co-translational aggregation model. This model influenced us to ask if there were other possible mechanisms by which ribosomes could be sequestered into inactive complexes. We then became aware of the work of Paul Anderson and Nancy Kedersha and Massachusetts General Hospital, showing that phosphorylation of eIF2α was linked to the formation of stress granules.

9.7.2 Stress Granules

Stress granules (SGs) had been shown to contain mRNA molecules, inactive 40S subunits, and a list of other mRNA-binding proteins. The protein TIA-1 (T cell internal antigen 1) was shown by Anderson and Kedersha to be a critical component of SGs. TIA-

1 is normally a nuclear-localized protein, involved in processing nascent mRNA transcripts.[2] Anderson and Kedersha showed that, upon administration of cell stress (for example, AsO_4^-), TIA-1 exported from the nucleus and, in the cytoplasm, aggregated to form SGs.[22] Their observations led to the following model:

Cell stress → eIF2α phosphorylation → polysome dissociation → free mRNA/40S → TIA-1 nuclear export → TIA-1/40S/mRNA binding → SG formation

In this model, cell stress activates an eIF2α kinase, which phosphorylates the alpha subunit of eIF2, leading directly to polysome dissociation by inhibiting translation initiation. Polysomes dissociate into free 60S subunits, and at least some 40S subunits bound to mRNA (in a structure perhaps similar to the 48S preinitiation complex).[94] TIA-1 is predominantly nuclear, but maintains equilibrium with some percent localized to the cytoplasm; this is a common profile for mRNA-binding proteins.[95] Upon polysome dissociation, cytoplasmic TIA-1 binds to the modified 48S complexes via a TIA-1 · mRNA interaction. In addition, TIA-1 has the ability to bind itself, and this was shown to be the basis for TIA-1 aggregates.[96] Thus, following polysome dissociation, cytoplasmic TIA-1 self-assembles into complexes bound to mRNA and 40S subunits. This sets up a mass action effect, promoting TIA-1 exit from the nucleus, where the additional TIA-1 forms more and/or larger aggregates containing 40S subunits and mRNA, and providing a scaffolding structure that generates SGs.

Over the past decade, there has been exponential growth of the understanding of SGs. They are now known to contain, in addition to TIA-1, 40S subunits, and mRNA molecules, over two dozen other proteins, many of which are involved in mRNA-related functions.[2] In terms of SG function, it was shown early on that SGs are highly dynamic. TIA-1 flux into and out of SGs, as assessed by fluorescent recovery after photo-bleaching, was on the order of milliseconds.[97] Similarly, mRNAs were only transiently retained in SGs before being passed on to other structures such as P bodies.[97] Kedersha and Anderson interpreted the dynamic nature of SGs as "triage centers" for mRNA handling. mRNAs would only transiently be held in SGs before being routed to some other mRNP. Thus, SGs are currently understood to be mRNPs that mediate mRNA transport, or mRNA flux.

9.7.3   Detecting and Quantifying Stress Granules in Brain

Our initial studies of SGs following brain I/R were not pursued by considering SGs as mRNA regulatory structures. Instead, we were guided by the idea that SGs could potentially serve as a second site of ribosomal sequestration, in addition to ribosomal subunits being sequestered in PAs, as shown by the Hu laboratory. SGs contain inactive 40S subunits. Thus, we tested the hypothesis that 40S subunits became sequestered in SGs. If this hypothesis was true, it would provide an additional mechanism of TA in post-ischemic neurons. Additionally, we hypothesized that SGs would persist in vulnerable neurons and be reversible in resistant neurons.

Our method for testing these hypotheses was simple. We used double labeling immunofluorescence histochemistry (IF) with antisera directed against TIA-1 and the small ribosomal protein S6. We then sought to determine how much of the cytoplasmic S6 signal colocalized with the cytoplasmic TIA-1 signal.[98] Figure 9.5 shows what we expected to see from these studies. At early reperfusion durations, when TA occurs in all post-ischemic neurons, we expected to see SGs in both vulnerable and resistant neuron populations. However, at later reperfusion we expected to see SGs disappear from resistant neurons (correlating with the recovery of translation in these neurons), but persist in the vulnerable neurons.

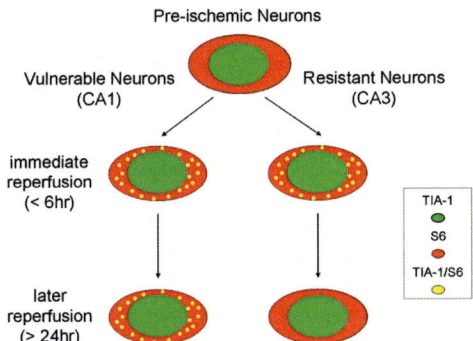

Figure 9.5   Diagram of hypothesis driving stress granule (SGs) experiments. SGs would accumulate in all neurons early in reperfusion. At later reperfusion duration, SGs would clear from resistant neurons (e.g., hippocampal CA3), but persist in vulnerable neurons (e.g., hippocampal CA1). SGs detected by costaining TIA-1 and S6 and looking for yellow punctate cytoplasmic colocalization.

We developed a simple semi-quantitative method for counting SGs in photomicrographs. The typical merged red and green channels used in double-labeling IF can be thought of as mathematical sets (Figs. 9.6a,b). Then, the typical merge operation that shows both the green and red signals plus the overlapping yellow signals is

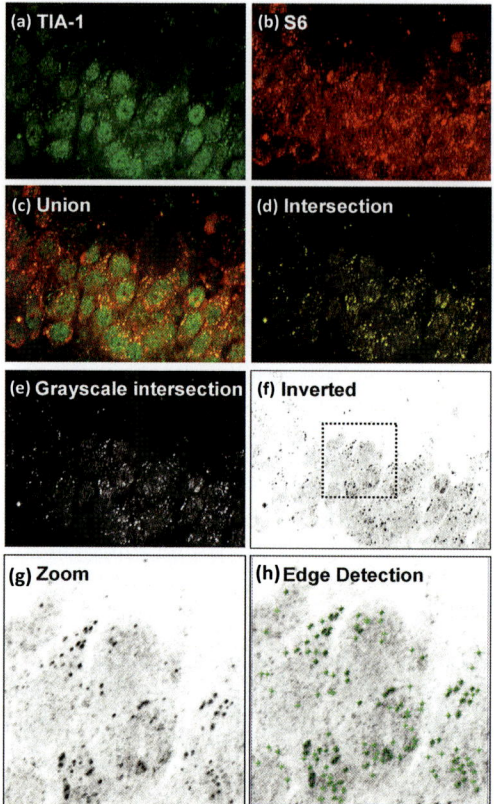

Figure 9.6   Method for estimating the number of SGs. Input channels: (a) TIA-1 immunofluorescence (IF), (b) S6 IF can be thought of as sets. (c) A standard merge of the two channels is actually the mathematical operation of the union of the two sets. (d) A "yellow channel" is produced by taking the intersection of the two sets. (e) The yellow channel is converted to gray scale, and then (f) inverted. (g) Blow up of inverted yellow channel shows SGs as punctate black spots. (h) Black spots are detected by 2D-PAGE spot detection software and counted to estimate number of spots, which are then normalized to the number of nuclei visible in the TIA-1 channel.

equivalent to a *union* of the two sets (Fig. 9.6c). There is no reason that the red and green channels cannot be subjected to an *intersection* operation that outputs only the overlap between the two channels, consisting only of the yellow signal. When we treat IF data in this fashion, we call the intersection output channel a "yellow channel" (Fig. 9.6d). This approach was feasible in the present instance because TIA-1 is mostly localized in the nucleus, and S6 is mostly localized in the cytoplasm (compare Figs. 9.6a,b). Hence, in the uninjured state, there is very little overlap between the two antigens. The formation of SGs results in intense yellow signal only in the cytoplasm. The yellow channel is converted to grey scale (Fig. 9.6e), and inverted (Fig. 9.6f). In the blow up of the inverted image, the SGs appear as intense black spots (Fig. 9.6g), which are then detected and counted using 2D PAGE software (Fig. 9.6h). The total number of detected spots are divide by the number of nuclei visible in the TIA-1 channel (Fig. 9.6a), and this provides an estimate of the number of SGs per neuron.

9.7.4 Stress Granules in Brain I/R

With the above tools and methods, we evaluated SGs in vulnerable and resistant neurons. We evaluated SGs in three different models of brain ischemia: (1) cardiac arrest-induced global brain ischemia and resuscitation (CA/R),[98] (2) bilateral carotid artery occlusion with hypovolemic hypotension global forebrain ischemia (2VO/HT),[76] and (3) a focal ischemia model of middle cerebral artery occlusion (MCAO). For CA/R and 2VO/HT, we followed the design in Fig. 9.5 and compared resistant CA3 hippocampal neurons to vulnerable CA1 neurons. In the stroke model, we compared core and penumbral neurons. We now concisely summarize our findings in studies of SGs following brain I/R.

Our first significant observation was that SGs are constitutive to normal, control brain neurons. We tested whether this was an effect of the anesthetic used, the means of fixing the brain, or background due to our primary or secondary antisera. We ruled out all of these and have concluded that SGs are constitutively present in brain neurons. This raises the question: why are they there? However, we have not pursued this basic neuroscience question. Instead, the practical consequence for our studies was that the control state was not a baseline of zero SGs, but a baseline of a minimum number of SGs.

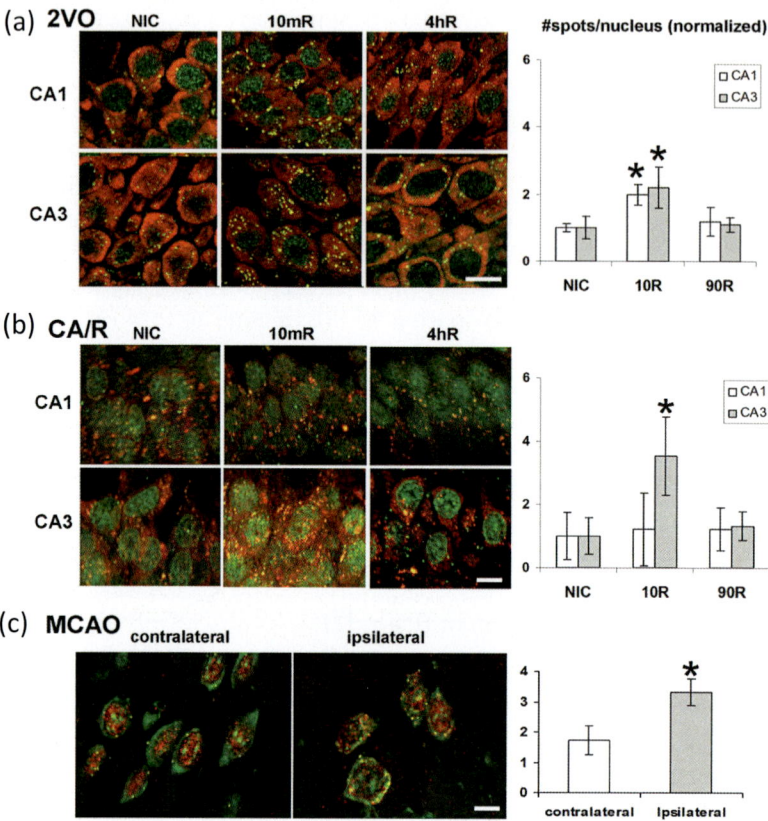

Figure 9.7   Quantitative estimation of SGs following (a) 2VO/HT, (b) CA/R and (c) pMCAO ischemia. In (a) and (b), representative photos of CA1 and CA3 showing merged TIA-1 (green) and S6 (red) channels in non-ischemic controls (NIC) and at 10 min (10mR) and 4 h (4hR) reperfusion. In (c) representative micrographs of merged TIA-1 IF (red) and poly(A) FISH in the contralateral and ipsilateral striatum of an animal subjected to 2 h pMCAO. Graphs are mean number of spots as estimated by yellow channel counting, normalized to the number of nuclei plus or minus standard deviations. Asterisks (*) indicate Tukey post hoc $p < 0.05$ for plots in (a) and (b) compared to NIC, or Student's t test for plot in (c) compared to contralateral. All scale bars are 10 microns; scale bars in (a), (b), (c), apply to all panels in (a), (b), (c), respectively.

With respect to SG behavior following brain I/R, the 2VO/HT model presented what we now take as the "standard" case of SG behavior. Rats were subjected to 10 min of 2VO/HT ischemia and evaluated out to 48 h reperfusion. We observed the number of SGs to increase at very short reperfusion durations (10 min), but then decline rapidly so that SGs were at baseline by 1 h reperfusion. There was no difference between CA1 and CA3 in the 2VO/HT model (Fig. 9.7a). This result falsified the hypothesis in Fig. 9.5.

In the CA/R model, we observed the same transient increase in SGs in CA3 and other resistant neuron populations (i.e., dentate gyrus and hilus of hippocampus).[98] However, in CA1, we saw no quantitative change in SGs. Instead, we observed a loss of S6 staining in the cytoplasm, but not in cytoplasmic SGs (Fig. 9.7b, 4hR CA1). We initially interpreted this result incorrectly to indicate that SGs were quantitatively sequestering 40S subunits. Our subsequent work on focal ischemia clarified the situation and allowed us to reinterpret the CA/R results.

In the MCAO model we compared the penumbra of the stroked ipsilateral side to the non-ischemic contralateral side for different brain regions. In striatum (Fig. 9.7c), we observed an increase in the number of SGs. Other regions showed no change in SGs, such as layer V pyramidal neurons (data not shown).

There was an additional interesting phenomenon in the MCAO model. In the core region we observed a wholesale loss protein antigen staining. We tested 14 antigens, and all of them showed significant declines in signal in the core region of the stroke; three of these are shown in Figs. 9.8a–c; PABP, ribosomal protein S6, and HuR. This observation is reminiscent of the loss of S6 staining in CA1 neurons at 4 h reperfusion after 10 min CA/R (Fig. 9.8d).

The DND discussed above in Section 9.3.2 is to be contrasted to **necrosis**, which occurs if ischemia is too intense or prolonged.[43] Necrotic cell death is characteristic of core, and DND occurs in penumbra following focal ischemia. That we observed loss of signal of 14 antigens in core neurons indicates there is wholesale loss of proteins in the necrotic core neurons. Although we did not test so many antigens in the CA/R model, the wholesale loss of S6 is more similar to the behavior of core, as opposed to penumbral neurons. On the other hand, the core neurons following MCAO are clearly necrotic, but those in CA1 at 4 h reperfusion after CA/R do not have an overt necrotic phenotype. Nonetheless, we now reinterpret the CA/R data to indicate that CA1 neurons in that model are undergoing a process with features similar to necrosis. Although CA1 neurons following CA/R die after a time delay, our data suggest that CA/R is inducing a hybrid of DND and necrosis. Some additional support for this notion is that the level of eIF2α(P) is higher by a factor of ~5 in CA1 neurons after CA/R than after 2VO/HT.[1] Also, the increase in stress granules in CA3 was ~4× after CA/R, but only ~2× after 2VO/HT (plots, Fig. 9.7). The increased quantities of stress markers indicate that 10 min CA/R is a greater stress on the neurons than 10 min 2VO/HT. This is a sensible conclusion because it is known that peripheral organ ischemia following CA/R, but not 2VO/HT, damages peripheral organs, which in turn enhances brain damage.[40] In general, many other workers have come to the conclusion that DND shares features with necrosis as well as apoptosis.[99]

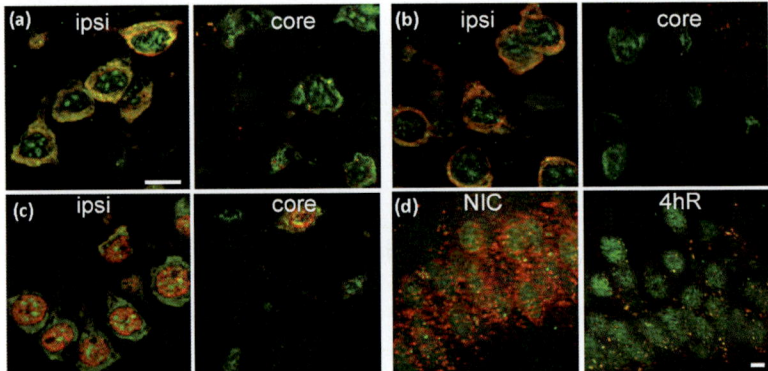

Figure 9.8   Comparison of loss of protein antigenicity in focal and global ischemia. Costaining of poly(A) FISH (green) with (a) PABP, (b), ribosomal protein S6, and (c) HuR (all proteins red) in ipsilateral penumbral and core neurons following 6 h pMCAO. (d) Merged images of TIA-1 (green) and S6 (red) in CA1 neurons at 4 h reperfusion following 10 min CA/R. The loss of protein antigenicity in core is similar to that in CA1 after CA/R; however, core neurons are shrunken and distorted and undergoing necrosis, whereas the CA1 neurons retain their morphology at 4 h reperfusion. Scale bar in the first panel of (a) is 10 microns and applies to all panels in A–C. Scale bar in (d) is 10 microns and applies only to panels in (d).

With respect to SGs, from the above data we were able to conclude that SGs do not sequester 40S subunits and therefore do not mediate prolonged TA in vulnerable neurons. In spite of falsifying our initial idea, we made a number of important observations that advanced our knowledge. We showed that the increases in the number of SGs correlated with eIF2α(P) levels, suggesting a pathway such as described in Section 9.7.2 occurs in reperfused neurons. However, such a pathway is transient and does not correlate with persistent TA or cell death. The transient increase in SGs is consistent with their dynamic nature as structures that route mRNA to other sites. In Section 9.8 we discuss to where the SGs may be routing the mRNAs.

9.7.5 Other SG Antigens

Our initial studies of SGs required we convince grant and manuscript reviewers that the objects in our photomicrographs were indeed SGs. Hence, we developed the ability to stain for a variety of proteins known to be in SGs, including eIF4E, eIF4G, poly(A)-binding protein (PABP), HuR, and TTP, in addition to S6 and TIA-1. We also developed stains for a number of antigens known *not* to be in SGs to prove that the colocalizations we observed were specific and not nonspecific artifacts. These included markers of the large 60S ribosomal subunit (ribosome P antigen),[100] markers of endoplasmic reticulum (PDI, erp29), Golgi (GM130, TGN38), mitochondria (COXIV) and cytoskeleton (α-tubulin, neurofilament H/M). We have shown that the proteins known to be in SGs colocalized with the punctate cytoplasmic TIA-1, and those known not to be in SGs did not colocalize with the punctate cytoplasmic TIA-1 in brain neurons.[76,98,101] Thus, we have a high degree of confidence in our SG staining.

In the course of these extensive double-labeling studies, in retrospect, one turned out to be quite significant. We noted above the observation that eIF4G is degraded as a function of ischemic duration. Having developed the ability to stain eIF4G in brain slices, we proceeded to assess it by IF. eIF4G staining was studied in hippocampal neurons following CA/R. We confirmed in this study that SGs, as measured by colocalizing eIF4G with TIA-1, transiently increased in CA3 but not CA1 during reperfusion.[102] However, eIF4G staining in reperfused neurons generated "large areas of increased eIF4G intensity." In controls, eIF4G staining was relatively homogeneous through the cell body cytoplasm (Fig. 9.9a). By 90 min reperfusion eIF4G appeared highly granular (Fig. 9.9b). By 4 h reperfusion, the smaller granules had been replaced by larger areas of increased eIF4G intensity (Fig. 9.9c, arrows). The granulation of eIF4G staining was not detected until 90 min reperfusion. The eIF4G granules were observed in both CA1 and CA3. Our initial impression of these results was that the granular eIF4G staining might have represented its inclusion in PAs. However, this granular staining pattern was to emerge as a most important discovery.

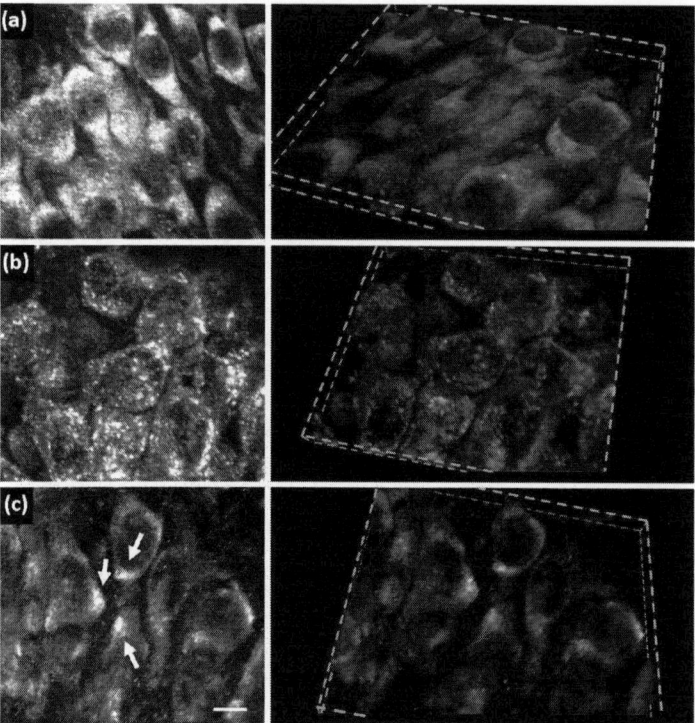

Figure 9.9 IF staining of eIF4G in CA1 neurons following CA/R. (a) Non-ischemic controls, (b) 10 min CA/R and 90 min reperfusion, and (c) 10 min CA/R and 4 h reperfusion. Left panels are maximum intensity orthographic projections constructed from 10 sequential z-slices separated by 0.75 microns with x:y:z pixel ratios of 1:1:7.5. Right panels are three-dimensional volumetric reconstructions made using ImageJ 3D viewer (ImageJ 1.44). Scale bar in (c) is 10 microns and applies to left panels only. Arrows in (c) point to "areas of increased intensity" of eIF4G staining.

9.8　mRNA Granules in Reperfused Neurons

In the effort to redundantly confirm that we were observing SGs, it was essential to develop a marker for one of the preeminent components of SGs: mRNA. To this end, we developed methods to colocalize an mRNA species and a protein via double staining.[76] To detect mRNA we used fluorescent in situ hybridization (FISH) in tandem with single labeling IF. To obtain the strongest possible FISH signal, we used as a probe a poly(T) 50-mer to detect the poly(A) tails of mRNAs. With suitable controls and validation, we were able to demonstrate specificity of poly(A) mRNA staining via FISH (Fig. 9.10). We then proceeded to study the colocalization of poly(A) mRNAs with different proteins, and under a variety of ischemic conditions including global and focal ischemia and ischemic preconditioning. We were now in a position to investigate the question asked at the beginning of Section 9.7: "What happens to the mRNA after polysomes dissociate?"

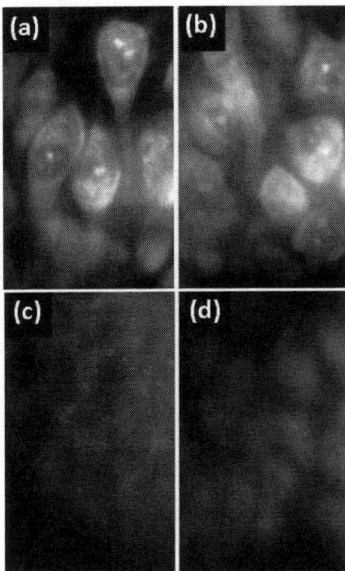

Figure 9.10　Staining controls for poly(A) FISH. (a) Regular FISH staining with no treatment. (b) Slices pretreated with 10 U/mL DNAase prior to FISH, (c) Slices treated with 0.1 M NaOH prior to FISH (NaOH hastens RNA but not DNA degradation), (d) Slices treated with 10 U/mL RNAse A prior to FISH. NaOH and RNAse A treatment abolished FISH signal.

In this section we describe several significant findings that emerged from studying global mRNA staining patterns in post-ischemic neurons. We showed that (1) mRNA globally transforms into a granular cytoplasmic distribution in reperfusion neurons, into structures we call "mRNA granules," (2) the presence of the mRNA granules correlates perfectly with TA, (3) the mRNA granules do *not* colocalize with ribosomal subunits, thus showing why mRNA granules correlate with TA (4) colocalization of the mRNA-binding protein, HuR (discussed in Section 9.2.2) with the mRNA granules correlated with translation of HSP70 protein, and (5) the punctate TIA-1 cytoplasmic granules contained mRNA, reinforcing their identity as SGs, which, in light of the other observations, made this a relatively incidental finding. In net, these observations have significantly advanced the discussion, if not having actually solved, the persistent TA problem, and advanced the discussion of why stress proteins are translated in resistant but not vulnerable neurons by shifting focus from ribosome to mRNA regulation. We now summarize these findings.

9.8.1 mRNA Granules

Our initial studies of poly(A) mRNA staining were conducted in the 2VO/HT model using 10 min ischemia. When we stained 1 h reperfused samples via poly(A) FISH, we were surprised to see the mRNA staining pattern was radically different from non-ischemic control neuronal mRNA staining patterns. In non-ischemic controls (NIC), the mRNA staining was relatively smooth and homogeneous throughout the cytoplasm, as illustrated for hippocampal CA1 and CA3 (Figs. 9.11a and 9.12a). However, at 1 h reperfusion, the mRNA staining was highly granular (Figs. 11b and 12b). 3D reconstructions of the NIC and 1hR poly(A) staining patterns (Figs. 9.11 and 9.12) demonstrate how the poly(A) morphology distributed throughout the cell bodies.

We thus called the granular mRNA staining pattern "mRNA granules," which is meant only as a descriptive term with no mechanistic connotation because, at this stage in our investigations, we have not definitively identified the molecular basis of these structures. Instead we have gone on to characterize the phenomenology of the mRNA granules: (1) their time course and correlation with TA, (2) their colocalization pattern with a number of proteins, and (3) how the mRNA granules behaved in neurons under a variety of ischemic conditions.

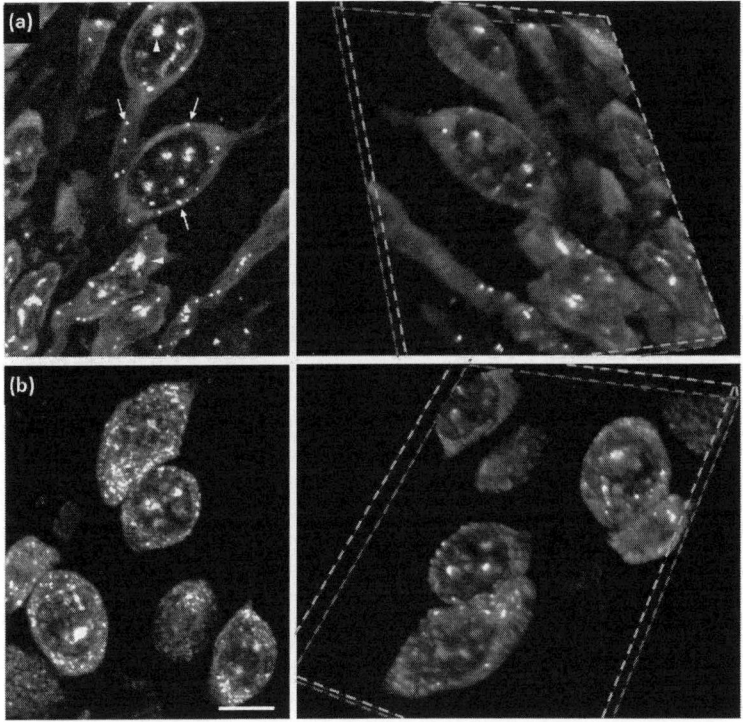

Figure 9.11   mRNA granules in CA1 at 1 h reperfusion. (a) Non-ischemic controls, and (b) 1hR reperfusion. Left panels are orthographic projections as described in the legend of Fig. 9.8, but with *x:y:z* pixel ratio = 1:1:3.5. Right panels are volumetric renderings as described in the legend of Fig. 9.8. Arrows in left panel, A, point to punctate cytoplasmic poly(A) staining that colocalize with components known to be in stress granules and illustrates presence of SGs in control rat brain neurons; volumetric rendering (right) shows these embedded in the cell body cytoplasm. Arrowheads point to nuclear poly(A) speckles that are routinely observed in neurons following poly(A) FISH staining. The mRNA granules are apparent at 1 h reperfusion (b). Scale bar on left, (b) applies to left panels only and is 10 microns.

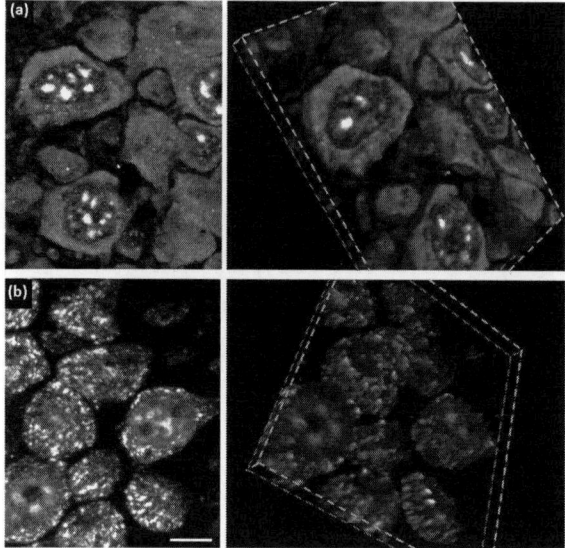

Figure 9.12 mRNA granules in CA3 at 1 h reperfusion. (a) Non-ischemic controls, and (b) 1hR reperfusion. Left panels are orthographic projections and right panels are volumetric renderings as described in the legend of Fig. 9.11. (b) Scale bar on left, (b) applies to left panels only and is 10 microns.

9.8.2 mRNA Granules Time Course and Correlation with Translation Arrest in 2VO/HT

From 1–48 h reperfusion after 10 min ischemia in the 2VO/HT model, in vivo translation was assessed 1 h after IV administration of a bolus of 1 mCi of ^{35}S-labeled cysteine and methionine. We worked in this time frame to avoid having our results confounded by overt cell loss in vulnerable hippocampal CA1 neurons, which occurs ~72 h reperfusion. Hippocampal CA1 and CA3 were microdissected, homogenized, and aliquots quantified for radioactive incorporation into TCA-precipitable proteins; aliquots of the same homogenates were also Western blotted for eIF2α(P). The same time course of reperfusion was repeated, brain slices obtained, and stained by FISH to detect poly(A) mRNAs.

The in vivo translation study confirmed the well-established fact that TA persists in CA1, but is transient in CA3 where translation fully recovered after 36 h reperfusion (Fig. 9.13a). The only statistically significant increase in eIF2α(P) occurred at 10 min reperfusion and returned to control levels for the remainder of the time course in both CA1 and CA3 (Fig. 9.13b); this result confirmed that levels of phosphorylated eIF2 alpha subunit do not correlate with TA in either CA1 or CA3. Significantly, in every sample where translation was below control levels, at those time points neurons stained by poly(A) FISH showed the presence of the mRNA granules (Figs. 9.13c,d). Thus, there was a direct correlation between mRNA granules and TA.

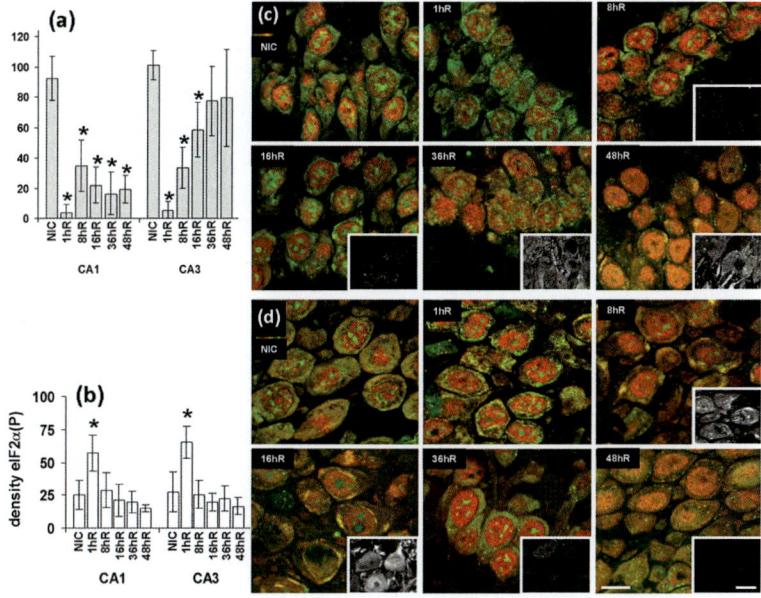

Figure 9.13   Correlation of post-ischemic TA with mRNA granules. (a) In vivo translation in CA1 and CA3 measured by radioactive amino acid incorporation (mean ± standard deviation) out to 48 h reperfusion. Asterisks denote Tukey post hoc $p < 0.05$ compared to NICs. (b) Densitometry of eIF2α(P), normalized to GAPDH, western blots over the same time course; *, Tukey post hoc $p < 0.05$ compared to NICs. (c) Double IF/FISH for HuR (red) and poly(A) green over same time course. mRNA granules were present at the same times at which translation rates were below controls. Insets in C are IF for HSP70 for corresponding experimental groups. Scale bar in last panel of D is 10 microns and applies to all HuR/poly(A) panels. Scale bar in insets is also 10 microns and applies to all inset images. All photomicrographs show under X63 oil immersion, and are orthographic projections from $z = 10$ with pixel $x{:}y{:}z{:} = 1{:}1{:}3.5$.

In both CA1 and CA3, we observed the mRNA granules to change their appearance over the reperfusion time course. At the earliest reperfusion times, the mRNA granules were smaller and appeared distributed throughout the whole of the cell body cytoplasm of the neurons (Fig. 9.14b). At later reperfusion durations, the mRNA granules became larger, less numerous, and tended to be associated with the plasma and/or nuclear membranes (Fig. 9.14c).

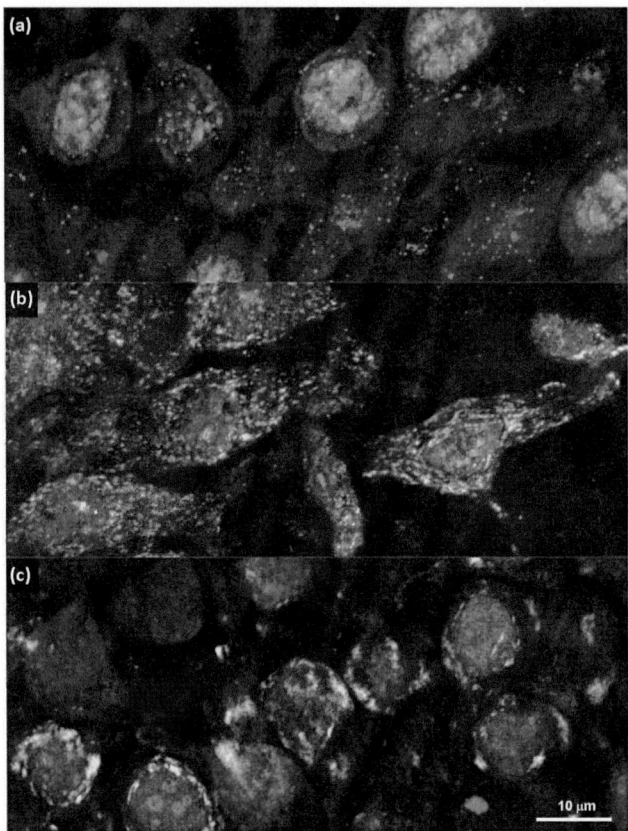

Figure 9.14 Illustration of change in morphology of mRNA granules with reperfusion duration in CA3. Staining for HuR and poly(A) in (a) Non-ischemic controls, (b) 1 h reperfusion, and (c) 16 h reperfusion. Each panel is an orthographic projection derived from X63 oil immersion z-stack photomicrographs with pixel x:y:z = 1:1:3.5, where (a) z = 49, (b) z = 50, and (c) z = 36. Scale bar in C is 10 microns and applies to panels.

9.8.3 Colocalization of Proteins with mRNA Granules

The spatial pattern of the mRNA staining following 2VO/HT was similar to that of eIF4G staining following CA/R (Fig. 9.14 vs Fig. 9.9, respectively). It was natural to ask if these were the same structures. It was also natural to consider what other proteins, if any, would colocalize with the mRNA granules. Hence colocalization studies were undertaken between poly(A) mRNAs and a variety of protein antigens, using FISH and IF double-labeling.[101]

We observed that the mRNA-binding proteins eIF4G (not shown) and PABP (Fig. 9.15a) colocalized with poly(A) mRNAs. PABP, like eIF4G (Fig. 9.9), exhibited a granular pattern in reperfused neurons if stained only by IF without co-labeling with FISH (not shown). Staining specificity was confirmed by the fact that the mRBPs TIA-1 (Fig. 9.15b) and TTP (not shown) did not colocalize with the mRNA granular staining. A number of organelle markers were also tested to determine colocalization with the mRNA granules. Shown are α-tubulin, the mitochondria marker cytochrome C oxidase subunit IV(COXIV), the endoplasmic reticulum marker protein disulfide isomerase (PDI) and the trans-Golgi network marker TGN38 (Figs. 9.15c–f, respectively). We also tested the neuronal nucleus marker NeuN, the Golgi apparatus marker GM130 and neurofilament H/M. None of the organelle markers colocalized with the mRNA granules, indicating the latter are not associated with the respective organelles, and suggesting the mRNA granules are cytoplasmic structures.

Most significantly, markers of the 40S and 60S subunits did not colocalize with the mRNA granules (Fig. 9.16). This observation suggests that mRNAs granules are a structure that partitions mRNA from the ribosomal subunits. This reinforces the idea of sequestration initially advocated by Hu and colleagues, but with a twist: it is not that ribosomal subunits are completely sequestered in non-functional complexes, but that mRNA appears to be quantitatively sequestered away from the ribosomal subunits. Physically separating ribosomal subunits and mRNAs will clearly result in a lack of translation. Hence, we feel these observations solve the persistent TA problem in vulnerable post-ischemic neurons:

persistent TA is caused by the physical separation of mRNA and ribosomal subunits.

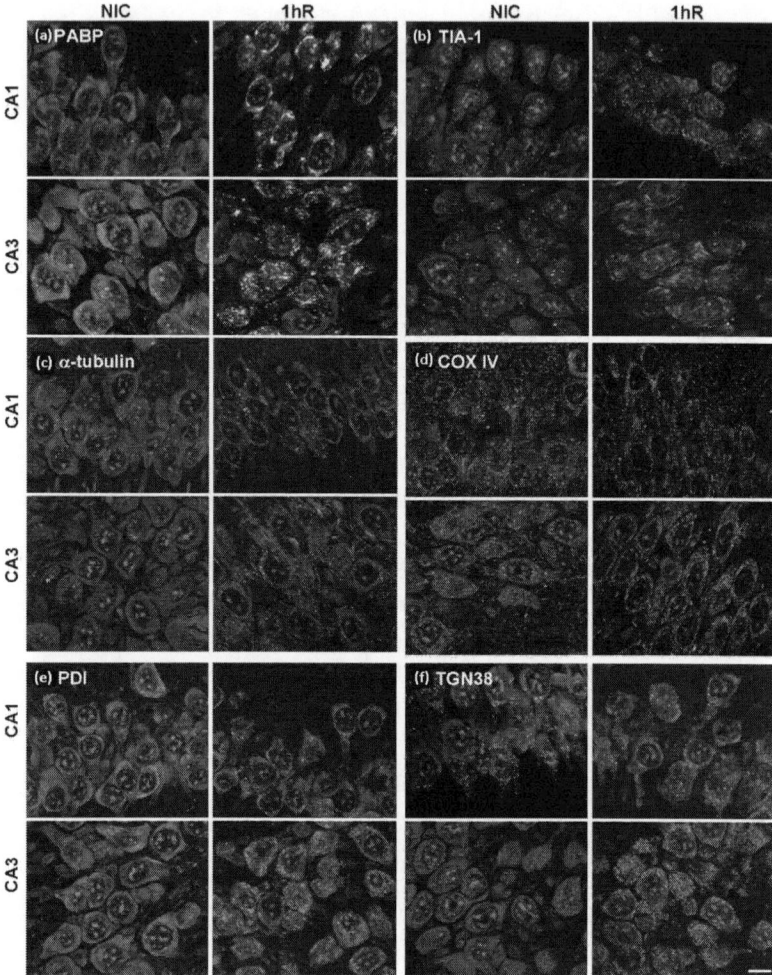

Figure 9.15 Double IF/FISH to test colocalization of proteins and organelle markers with mRNA granules. Antigens as indicated in a–f. Each staining pair shows CA1 and CA3 in non-ischemic controls and at 1 h reperfusion, as indicated. Scale bar in lower right image is 10 microns and applies to all images. All images are orthographic projections of z = 10 and pixel ratio = 1:1:3.5, obtained under X63 oil immersion.

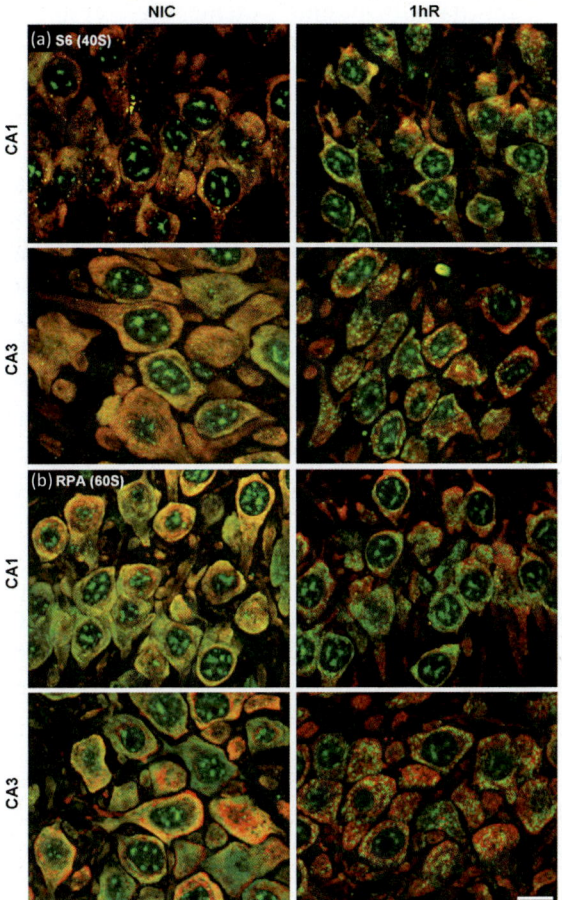

Figure 9.16 Ribosomal subunits do not colocalize with mRNA granules. (a) S6 IF (red) and poly(A) FISH (green); S6 is a marker of the 40S subunit. (b) Ribosomal P antigen (RPA) IF (red) and poly(A) FISH (green); RPA is a marker of the 60S subunit. Each staining pair shows CA1 and CA3 in non-ischemic controls and at 1 h reperfusion, as indicated. Scale bar in lower right image is 10 microns and applies to all images. All images obtained as described in the legend of Fig. 9.14.

9.8.4 HuR Colocalization with mRNA Granules

The colocalization behavior of HuR showed differential staining between CA1 and CA3. At all time points tested, HuR always colocalized with the mRNA granules when present in CA3 neurons (Fig. 9.13d). However, in CA1, HuR did not colocalize with the mRNA granules until 36 or 48 h reperfusion (Fig. 9.13c). Thus, unlike any other antigen we tested, HuR colocalization with the mRNA granules distinguished vulnerable and resistant neurons.

In addition, we observed that, whenever HuR colocalized with the mRNA granules, those same cells translated HSP70 protein (insets, Figs. 9.13c,d). This correlation held for all but the 1 h reperfusion time point, at which time there is no transcription of the hsp70 Mrna.[87] We have extended the microscope observations by showing the immunoprecipitation of HuR protein co-precipitates hsp70 mRNA, but not a non-ARE containing protein.[101] The correlation between HuR colocalization in mRNA granules and translation of HSP70 protein suggests these events are linked. There is evidence that HuR binds to hsp70 mRNA.[103] It is therefore reasonable to hypothesize that HuR plays some direct or indirect role in mediating the selective translation of hsp70 mRNA. This hypothesis is currently driving our ongoing work and we have not yet resolved this issue. However, this represents an advance in understanding the differential translation of HSP70 between vulnerable and resistant neurons, and shifts the research focus to the molecular biology of mRNA regulation.

9.9 Behavior of mRNA Granules Under Other Conditions

The pattern of mRNA granule behavior described above is that following global forebrain ischemia in the 2VO/HT model. In this section, we describe several other studies that evaluated mRNA granules under different conditions of ischemia or following pharmacologic treatments.

9.9.1 2VO/HT Model: Effect of Ischemia Duration on mRNA Granules

The relationship between ischemia duration and outcome (survival or death) is nonlinear.[104–107] The threshold to induce CA1 neuronal death with the 2VO/HT model is between 9 and 10 min of normothermic ischemia. At 8 min, there is no brain cell death, thus

8 min is a sublethal insult. Lesser ischemia durations are potent inducers of ischemic preconditioning (IPC, see Section 9.3.3): 2 min of 2VO/HT ischemia can induce essentially 100% neuroprotection if a second, lethal 10 min ischemia is induced exactly 2 days after the sublethal ischemia.[51] For durations longer than 10 min, neuronal populations in addition to CA1 are drawn into DND. For ischemia > ~25 min, immediate necrosis results. Thus, it was of interest to examine the effects of ischemia duration on mRNA granule formation.

For sublethal 2VO/HT ischemia of <10 min, we observed the frequency of mRNA granule formation to decline with decreasing duration. At 1 h reperfusion following 7 and 8 min of ischemia >75% of animals showed mRNA granules in hippocampal neurons. At 6 min, ~50% of animals showed mRNA granules. For ischemia <5 min, <25% of animals showed mRNA granules in hippocampal neurons. In six animals given 2 min 2VO/HT ischemia, none of the animals showed mRNA granules. From these data, we can surmise that the threshold for formation of mRNA granules is less than that which induces CA1 cell death, but is greater than that required for an optimal IPC response. Thus, mRNA granule formation is not required for IPC. This leads us to the tentative conclusion that mRNA granules represent a relatively intense aspect of the post-ischemic stress responses that are not induced by weak sublethal insults ~2 min in duration.

For supralethal ischemia doses, we studied 15 min ischemia and 1 h reperfusion.[108] Similar to 10 min ischemia, all of the post-ischemic neurons showed mRNA granules that did not colocalize with ribosomes. The main result of this study was the observation of some cortical neurons that, in addition to forming mRNA granules, lost the nuclear poly(A) speckles, as described below for endothelin-1 (ET1) treatment.

9.9.2　Focal Cerebral Ischemia Effect on mRNA Granules

We used the MCAO model of focal brain ischemia to study the effect of the duration of focal ischemia on mRNA granule formation. Focal ischemia is more complex than global ischemia because there is a gradation of blood flow decrements in the brain of the same animal, where blood flow is zero at the site of occlusion, and 100% some distance from the occlusion. Hence, we studied only the effect of ischemia duration, and did not study reperfusion. When only ischemia, and not reperfusion, is used in the MCAO model, this is known as "permanent" MCAO (pMCAO), and is a model of stroke where spontaneous reperfusion does not occur, leading to frank necrosis of the brain tissue. In addition, we studied the effect of streptozocin-induced diabetes superimposed on the pMCAO condition.

In the pMCAO model in rat, it is well known that the full extent of necrotic brain tissue, the core, forms by 8 h after induction of pMCAO.[46] We therefore studied pMCAO durations of 2, 4, 6, and 8 h. In these studies we evaluated (1) which neurons in the MCA territory formed mRNA granules, (2) how the distribution of mRNA granule–containing neurons changed with pMCAO duration, and (3) the colocalization properties of the mRNA granules. With regard the latter, the mRNA granules that formed following pMCAO had the same colocalization characteristics as those identified during 2VO/HT ischemia (Section 9.8.3).

MCAO affects forebrain at the coronal levels of the parietal and caudal frontal lobes, and the underlying subcortical structures, which is mainly striatum in the rat brain. We systematically mapped the distribution of mRNA granule–containing neurons through this volume of brain (Fig. 9.17a). In these studies we identified two broad types of response to the pMCAO: (1) what we termed "granular" tissue, which referred to regions in the cortex and striatum where the neurons clearly showed mRNA granules in their cytoplasm (Fig. 9.17b) and (2) core tissue where neurons had a distorted and shrunken morphology, and were clearly undergoing necrosis (Fig. 9.17c). We called the "affected volume" the sum of the volumes containing core + "granular" brain tissue. In any particular animal, the line between these regions was relatively discreet. Outside of the affected volume, the neurons were indistinguishable from non-ischemic control tissue with respect to poly(A) staining by FISH.

Our main findings were as follows. The relative proportion of "granular" and core tissue changed. At 2 h pMCAO, ~75% of the affected volume was occupied by "granular" tissue. By 8 h, 100% of the affected volume was core (Fig. 9.17d). An exponential fit of the percentage of total affected volume that was granular verses pMCAO duration gave a correlation coefficient of 0.91 (a linear fit gave a correlation coefficient of 0.77). The total affected volume did not change as a function of pMCAO duration (Fig. 9.17e).

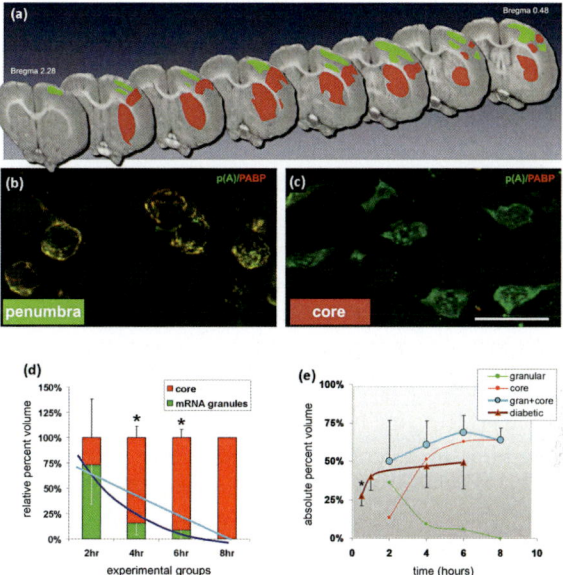

Figure 9.17 Distribution of mRNA granule–containing ("granular") and core neurons following pMCAO in rat. (a) Individual coronal sections taken from the Laboratory of Neuroimaging (LONI) at the University of California, Los Angeles rat brain dataset serving as a backdrop to specify areas with mRNA granule–containing neurons ("granular," green) or shrunken and distorted neurons ("core," red). (b) Representative "granular" neurons detected using PABP/poly(A) IF/FISH. (c) Representative core neurons detected using PABP/poly(A) IF/FISH. Scale bar in C is 20 microns and applies to b and c. (d) Averages ± standard deviations of the relative percent volumes of mRNA granule–containing neurons (green bars) and shrunken and distorted neurons (red bars) following durations of pMCAO. Blue and cyan curves are exponential (correlation coefficient = 0.91) and linear (correlation coefficient = 0.77) fits through the mRNA granule–containing neurons volume averages, respectively. Asterisks (*) indicate Tukey post hoc $p < 0.05$ between core and mRNA granule volumes at the respective pMCAO durations. (e) Total affected volumes (average ± standard deviation) of diabetic (brown curve) and non-diabetic (cyan) Long Evans as a percentage of one hemisphere of the rendered LONI dataset volume. Average total volumes of mRNA granule–containing (green curves) and shrunken and distorted neurons (red curve); error bars are omitted from these curves for clarity but are proportionally the same as those shown in the plot in d. Asterisk (*) indicates Tukey post hoc $p < 0.05$ for 30 min pMCAO diabetic group compared to all other experimental groups.

When the same experiment was repeated in diabetic animals, we were surprised to find that no pMCAO duration showed "granular" tissue, and all of it was core (Fig. 9.17e). The total core volume in the diabetic animals was statistically indistinguishable from the total affected volume in the non-diabetics.

Although these were only initial descriptive studies, the results suggest the following tentative conclusion. Even at 2 h pMCAO, the large volume of "granular" tissue suggests these regions were under intense enough injury to induce mRNA granules. Hence, these regions behave analogous to the 8–10 min duration in the 2VO/HT global ischemia model. However, as pMCAO duration increased, the tissue became increasingly necrotic, which is equivalent to > ~25 min 2VO/HT ischemia. Thus, characterizing the mRNA morphology in the neurons provided a basis to compare what are otherwise difficult models to compare. Diabetes represented an intrinsic injury to the brain tissue prior to application of the pMCAO: when the pMCAO was administered, even the 2 h pMCAO was enough to tip the neurons to the necrotic phenotype, and this suggests that the effects of the two injuries—diabetes and pMCAO—were additive on outcome.

9.9.3   Effect of Endothelin-1 on mRNA Granules

Endothelin-1 (ET-1) is a highly potent vasoconstrictor. We studied the effect of direct injection of ET-1 bilaterally it into the third ventricle.[108] Our rationale was to attempt to cause a graded decrease in forebrain blood flow as a function of ET-1 does, and then determine the threshold decrement in blood flow required to induce mRNA granules in the neurons. It is well known that TA occurs in neurons at ~50% decrement in blood flow rate. Therefore, if we could demonstrate that mRNA granules formed at ~50% decrement in blood flow, that would be additional evidence linking mRNA granules and TA.

However, these studies did not work as anticipated. ET-1 is such a potent vasoconstrictor that it fully constricted individual vessels, causing necrotic damage in localized regions of cortex around the third ventricle (Fig. 9.18a). With increasing ET-1 dose, the area of necrotic tissue increased. At no dose of ET-1 did we observe mRNA granules in the neurons. Instead, we observed loss of nuclear poly(A) speckles, similar to our finding at 15 min 2VO/HT ischemia (Figs. 9.18b,c). That we observed loss of nuclear speckles in the absence of mRNA granules indicates these two phenomena are not causally linked. Our overall conclusion of these studies was that ET-1 is not a good means to obtain graded decreases in brain blood flow, and it did not allow us to determine the threshold of blood flow decrement that induced mRNA granule formation.

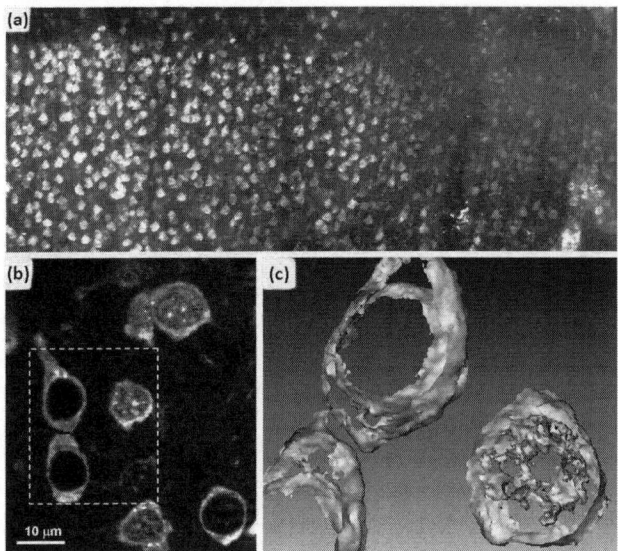

Figure 9.18 Effect of endothelin-1 (ET1) on poly(A) staining in cerebral cortex. (a) 200 pg of ET-1 causes discreet lesions in cerebral cortex. Shown is layer II at cusp of an unaffected and affected area. From left to right, the amount of neuronal death increases precipitously. (b) High magnification (X63 oil) orthographic projection ($z = 10$) of layer II neurons showing loss of nuclear poly(A) speckles in some neurons but not others. Scale bar 10 microns. (c) Loss of poly(A) nuclear speckles is not a probe penetration artifact as 3D surface reconstructions from the z-stack in B show that one neuron lacks poly(A) speckles and a second neurons only microns from the first still contains poly(A) speckles, at the exact same z-levels.

9.9.4 Effect of Cycloheximide and Puromycin on mRNA Granules

The final study we describe was aimed at using pharmacologics to learn more about the molecular biology of the mRNA granules. Kedersha and Anderson used agents well known to induce TA to study the relationship between translation and SG formation.[97] Specifically, emetine halts elongation, preventing polysomes disassembly.[109] Puromycin halts translation by mimicking the translation termination factor,[110] thereby halting elongation, and causing premature termination of polysomes and polysome dissociation. Anderson and Kedersha showed that emetine prevented SG formation by maintaining mRNA and 40S subunits in the inactive polysomes, and that puromycin administration alone could induce SG formation by fostering polysome dissociation as occurs when translation initiation is halted (as, for example, when eIF2α becomes phosphorylated). We adapted this strategy to assess if a similar link held between polysomes and mRNA granules in rat brain.[111]

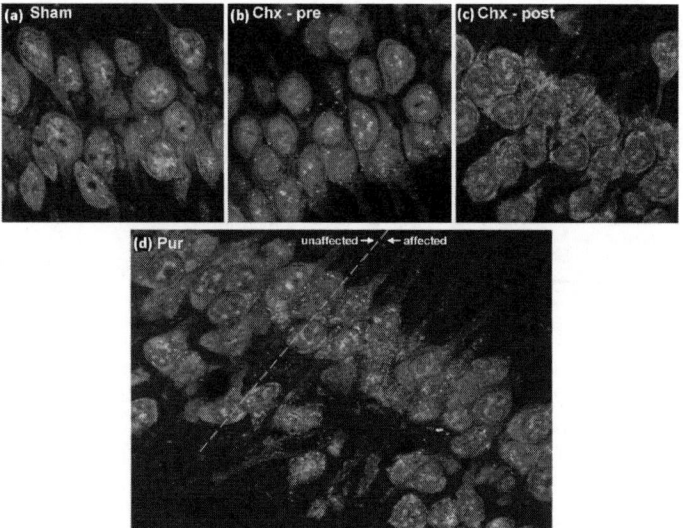

Figure 9.19 Effects of cycloheximide (Chx) and puromycin (Pur) on mRNA granules. (a) Vehicle-treated control, (b) Chx administered 15 min before 10 min ischemia and 1 h reperfusion; Chx inhibited mRNA granule formation. (c) Chx administered 15 min after 10 min ischemia and reperfusion for 1 h; Chx did not affect mRNA granule formation because polysomes were already dissociated. (d) Pur administered intraventricularly (third ventricle) to a non-ischemic control induced formation of mRNA granules. Photo taken at the cusp of affect and unaffected area in CA1; cyan line divides these. Scale bar in d is 10 microns and applies to all images

In place of emetine we used cycloheximide, which works by a similar mechanism and "freezes" polysomes into an inactive state.[110] When cycloheximide was administered 15 min before ischemia, mRNA granule formation was inhibited at 1 h reperfusion (Fig. 9.19b). If cycloheximide was administered at 10 min reperfusion, after 10 min ischemia, then it had no effect, and mRNA granules formed in a manner indistinguishable from vehicle-treated reperfused animals (Figs. 9.19a,c, respectively). Intraventricular administration of puromycin induced mRNA granule formation in non-ischemic control rat brain neurons (Fig. 9.19d). These results indicate that polysome dissociation is upstream of, and a prerequisite to, mRNA granule formation.

9.10   Summary of mRNA Granules

The above summarizes our current findings on mRNA regulation in the context of post-ischemic TA. The most important finding at this stage is that TA is caused by a partitioning of mRNA away from ribosomal subunits in the form of granular structures, the mRNA granules. Further, a link between HuR and mRNA granules suggests they play an intimate role in post-ischemic stress response. We summarize these findings with the following model depicting the role of TA and mRNA regulation in the execution of stress responses induced in neurons by I/R injury (Fig. 9.20).

I/R-induced cellular stress activates the eIF2α kinase PERK,[90] which phosphorylates eIF2α,[81,82] thereby inhibiting eIF2B, leading to polysome dissociation,[77] Our data show that SGs increase rapidly and transiently.[98,76] It is reasonable to presume the transient SG increase occurs in response to the increased concentration of free mRNA following polysome dissociation. We do not see mRNA granules form much before 45 min of reperfusion, indicating that SGs precede mRNA granule formation. Thus, we posit that SGs serve as a transient vehicle to transport mRNAs such that mRNA granules can form. Formation of mRNA granules is then the definitive marker of TA because poly(A) mRNAs are kept sequestered away from the ribosomal subunits.[76,101] Functionally, the mRNA granules in the first hours of reperfusion act as mRNA silencing complexes, holding constitutive mRNAs in abeyance while stress-induced mRNAs build up in the nucleus and are transported into the cytoplasm. The link between HuR and the mRNA granules suggest they also play a role in the selective translation of the stress-induced mRNAs.

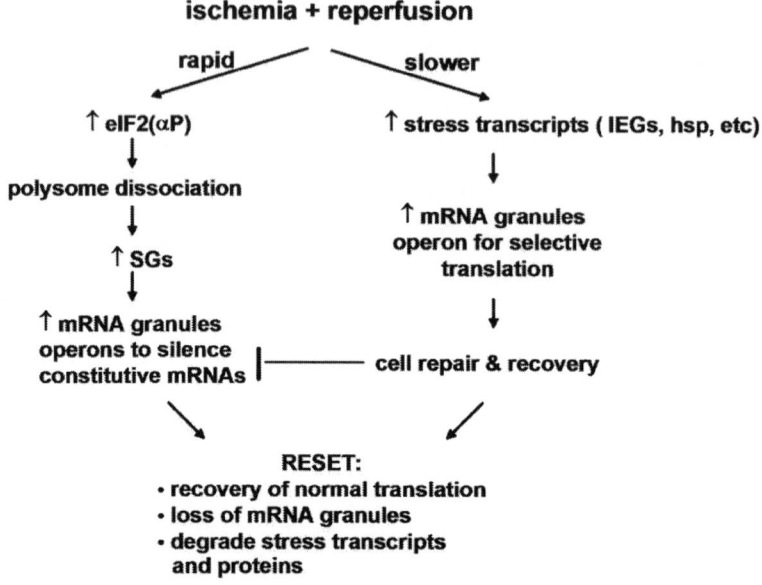

Figure 9.20 Summary of changes in ribosomal and mRNA regulation and their role in stress-induced translation in reperfused neurons. The processes depicted here are instantiations of the general algorithm shown in Fig. 9.2.

Successful translation of stress proteins is a watershed event in the intracellular stress response, providing essential machinery to halt damage, mechanisms, clear damaged cell components, and rebuild the cellular machinery. As these "damage control" processes come to a completion, the signals maintaining the stress responses are diminished and stress responses eventually shut off. Then, only the stress-induced mRNAs and proteins need to be degraded because they are no longer needs. At this point, the cell has successfully coped with the injury and reset itself to its pre-injury state. Such a scheme is particularly important for terminally differentiated cells such as neurons, which do not have the option to increase mitosis and cell division as a way to repair the damaged tissue.

Thus, it can be seen with this model that ischemic vulnerability will readily result if the cell is unable to express any of these steps in the successful execution of the stress responses. Thereby ischemia does not kill cells solely because of accumulation of cellular damage, but instead by overwhelming the cell's intrinsic defenses against injury. In this regard then, the prolonged TA of the vulnerable neurons is a marker of the inability of the neuron to successfully execute its endogenous stress response. In the final section, we further abstract the above qualitative model to an abstract and highly general model that posits that cell death results from a competition between the total amount of damage experienced by the cell and the total amount of stress responses induced by the cell.

9.10.1 Weakness and Limitation

The above model must be qualified by noting the limitations of the studies on which it is based. At present there are simply no deep molecular studies related to the following issues:

(1) What molecular mechanisms cause the mRNA granules form?
(2) What is their full molecular composition?
(3) By what mechanisms do they keep the mRNA away from ribosomal subunits?
(4) What mechanism, if any, allow them to participate in selective translation of stress-induced mRNAs?
(5) By what means are they dissolved in cells that recover from ischemia?

Thus, without a deeper molecular understanding, the model above can only be taken to be a set of phenomenological hypotheses consistent with the studies described above. As with any model in science, further evidence may completely alter the picture we have generated here.

In addition to the need for molecular studies, further types of microscope studies would be of use. To date we have tried unsuccessfully to use electron microscopy (EM) with immunogold labeling of proteins and poly(A) mRNAs to see the mRNA granules at the ultrastructural level. There have been many prior and extensive studies of brain ischemia at the EM level, and nothing resembling the morphology of mRNA granules has been described. Thus, one must conclude the mRNA granules do not bind OsO_4 and are electron transparent. We have also tried to develop FISH using specific probes for specific nucleotide sequences. These efforts have been stymied by the low signal obtained, which are orders of magnitude lower than the poly(A) signal.

Finally, it is important to expand the phenomenology. How general are mRNA granules? Do they only occur in neurons? Are they present in other neurological disease conditions? Do they occur in other cell types and other non-neurological diseases?

9.11 Bringing It Altogether: A Unified Model of Cell Injury

The limitations just mentioned pertain to the specific molecular details of I/R-induced cellular damage and I/R-induced stress responses. Such details are of potential importance for developing therapeutic strategies that target ischemic brain damage. However, there is a different direction from which to consider the above findings and that is in the direction away from the specific and toward the general. Many other forms of injury bear a formal similarity to the how neurons are injured by brain ischemia. The pattern is this:

(1) Cell is injured by some mechanism, X; X can be any form of injury: chemical, thermal, mechanical, biologically based (as is ischemia), etc.

(2) X causes the accumulation of damage products inside the cell. If X is uniquely specific, there will be only one form of cell damage, call it d_X. However, ischemia is more representative of most forms of cell injury by inducing, through a variety of parallel and simultaneous mechanism, a number of different forms of cell damage. Thus, a general injury, I will induced many forms of cell damage products $d_1, d_2, d_3, \ldots d_N$.

(3) The accumulation of $d_1, d_2, d_3, \ldots d_N$ will induce whatever intrinsic mechanisms the cell possesses for coping with the damage products. These we can label $s_1, s_2, s_3, \ldots s_N$. We do not expect there to be an exact 1 to 1 correspondence between the d_i and s_i, but we do expect enumerable relationships.[112]

(4) Then, the various damage mechanisms and various stress responses become involved in a competition. Stress responses exist to inhibit damage. Damage, by its very nature destroys cell components, and potentially, even the mediators of the stress responses. Hence damage and stress responses are mutually antagonistic. The term "competition" is apropos because the outcome of this competition is the very life of the cell.

9.11.1 Nonlinear Dynamical Model of Cell Injury

We have recently devised a simple mathematical model that captures this qualitative picture.[113] The model is based on the logic that underlies the use of Hill functions in pharmacology. The Hill function defines a threshold, Θ, which is the concentration of a drug that produces a 50% effect. We have carried this logic over to the competition between damage and stress responses induced by injury. The key idea to the whole model is not to consider any specific form of damage or stress response, but to imagine the combined action of all damage mechanisms, which we call the **total damage**, D, and the combined action of all induced stress response, which is the **total induced stress response**s, S. The competition then is between D and S.

The most simplified form of the model is as follows. We say that there is some threshold amount of total damage, Θ_D, which exerts a 50% inhibition on total stress responses, S. Similarly, there is some threshold amount of total induced stress responses, Θ_S, which exerts a 50% inhibition on total damage, D. We then express the competition as a nonlinear system of two differential equations coupled in D and S:

$$\frac{dD}{dt} = \frac{\Theta_D^{n_D}}{\Theta_D^{n_D} + S^{n_D}} - D$$

$$\frac{dS}{dt} = \frac{\Theta_S^{n_S}}{\Theta_S^{n_S} + D^{n_S}} - S$$

$$(9.1)$$

Equation (9.1) describes that the rate of accumulation of D is inversely proportion to S, where S is limited by Θ_D. Simultaneously, the rate of accumulation of S is inversely proportional to D, where D is limited by Θ_S. The parameters n_D and n_S are Hill coefficients representing the coupling strengths amongst the individual d_i and s_i, respectively. We can link injury magnitude to the threshold terms by the following expressions:

$$
\begin{aligned}
\Theta_D &= c_D I e^{\lambda_D I} \\
\Theta_S &= c_S I e^{-\lambda_S I}
\end{aligned}
$$

(9.2)

In Eq. (9.2), the terms c_D and λ_D are parameters that determine the relative toxicity of the applied injury. I is the quantitative amount of the injury applied. The terms c_S and λ_S are parameters that set the relative strength of a cell's intrinsic stress responses. Equation (9.2) is the core supposition of the model. Technically it posits that Θ_D will increase exponentially as Ie^I and Θ_S will change as Ie^{-I}. Intuitively it says that Θ_D will be small at low injury magnitudes and grow with increasing injury. On the other hand, Θ_S will increase for low injury magnitudes but decrease as injury magnitude increases. Substituting Eq. (9.2) into Eq. (9.1) gives our nonlinear dynamical model of cell injury:

$$
\begin{aligned}
\frac{dD}{dt} &= \frac{(c_D I e^{\lambda_D I})^{n_D}}{(c_D I e^{\lambda_D I})^{n_D} + S^{n_D}} - D \\
\frac{dS}{dt} &= \frac{(c_S I e^{-\lambda_S I})^{n_S}}{(c_S I e^{-\ddot{e}_S I})^{n_S} + D^{n_S}} - S
\end{aligned}
$$

(9.3)

When conventional bifurcation analysis is performed on Eq. (9.3), it producing output that is qualitatively similar to real biological injury systems. Equation (9.3) predicts that four qualitatively different types of injury exist, each type representing a class of solutions to Eq. (9.3) under different parameters limits (Figs. 9.21a–d; further explanation is provided in the legend of Fig. 9.21). Three of the four classes of injury are characterized by the property of **bistability**, where there are two simultaneous solutions to Eq. (9.3).

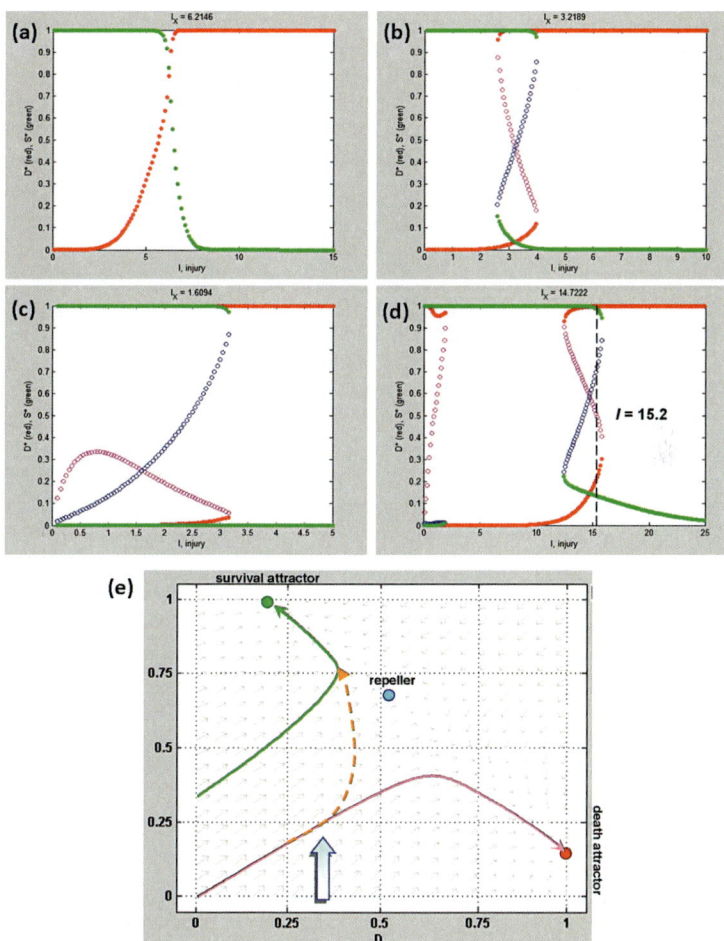

Figure 9.21 Example output from the nonlinear model of cellular injury. The model gives rise to four different types of injury courses: (a) monostable injury course where there is a single attractor at each *I*. (b) Bistable type A injury courses, where the system bifurcates over a subrange of the injury course, producing two attractors and their associated repeller (*D* repeller magenta, *S* repeller blue). (c) Bistable type B injury courses where the system bifurcates beginning at *I* = 0 and extending into the injury course range. (d) Double bistable injury courses where both type B and type A are present across the injury course. The injury courses are simply bifurcation diagrams of *D** and *S** plotted on the same axes and using *I* as the control

parameter. Because this version of the model is normalized, D^* and S^* are confined to vary only between 0 and 1. (e) A phase plane taken from the bifurcation curve in (d) at $I = 15.2$ [position and fixed points indicated by black line on (d)]. For this system, $I = 15.2$ is lethal because $I > I_X$ (e.g., 15.2 > 14.72). The phase plane shows all possible trajectories (e.g., time courses) of D and S at $I = 15.2$. In particular it reveals the therapeutic potential of bistability in an injured cell system. If the system is injured from initial conditions $D_0 = 0$ and $S_0 = 0$, the system will inevitably die. This is indicated by the pink trajectory that begins at (0, 0) and ends at the death attractor for this phase plane (red circle). However, because the system is bistable, there are possible paths to the survival attractor (green circle). Thus, in a bistable phase plane, the potential exist for the system to either die or recover. The green trajectory that beings at (0, 0.3) shows a potential path to the survival attractor. The blue arrow indicates that a system could be traveling along the red attractor, but if subjected to a "sufficient force" could be diverted from the death trajectory to the survival trajector. The deflection of the system from the death to the survival trajectory is indicated by the orange dashed curve. While it is unclear what a "sufficient force" is, the model gives a clear-cut theoretical basis to realize that therapy is the diverting of a system on a pro-death to a pro-survival trajectory. This general insight is superior to the mainstream inductive thinking that believes intervention in a single molecular pathway can serve as a basis for therapy. In the stroke field, such think has a 100% failure rate in stroke clinical trials. Thus, some alternative means of thinking is required and the nonlinear model of cell injury offers one such alternative. The parameter values for a–d, expressed as the array $(c_D, \lambda_D, n_D, c_S, \lambda_S, n_S)$ are (a) (0.1, 0.1, 4, 50, 0.9, 4); (b) (0.1, 0.1, 4, 2.5, 0.9, 4); (c) (0.1, 0.1, 4, 0.5, 0.9, 4); (d) (0.01, 0.1, 4, 0.19, 0.1, 4).

To understand what bistability represents in Eq. (9.3), we consider the case where $\Theta_D = \Theta_S$. Setting these equal from Eq. (9.2) and solving for I gives

$$I_x = \frac{\ln c_S - \ln c_D}{\lambda_D + \lambda_S} \tag{9.4}$$

Equation (9.4) defines that value of I where $\Theta_D = \Theta_S$. When I_X is used in Eq. (9.3), it always produces a solution such that the equilibrium or fixed point solutions to Eq. (9.3), notated D^* and S^*, are equal. But what does this mean? The value I_X is the *tipping point magnitude of injury between life and death of the injured system*. For all $I < I_X$,

$S^* > D^*$, meaning the total induced stress response will win the competition and the system will recover. For all $I > I_X$, $D^* > S^*$, total damage will win the competition, and the system will die.

In bistable solutions to Eq. (9.3), one solution to Eq. (9.3) is the state $S^* > D^*$, and the other solution is $D^* > S^*$. The significance of this is that a system that is fated to die, but is bistable, possess possible paths to recovery (Fig. 9.21e). Thus, Eq. (9.3) suggests that for therapeutic purposes, one must seek the bistable regime of the injury and find means to induce trajectories toward recovery in systems that otherwise left alone would die. The implications of this insight for biomedical research are very significant.

9.11.2 Brief Comments on the Qualitative Interpretation of this Model

The full interpretation of Eq. (9.3) is beyond the scope of this chapter and the reader is referred to DeGracia, Huang, and Huang (2012) for a fuller description. Here we briefly comment on a qualitative interpretation of Eq. (9.3).

Equation (9.3) suggests that injury need be interpreted not in terms of individual molecular events, but as a network of interconnected and interdependent molecular changes. Equation (9.3) describes the dynamics of a network made up of antagonistic components whose interactions are driving by Eq. (9.2). Equation (9.3) suggests that understanding injury means understanding the global network dynamics, not just some specific piece of the underlying network. Understanding the network dynamics implies being able to manipulate these. Equation (9.3) provides a quantitative roadmap to determine the temporal network dynamics of any specific injury, and with such a roadmap, one can determine if the system displays bistability, and if so, how to manipulate lethal bistable injuries and convert them from death to recovery. Equation (9.3) provides a formal and precise roadmap for a systematic approach to therapies against such injuries as brain ischemia.

9.12 Conclusion and Future Prospects

Equation (9.3) is an abstraction of the work described throughout this chapter on ischemic brain injury. Forty plus years of detailed empirical study of brain ischemia has revealed that it consists of many parallel injury mechanisms. The work directed toward understanding TA in post-ischemic neurons, along with other related lines of research, has revealed the operation of intrinsic stress responses, removing any misconception that the cell is a passive victim of the injury, and demonstrating the active role the cell takes to combat the injury. Our recent work on mRNA regulation and its central contribution to cell recovery or demise shows clearly one example of how the competition between total damage and total stress responses manifests in one specific injury system. The difference between hippocampal CA1 and CA3 is the difference between $D^* > S^*$ and $S^* > D^*$, respectively. The nonlinear nature of injury dose (or magnitude) to cell outcome of recovery or death is captured by Eq. (9.3).

From a nanotechnology point of view, the work on ischemic brain injury, and its culmination in the abstractions captured in Eq. (9.3) reveals that cells are not formed from pieces of linearly independent components. To understand the nature of a cell, and hence emulate it, we must recognize that all of the pieces inside the cell act in concert, as a vast molecular network. Further, as Eq. (9.3) illustrates, we do not need to know every detail of every molecule in the network to understand the global dynamics of the system. All that is required is a concise definition of the functions sought, so that an appropriate nonlinear expression can be written such that the attractor states model the desired function. Then, it is expected that any number of physical instantiations of the equation can emulate the network dynamics. Successfully modeling and technologically emulating biological systems requires recognition that a cell is a physical manifestation of network dynamics. At first glance, one might not expect a traditional biomedical area such as brain ischemia to lead to such as conclusion, but as this chapter illustrates, such logic applies even to the decision between life and death.

Acknowledgemnts

The work summarized here spans two decades of research and it is not possible to thank all of my colleagues and students who have contributed to the insights expressed. It is hoped they know of our respect and appreciation of their contributions. We would like to thank key contributors. We thank Blaine C. White for many years of support; Peter Hossmann, who is our intellectual grandfather and whose work trail blazed paths; Sui Huang for his insights into biological dynamics. We also thank Foaz Kayali, PhD, MD, who performed many of the stress granule studies, and perfected poly(A) FISH. We thank Jie Wang our meticulous laboratory surgeon. We thank the granting agencies that have supported this work: National Institute of Neurological Disorders and Stroke, NS 057167 (D. J. D.); a Ruth L. Kirschstein National Research Service Award, NS063651 (J. J. S.); an NINDS Minority Supplement on NS057167 (M. K. L.); and a Thomas C. Rumble Fellowship, Wayne State University (J. T. J.).

References

1. DeGracia, D. J., Jamison, J. T., Szymanski, J. J., Lewis, M. K. (2008). Translation arrest and ribonomics in post-ischemic brain, layers and layers of players. *J Neurochem.*, **106**(6), 2288–2301.

2. Anderson, P., Kedersha, N. (2009). RNA granules, post-transcriptional and epigenetic modulators of gene expression. *Nat Rev Mol Cell Biol*, **10**(6), 430–436.

3. Ivanov, P., Kedersha, N., Anderson, P. (2011). Stress puts TIA on TOP. *Genes Dev*, **25**(20), 2119–2124.

4. Hossmann, K. A. (1993). Disturbances of cerebral protein synthesis and ischemic cell death. *Prog Brain Res*, **96**, 161–177.

5. Lee, M.-H., Schedl, T. (2006). RNA-binding proteins (April 18, 2006), WormBook, ed. The C. elegans Research Community, WormBook, doi/10.1895/wormbook.1.79.1, http, //www.wormbook.org.

6. Claverie, J. M. (2001). What if there are only 30,000 human genes? *Science*, **291**(5507), 1255–1257.

7. Paik, Y., Jeong, S., Omenn, G. S., Uhlen, M., Hanash, S., Cho, S., Lee, H., Na, K., Choi, E., Yan, F., Zhang, F., Zhang, Y., Snyder, M., Cheng, Y., Chen, R., Marko-Varga, G., Deutsch, E., Kim, H., Kwon, J., Aebersold, R., Bairoch, A., Taylor, A., Kim, K., Lee, E., Hochstrasser, D., Legrain, P., Hancock, W. (2012). The chromosome-centric human proteome project for cataloging proteins encoded in the genome. *Nat Biotechnol*, **30**(3), 221–223.

8. Sanchez-Diaz, P., Penalva, L. (2006). Post-transcription meets post-genomic, the saga of RNA binding proteins in a new era. *RNA Biol*, **3**(3), 101–109.

9. Lunde, B. M., Moore, C., Varani, G. (2007). RNA-binding proteins, modular design for efficient function. *Nat Rev Mol Cell Biol*, **8**(6), 479–490.

10. Hamilton, T. L., Stoneley, M., Spriggs, K. A., Bushell, M. (2006). TOPs and their regulation. *Biochem Soc Trans*, **34**(Pt 1), 12–16.

11. Barreau, C., Paillard, L., Osborne H. (2006). AU-rich elements and associated factors, are there unifying principles? *Nucleic Acids Res*, **33**(22), 7138–7150.

12. Hollams, E. M., Giles, K. M., Thomson, A. M., Leedman, P. J. (2002). MRNA stability and the control of gene expression, implications for human disease. *Neurochem Res*, **27**(10), 957–980.

13. Bakheet, T., Williams B., Khabar K. (2003). ARED 2.0, an update of AU-rich element mRNA database. *Nucleic Acids Res*, **31**(1), 421–423.

14. Villalba, A., Coll, O., Gebauer, F. (2011). Cytoplasmic polyadenylation and translational control. *Curr Opin Genet Dev*, **21**(4), 452–457.

15. Keene, J. D., Lager, P. J. (2005). Post-transcriptional operons and regulons co-ordinating gene expression. *Chromosome Res*, **3**(3), 327–337.

16. Erickson, S. L., Lykke-Andersen, J. (2011). Cytoplasmic mRNP granules at a glance. *J Cell Sci*, **124**(Pt 3), 293–297.

17. Carpenter, B., MacKay, C., Alnabulsi, A., MacKay, M., Telfer, C., Melvin, W. T, Murray, G. I. (2006). The roles of heterogeneous nuclear ribonucleoproteins in tumour development and progression. *Biochim Biophys Acta*, **1765**(2), 85–100.

18. Stewart, M. (2007). Ratcheting mRNA out of the nucleus. *Mol Cell*, **25**(3), 327–330.

19. Balagopal, V., Parker, R. (2009). Polysomes, P. bodies and stress granules, states and fates of eukaryotic mRNAs. *Curr Opin Cell Biol*, **21**(3), 403–408.

20. Simpson, R., Lim, J., Moritz, R., Mathivanan, S. (2009). Exosomes, proteomic insights and diagnostic potential. *Expert Rev Proteomics*, **6**(3), 267–283.

21. Ceman, S., Saugstad, J. (2011). MicroRNAs, Meta-controllers of gene expression in synaptic activity emerge as genetic and diagnostic markers of human disease. *Pharmacol Ther*, **130**(1), 26–37.

22. Kedersha, N., Anderson, P. (2009). Regulation of translation by stress granules and processing bodies. *Prog Mol Biol Transl Sci*, **90**, 155–185.

23. Kang, M. K., Han, S. J. (2011). Post-transcriptional and post-translational regulation during mouse oocyte maturation. *BMB Rep*, **44**(3), 147–157.

24. Krichevsky, A. M., Kosik, K. S. (2001). Neuronal RNA granules, a link between RNA localization and stimulation-dependent translation. *Neuron*, **32**(4), 683–696.

25. Burry, R. W., Smith, C. L. (2006). HuD distribution changes in response to heat shock but not neurotrophic stimulation. *J Histochem Cytochem*, **54**(10), 1129–1138.

26. Hinman, M. N., Lou, H. (2008). Diverse molecular functions of Hu proteins. *Cell Mol Life Sci*, **65**(20), 3168–3181.

27. von Roretz, C., Di Marco, S., Mazroui, R., Gallouzi, I. E. (2011). Turnover of AU-rich-containing mRNAs during stress, a matter of survival. *Wiley Interdiscip Rev RNA*, **2**(3), 336–347.

28. Keene, J. D. (2007). RNA regulons, coordination of post-transcriptional events. *Nat Rev Genet*, **8**(7), 533–543.

29. Güell, M., Yus, E., Lluch-Senar, M., Serrano, L. (2011). Bacterial transcriptomics, what is beyond the RNA horiz-ome? *Nat Rev Microbiol*, **9**(9), 658–669.

30. Morris, A. R., Mukherjee, N., Keene, J. D. (2010). Systematic analysis of posttranscriptional gene expression. *Wiley Interdiscip Rev Syst Biol Med*, **2**(2), 162–180.

31. Kapp, L. D., Lorsch, J. R. (2004). The molecular mechanics of eukaryotic translation. *Annu Rev Biochem*, **73**, 657–704.

32. Pesole, G., Mignone, F., Gissi, C., Grillo, G., Licciulli, F., Liuni, S. (2001). Structural and functional features of eukaryotic mRNA untranslated regions. *Gene*, **276**(1–2), 73–81.

33. Maier, T., Güell, M., Serrano, L. (2009). Correlation of mRNA and protein in complex biological samples. *FEBS Lett*, **583**(24), 3966–3973.

34. Saugstad, J. (2010). MicroRNAs as effectors of brain function with roles in ischemia and injury, neuroprotection, and neurodegeneration. *J Cereb Blood Flow Metab*, **30**(9), 1564–1576.

35. Adeli, K. (2011). Translational control mechanisms in metabolic regulation, critical role of RNA binding proteins, microRNAs, and cytoplasmic RNA granules. *Am J Physiol Endocrinol Metab*, **301**(6), E1051–E1064.

36. Kushmerick, M. J., Conley, K. E. (2002). Energetics of muscle contraction, the whole is less than the sum of its parts. *Biochem Soc Trans*, **30**(2), 227–231.

37. Hertz, L., Dienel, G. A. (2002). Energy metabolism in the brain. *Int Rev Neurobiol*, **51**, 1–102.

38. Sabido, F., Milazzo, V., Hobson, R. 2nd, Duran, W. N. (1994). Skeletal muscle ischemia-reperfusion injury, a review of endothelial cell-leukocyte interactions. *J Invest Surg*, **7**(1), 39–47.

39. Kirino, T. (2000). Delayed neuronal death. *Neuropathology*, **20** Suppl, S95–S97.

40. Montie, H., Haezebrouck, A., Gutwald, J., DeGracia, D. J. (2005). PERK is activated differentially in peripheral organs following cardiac arrest and resuscitation. *Resuscitation*, **66**(3), 379–389.

41. Dienel, G. A. (2009). Energy metabolism in the brain. In *From Molecules to Networks, Second Edition, An Introduction to Cellular and Molecular Neuroscience*, (Byrne, Byrne and Roberts, eds.), Academic Press, Burlington, MA. P. 49–110.

42. Lipton, P. (1999). Ischemic cell death in brain neurons. *Physiol Rev*, **79**(4), 1431–1568.

43. Hossmann, K. A., Traystman, R. J. (2009). Cerebral blood flow and the ischemic penumbra. *Handb Clin Neurol*, **92**, 67–92.

44. Hess, E. P., White, R. D. (2010). Optimizing survival from out-of-hospital cardiac arrest. *J Cardiovasc Electrophysiol*, **21**(5), 590–595.

45. White, B. C., Grossman, L., O.'Neil, B., DeGracia, D. J., Neumar, R., Rafols, J., Krause, G. S. (1996). Global brain ischemia and reperfusion. *Ann Emerg Med*, **27**(5), 588–594.

46. Hossmann, K. A. (2009). Pathophysiological basis of translational stroke research. *Folia Neuropathol*, **47**(3), 213–227.

47. O.'Collins, V., Macleod, M., Donnan, G., Horky, L., van der Worp, B., Howells, D. (2006). 1,026 experimental treatments in acute stroke. *Ann Neurol*, **59**(3), 467–477.

48. Vivien, D., Gauberti, M., Montagne, A., Defer, G., Touzé, E. (2011). Impact of tissue plasminogen activator on the neurovascular unit, from clinical data to experimental evidence. *J Cereb Blood Flow Metab*, **31**(11), 2119–2134.

49. Johnson, M., Bakas, T. (2010). A review of barriers to thrombolytic therapy, implications for nursing care in the emergency department. *J Neurosci Nurs*, **42**(2), 88–94.

50. Delhaye, C., Mahmoudi, M., Waksman, R. (2012). Hypothermia therapy, neurological and cardiac benefits. *J Am Coll Cardiol.*, **59**(3), 197–210.

51. Kirino, T. (2002). Ischemic tolerance. *J Cereb Blood Flow Metab*, **22**(11), 1283–1296.

52. Dirnagl, U., Simon, R. P., Hallenbeck, J. M. (2003). Ischemic tolerance and endogenous neuroprotection. *Trends Neurosci*, **26**(5), 248–254.

53. Silver, I., Deas, J., Erecińska, M. (1997). Ion homeostasis in brain cells, differences in intracellular ion responses to energy limitation between cultured neurons and glial cells. *Neuroscience*, **78**(2), 589–601.

54. Lau, A., Tymianski, M. (2010). Glutamate receptors, neurotoxicity and neurodegeneration. *Pflugers Arch*, **460**(2), 525–542.

55. Murphy, S., Gibson, C. L. (2007). Nitric oxide, ischaemia and brain inflammation. *Biochem Soc Trans*, **35**(Pt 5), 1133–1137.

56. White, B. C., Sullivan, J., DeGracia, D. J., O.'Neil, B., Neumar, R., Grossman, L., Rafols, J., Krause, G. S. (2000). Brain ischemia and reperfusion, molecular mechanisms of neuronal injury. *J Neurol Sci*, **179**(S1–2), 1–33.

57. Neumar, R. W., DeGracia, D. J., Konkoly, L., Khoury, J., White, B. C., Krause, G. S. (1998). Calpain mediates eukaryotic initiation factor 4G degradation during global brain ischemia. *J Cereb Blood Flow Metab*, **18**(8), 876–881.

58. Muralikrishna Adibhatla, R., Hatcher, J. F. (2006). Phospholipase A_2, reactive oxygen species, and lipid peroxidation in cerebral ischemia. *Free Radic Biol Med*, **40**(3), 376–387.

59. Bazan, N., Marcheselli, V., Cole-Edwards K. (2005). Brain response to injury and neurodegeneration, endogenous neuroprotective signaling. *Ann N. Y. Acad Sci*, **1053**, 137–147.

60. Matsumoto, S., Shamloo, M., Matsumoto, E., Isshiki, A., Wieloch, T. (2004). Protein kinase C-gamma and calcium/calmodulin-dependent protein kinase II-alpha are persistently translocated to cell membranes of the rat brain during and after middle cerebral artery occlusion. *J Cereb Blood Flow Metab*, **24**(1), 54–61.

61. Phillis, J., Horrocks, L., Farooqui, A. (2006). Cyclooxygenases, lipoxygenases, and epoxygenases in CNS, their role and involvement in neurological disorders. *Brain Res Rev*, **52**(2), 201–243.

62. Phillis, J. (1994). A "radical" view of cerebral ischemic injury. *Prog Neurobiol*, **42**(4), 441–448.

63. Starkov, A., Chinopoulos, C., Fiskum, G. (2004). Mitochondrial calcium and oxidative stress as mediators of ischemic brain injury. *Cell Calcium*, **36**(3–4), 257–264.

64. Chan, P. H. (2004). Mitochondria and neuronal death/survival signaling pathways in cerebral ischemia. *Neurochem Res*, **29**(11), 1943–1949.

65. Chen, X., Kintner, D. B., Luo, J, Baba, A., Matsuda, T., Sun, D. (2008). Endoplasmic reticulum Ca^{2+} dysregulation and endoplasmic reticulum stress following in vitro neuronal ischemia, role of Na^+-K^+-Cl^- cotransporter. *J Neurochem*, **106**(4), 1563–1576.

66. DeGracia, D. J., Montie, H. L. (2004). Cerebral ischemia and the unfolded protein response. *J Neurochem*, **91**(1), 1–8.

67. Wieloch, T., Bergstedt, K., Hu, B. R. (1993). Protein phosphorylation and the regulation of mRNA translation following cerebral ischemia. *Prog Brain Res*, **96**, 179–191.

68. Wieloch, T., Hu, B. R., Boris-Möller, A., Cardell, M., Kamme, F., Kurihara, J., Sakata, K. (1996). Intracellular signal transduction in the postischemic brain. *Adv Neurol*, **71**, 371–387.

69. Hu, B. R., Martone, M., Jones, Y. Z., Liu, C. L. (2000). Protein aggregation after transient cerebral ischemia. *J Neurosci*, **20**(9), 3191–3199.

70. Hu, B. R., Janelidze, S., Ginsberg, M. D., Busto, R., Perez-Pinzon, M., Sick T. J., Siesjö, B. K., Liu, C. L. (2001). Protein aggregation after focal brain ischemia and reperfusion. *J Cereb Blood Flow Metab*, **21**(7), 865–875.

71. DeGracia, D. J., Hu, B. R. (2007). Irreversible translation arrest in the reperfused brain. *J Cereb Blood Flow Metab*, **27**(5), 875–893.

72. Abe, H., Nowak, T. S. Jr. (1996). The stress response and its role in cellular defense mechanisms after ischemia. *Adv Neurol*, **71**, 451–466.

73. Bazan, N. (2005). Lipid signaling in neural plasticity, brain repair, and neuroprotection. *Mol Neurobiol*, **32**(1), 89–103.

74. Hayashi, T., Abe, K. (2004). Ischemic neuronal cell death and organellae damage. *Neurol Res*, **26**(8), 827–834.

75. Büttner, F., Cordes, C., Gerlach, F., Heimann, A., Alessandri, B., Luxemburger, U., Türeci, O., Hankeln, T., Kempski, O., Burmester, T. (2009). Genomic response of the rat brain to global ischemia and reperfusion. *Brain Res*, **1252**, 1–14.

76. Jamison, J. T., Kayali, F., Rudolph, J, Marshall, M., Kimball, S. R., DeGracia, D. J. (2008). Persistent redistribution of poly-adenylated mRNAs correlates with translation arrest and cell death following global brain ischemia and reperfusion. *Neuroscience*, **154**(2), 504–520.

77. Kleihues, P., Hossmann, K. A. (1971). Protein synthesis in the cat brain after prolonged cerebral ischemia. *Brain Res*, **35**(2), 409–418.

78. Kaufman, R. J. (2002). Orchestrating the unfolded protein response in health and disease. *J Clin Invest*, **110**(10), 1389–1398.

79. Hu, B. R., Wieloch, T. (1993). Stress-induced inhibition of protein synthesis initiation, modulation of initiation factor 2 and guanine nucleotide exchange factor activities following transient cerebral ischemia in the rat. *J Neurosci*, **13**(5), 1830–1838.

80. Merrick, W. (1992). Mechanism and regulation of eukaryotic protein synthesis. *Microbiol Rev*, **56**(2), 291–315.

81. Burda, J., Martín M., García, A., Alcázar, A., Fando, J., Salinas, M. (1994). Phosphorylation of the alpha subunit of initiation factor 2 correlates with the inhibition of translation following transient cerebral ischaemia in the rat. *Biochem J*, **302**(Pt 2), 335–338.

82. DeGracia, D. J., Neumar, R. W., White, B. C., Krause, G. S. (1996). Global brain ischemia and reperfusion, modifications in eukaryotic initiation factors associated with inhibition of translation initiation. *J Neurochem*, **67**(5), 5–12.

83. DeGracia, D. J., Sullivan, J. M., Neumar, R. W., Alousi, S. S., Hikade, K. R., Pittman, J. E., White, B. C., Rafols, J. A., Krause, G. S. (1997). Effect of brain ischemia and reperfusion on the localization of phosphorylated eukaryotic initiation factor 2 alpha. *J Cereb Blood Flow Metab*, **17**(12), 1291–1302.

84. Kumar, R., Krause, G. S., Yoshida, H., Mori, K., DeGracia, D. J. (2003). Dysfunction of the unfolded protein response during global brain ischemia and reperfusion. *J Cereb Blood Flow Metab*, **23**(4), 462–471.

85. DeGracia, D. J. (2004). Acute and persistent protein synthesis inhibition following cerebral reperfusion. *J Neurosci Res*, **77**(6), 771–776.

86. Nowak, T. S. Jr. (1990). Protein synthesis and the heart shock/stress response after ischemia. *Cerebrovasc Brain Metab Rev*, **2**(4), 345–366.

87. Roberts, G., Di Loreto, M., Marshall, M., Wang, J., DeGracia, D. J. (2007). Hippocampal cellular stress responses after global brain ischemia and reperfusion. *Antioxid Redox Signal*, **9**(12), 2265–2275.

88. Brostrom, C., Brostrom, M. (1990). Calcium-dependent regulation of protein synthesis in intact mammalian cells. *Annu Rev Physiol*, **52**, 577–590.

89. Paschen, W. (1996). Disturbances of calcium homeostasis within the endoplasmic reticulum may contribute to the development of ischemic-cell damage. *Med Hypotheses*, **47**(4), 283–288.

90. Kumar, R., Azam, S., Sullivan, J. M., Owen, C., Cavener, D. R., Zhang, P., Ron, D., Harding, H. P., Chen, J. J., Han, A., White, B. C., Krause, G. S., DeGracia, D. J. (2001). Brain ischemia and reperfusion activates the

eukaryotic initiation factor 2-alpha kinase, PERK. *J Neurochem*, **77**(5), 1418–1421.

91. Truettner, J., Hu, K., Liu, C., Dietrich, W., Hu, B. (2009). Subcellular stress response and induction of molecular chaperones and folding proteins after transient global ischemia in rats. *Brain Res*, **1249**, 9–18.

92. Liu, C. L., Ge, P., Zhang, F., Hu, B. R. (2005). Co-translational protein aggregation after transient cerebral ischemia. *Neuroscience*, **134**(4), 1273–1284.

93. Zhang, F., Liu, C., Hu, B. R. (2006). Irreversible aggregation of protein synthesis machinery after focal brain ischemia. *J Neurochem*, **98**(1), 102–112.

94. Kedersha, N. L., Gupta, M., Li, W., Miller, I., Anderson, P. (1999). RNA-binding proteins TIA-1 and TIAR link the phosphorylation of eIF-2 alpha to the assembly of mammalian stress granules. *J Cell Biol*, **147**(7), 1431–1442.

95. Twyffels, L., Gueydan, C., Kruys, V. (2011). Shuttling SR proteins, more than splicing factors. *FEBS J*, **278**(18), 3246–3255.

96. Gilks, N., Kedersha, N., Ayodele, M., Shen, L., Stoecklin, G., Dember, L. M., Anderson, P. (2004). Stress granule assembly is mediated by prion-like aggregation of TIA-1. *Mol Biol Cell*, **15**(12), 5383–5398.

97. Kedersha, N., Cho, M. R., Li, W., Yacono, P. W., Chen, S., Gilks, N., Golan, D. E., Anderson, P. (2000). Dynamic shuttling of TIA-1 accompanies the recruitment of mRNA to mammalian stress granules. *J Cell Biol*, **151**(6), 1257–1268.

98. Kayali, F., Montie, H., Rafols, J., DeGracia, D. J. (2005). Prolonged translation arrest in reperfused hippocampal cornu Ammonis 1 is mediated by stress granules. *Neuroscience*, **134**(4), 1223–1245.

99. Martin, L. J., Al-Abdulla, N. A., Brambrink, A., Kirsch, J., Sieber, F., Portera-Cailliau, C. (1998). Neurodegeneration in excitotoxicity, global cerebral ischemia, and target deprivation, A perspective on the contributions of apoptosis and necrosis. *Brain Res Bull*, **46**(4), 281–309.

100. Hines, J. J., Weissbach, H., Brot, N., Elkon, K. (1991). Anti-P autoantibody production requires P1/P2 as immunogens but is not driven by exogenous self-antigen in MRL mice. *J Immunol*, **146**(10), 3386–3395.

101. Jamison, J. T., Szymanski, J. J., DeGracia, D. J. (2011). Organelles do not colocalize with mRNA granules in post-ischemic neurons. *Neuroscience*, **199**, 394–400.

102. DeGracia, D. J., Rafols, J. A., Morley, S. J., Kayali, F. (2006). Immunohistochemical mapping of total and phosphorylated eukaryotic initiation

factor 4G in rat hippocampus following global brain ischemia and reperfusion. *Neuroscience*, **139**(4), 1235–1248.

103. Amadio, M., Scapagnini, G., Laforenza, U., Intrieri, M., Romeo, L., Govoni, S., Pascale, A. (2008). Post-transcriptional regulation of HSP70 expression following oxidative stress in SH-SY5Y cells, the potential involvement of the RNA-binding protein HuR. *Curr Pharm Des*, **14**(26), 2651–2658.

104. DeGracia, D. J. (2010). Towards a dynamical network view of brain ischemia and reperfusion. Part I., background and preliminaries. *J Exp Stroke Transl Med*, **3**(1), 59–71.

105. DeGracia, D. J. (2010). Towards a dynamical network view of brain ischemia and reperfusion. Part II, a post-ischemic neuronal state space. *J Exp Stroke Transl Med*, **3**(1), 72–89.

106. DeGracia, D. J. (2010). Towards a dynamical network view of brain ischemia and reperfusion. Part III, therapeutic implications. *J Exp Stroke Transl Med*, **3**(1), 90–103.

107. DeGracia, D. J. (2010). Towards a dynamical network view of brain ischemia and reperfusion. Part IV, additional considerations. *J Exp Stroke Transl Med*, **3**(1), 104–114.

108. Jamison, J. T., Lewis, M. K., Kreipke, C. W., Rafols, J. A., DeGracia, D. J. (2011). Polyadenylated mRNA staining reveals distinct neuronal phenotypes following endothelin-1, focal brain ischemia, and global brain ischemia/ reperfusion. *Neurol Res*, **33**(2), 145–161.

109. Ballestra, J. (1991). *Inhibitors of Eukaryotic Protein Synthesis*, (Trachseh, H. ed.), CRC Press, London.

110. Pestka, S. (1971). Inhibitors of ribosome functions. *Annu Rev Microbio*, **25**, 487–562.

111. Szymanski, J., Jamison, J., DeGracia, D. J. (2012). Texture analysis of poly-adenylated mRNA staining following global brain ischemia and reperfusion. *Comput Methods Programs Biomed*, **105**(1), 81–94.

112. DeGracia, D. J. (2008). Ischemic damage and neuronal stress responses, Towards a systematic approach with implications for therapeutic treatments. In *New Frontiers in Neurological Research*, (Wang D. Q., Ying, W., eds.), Research Signpost, Kerala, India. pp 235–264.

113. DeGracia, D. J., Huang, Z. F., Huang, S. (2012). A nonlinear dynamical theory of cell injury. *J Cereb Blood Flow Metab*, **32**, 1000–1013.

Chapter 10

Physical Properties and Biomedical Applications of Superparamagnetic Iron Oxide Nanoparticles

Gavin Lawes,[a] Ratna Naik,[a] and Prem Vaishnava[b]

[a]*Department of Physics and Astronomy, Wayne State University, Detroit MI 48201 USA*
[b]*Department of Physics, Kettering University, Flint MI 48504 USA*

glawes@wayne.edu

One of the central themes underlying the long-standing and continuing interest in nanomaterials is the recognition that the physical properties of nanoscale materials can be qualitatively different from the properties of bulk materials. Conceptually these differences can arise because some of the features that are negligible in bulk dominate the characteristics of nanoscale materials. One very well-known example of this type of effect is the surface-dominated behavior of nanoparticles as compared to bulk systems. The fraction of atoms at the surface of a nanoparticle scales as a/L, where a is the atomic spacing, typically on the order of 0.5 nm, and L is the size of the nanoparticle. For macroscopic values of L, the fraction of surface atoms is negligible. However, as L approaches the interatomic spacing, the fraction of surface atoms can become very large. The characteristic properties of these surface atoms has

NanoCellBiology: Multimodal Imaging in Biology and Medicine
Edited by Bhanu P. Jena and Douglas J. Taatjes
Copyright © 2014 Pan Stanford Publishing Pte. Ltd.
ISBN 978-981-4411-79-0 (Hardcover), 978-981-4411-80-6 (eBook)
www.panstanford.com

been associated with the large chemical reactivity of gold nano-particles,[1-4] with the development of magnetism in semiconduc-ting oxide nanoparticles,[5-8] and the very high catalytic activity of nanoparticles,[9-13] features that are absent in bulk materials.

The magnetic properties of nanoparticles are also distinctly different from those observed in bulk magnetic materials. One quantitative difference between bulk and nanoscale magnets is that the saturation magnetization is very typically reduced in small systems. This reduction in the saturation magnetization is routinely attributed to the much larger fraction of surface atoms in nanoscale systems.[14-16] These are expected to experience much weaker magnetic interactions than non-surface atoms, so these surface spins may not contribute to the net moment of the system. Furthermore, for sufficiently small particles, thermal fluctuations play a central role in determining how the system responds to external magnetic fields. From a practical point of view, these thermal fluctuations will disrupt any magnetic structure produced by applying an external field once this field is removed, leading to so-called "superparamagnetic" behavior.[17-20]

The physical properties, especially their magnetic properties, exhibited by magnetic nanoparticles have motivated considerable interest in using these systems to develop a variety of novel biomedical applications.[21-24] Magnetic nanoparticles have been proposed for applications ranging from targeted drug delivery and transfection,[25-35] to contrast agents for magnetic resonance imaging,[36-39] to local heating and hyperthermia,[40-49] to controlled drug release,[35,50-53] among a number of potential applications. Iron oxide nanoparticles in particular have been especially well investigated, since these are widely regarded to be minimally toxic and are expected to be suitable for clinical therapies.[24,54,55] There are a large number of review articles that address the potential applications for magnetic nanoparticles in medicine, including accessible general overviews of these applications,[22,24,56,57] along with more specialized reviews focusing on magnetic nanoparticles for drug delivery applications,[58-60] hyperthermia,[46,61-63] and MRI,[64] as well as reviews of nanoparticle synthesis.[65,66] The goal of this particular monograph is to offer an introduction to the physical properties of magnetic iron oxide nanoparticles in the context of these aforementioned biomedical applications in order to provide researchers in diverse fields with a working knowledge of the physics underlying these fascinating nanomaterials.

The chapter is structured as follows. First, we provide a short discussion of the origin of ferromagnetism in bulk systems along with a brief overview of the physical properties of magnets. Following this exposition, we discuss the important magnetic properties of nanoscale magnets, with an emphasis on providing a technical, but accessible, introduction to the physical mechanisms responsible for their behavior. In this context, we will discuss aspects of superparamagnetism and superparamagnetic blocking, the magnetohydrodynamics of magnetic nanoparticles in solution, magnetic relaxation and hyperthermia, and magnetic nanoparticles and dephasing in magnetic resonance imaging. The technical discussion will be accompanied by experimental results and some representative data to help clarify the underlying mechanisms. We hope that a firm understanding of the physical properties of iron oxide nanoparticles will allow for not only an appreciation of how these properties naturally suggest certain previously proposed applications, but may also lead to the identification of entirely new applications. This section will be followed with a brief detour to present a short review of some standard techniques to prepare iron oxide nanoparticles, including functionalization for biomedical applications. The remainder of the chapter will focus on providing illustrative examples of how the novel properties of magnetic nanoparticles can be exploited to develop potentially transformative applications in medicine.

10.1 Magnetism in Bulk Iron Oxide

At a fundamental level, magnetism arises from the quantum mechanical spin of electrons.[67,68] Unpaired, localized electrons on atoms tend to align their spins in the same direction to minimize the repulsive forces between them, which, when combined with contributions from the orbital angular momentum, leads to a local magnetic dipole moment on the atom. In magnetite, iron oxide having the composition Fe_3O_4, both Fe^{2+} and Fe^{3+} are present in an inverse spinel structure in a 1:2 ratio. The Fe^{2+} sites have 6 electrons in the d-shell, leading to an effective spin-only moment of 4.9 μ_B, which experimentally is found to increase to roughly 5.4 μ_B after orbital contributions are included.[67] Conversely, the Fe^{3+} ions have five electrons in the d-shell, producing an effective moment of 5.9 μ_B. The oxygen anions formally have filled electronic orbitals and do not contribute directly to the magnetism.

10.2 Ferromagnetic and Ferrimagnetic Ordering

Conceptually, the development of ferromagnetic order in a magnetic material can be understood by recognizing that a magnetic field will align these magnetic dipoles to reduce the free energy, and that the effective internal magnetic field produced by the moments increases as they align more completely. This behavior can be modeled microscopically by considering an *exchange energy* between adjacent moments, which is reduced for moments having a parallel alignment. This positive feedback loop for magnetic interactions, with a small alignment of the magnetic moments producing a small internal field producing greater alignment, will lead to all the spins being aligned in the same direction as long as this process is stable against thermal fluctuations. In real crystals, this simple model is modified by the magnetocrystalline anisotropy energy,[68] which favors the alignment of the magnetic moment along certain specific crystallographic directions because of the coupling between the spin and orbital components of the electron wave function. The critical temperature below which this ferromagnetic structure, having all the local moments aligned in the same direction, is stable is called the Curie temperature, and depends on parameters including the size of the local moments and their interaction energy. Since the ferromagnetic structure has all the moments aligned in the same direction, the Curie temperature is marked by the onset of a finite magnetization. The development of ferromagnetic order in $BiMnO_3$ is shown in Fig. 10.1. At high temperatures, the Mn spins are random, and the paramagnetic material shows only a negligible field-induced magnetic moment. As the sample is cooled below approximately 100 K, marking the Curie temperature, the magnetization shows a dramatic increase in the magnetization, associated with the magnetic ordering of the Mn local magnetic moments.

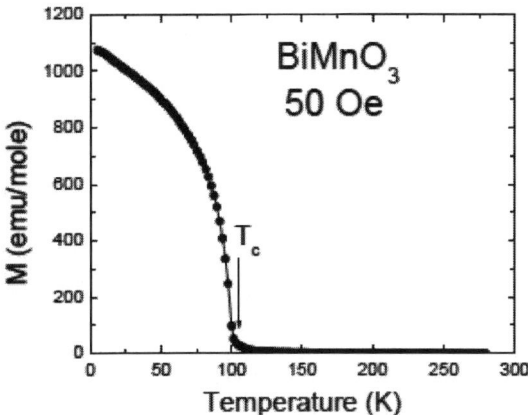

Figure 10.1 Temperature-dependent magnetization of $BiMnO_3$. The Curie temperature is indicated by an arrow.

The magnetic structure developing in iron oxide, both Fe_3O_4 and γ-Fe_2O_3, is somewhat more complicated. Rather than promoting a parallel arrangement of adjacent moments, such as occurs in ferromagnetic materials, the microscopic interactions in iron oxide favor antiparallel configurations for certain neighboring Fe ions.[69,70] In magnetite, which has the AB_2O_4 inverse spinel structure with Fe^{3+} on the A site (octahedral) and a mixture of Fe^{2+} and Fe^{3+} on the B site (octahedral), the microscopic interactions promote such antiparallel, or antiferromagnetic, interactions between A site and B site Fe ions. Thus, while Fe_3O_4 does develop long-range magnetic order, the partial cancellation of the magnetic moments on adjacent sites strongly reduces the saturation magnetization compared to what would be expected for a ferromagnetic structure consisting of the same ions. Thus, the fact that both Fe_3O_4 and γ-Fe_2O_3 are ferrimagnetic, with partial moment compensation due to antiferromagnetic order, rather than ferromagnetic, with fully aligned moments, means that their magnetic properties are not as strong as might be expected at first glance.

Two of the most important characteristics defining a magnetic material are the ordering temperature, called the *Curie temperature* for ferromagnets and *Neel temperature* for antiferromagnets and ferrimagnets, and saturation magnetization. The magnetic ordering temperatures for Fe_3O_4 and γ-Fe_2O_3 are both close to 600°C, implying that these materials are deep in the ordered phase for applications near room temperature. This is important for biomedical applications, since it means that the intrinsic bulk magnetic properties of these iron oxides are independent of temperature near physiological conditions. The saturation magnetization for Fe_3O_4 is approximately 90 emu/g, while for γ-Fe_2O_3 the value is close to 80 emu/g. This should be compared with the saturation magnetization of Fe metal, which is approximately a factor of three times larger at 221.7 emu/g.[71] However, as we will discuss later, there are a number of other materials properties that make iron oxide a much more desirable choice for biomedical applications than Fe metal, despite the latter's better magnetic properties. One other common form for iron oxide is α-Fe_2O_3, which is only very weakly magnetic, having a saturation magnetization of less than 0.1 emu/g.[72]

10.3 Magnetic Hysteresis

One of the other defining properties for ferromagnets is the presence of magnetic hysteresis.[68] The magnetization of a ferromagnet will increase and then saturate with the application of a sufficiently large external magnetic field, but the magnetization persists even after this external field is removed. This residual magnetization at zero field is called the *remnant* magnetization. In order to reduce the magnetization to zero, it is necessary to apply a sufficiently large magnetic field in the *opposite* direction, with the magnitude of this field called the *coercivity*. A representative magnetic hysteresis curve is shown in Fig. 10.2a with the saturation magnetization, remnant magnetization, and coercive field indicated. Because of this phenomenon, the zero field magnetization of a ferromagnet depends on the history of the sample, so that ferromagnetic materials can be used as memory storage devices, for example in conventional magnetic hard disks. Although magnetic iron oxides are ferrimagnetic, rather than ferromagnetic, similar hysteretic effects are present. For bulk Fe_3O_4, the saturation magnetization is close to 90 emu/g. The magnetization curves show some small anisotropy depending on the direction of the applied field,[73] with the remnant magnetization varying from approximately 82 emu/g for H along the [111] direction to 55 emu/g for H along [001][74] at room temperature, with the anisotropy becoming more pronounced at low temperature. The coercivity of bulk magnetite is relatively low, less than 100 Oe, but is found to depend strongly on magnetostriction.[75,76]

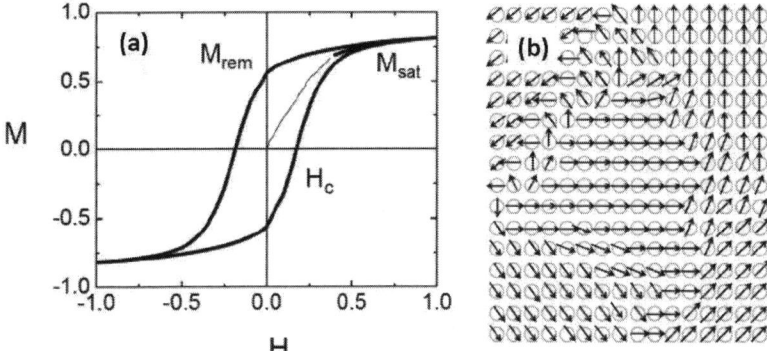

Figure 10.2 (a) Magnetization-magnetic field curve for a ferromagnet. (b) Schematic illustration of domains inside a ferromagnet.

Microscopically, hysteresis can be attributed to the presence of domains in a magnetic sample.[67,68] While the strong interactions between neighboring moments favor a uniform local alignment for the spins, over larger length scales other energy terms, including the magnetostatic and magnetocrystalline energies, can become significant and modify this spin structure. Experimentally, the magnetic moments in ferromagnets are aligned in small volumes of the sample, but different regions, or *domains*, of the magnet can have different directions for the local alignment. The interfacial regions between the magnetic domains are called *domain walls*, in which adjacent spins are not aligned, but show a smooth variation between the alignments of the neighboring magnetic domains. If the orientations of these magnetic domains are random, the net magnetization of the sample will be zero, even though each domain exhibits ferromagnetic order. This is schematically illustrated in Fig. 10.2b. Magnetic hysteresis can therefore be understood as a movement of domain walls in an external magnetic field. Domains with a magnetization in the direction of the applied field will grow in favor of other domains, producing a net magnetization for the sample. This process is not perfectly reversible as the applied field is removed leading to a remnant magnetization at zero field.

The characteristic size of these magnetic domains depends very strongly on the different contributions to the internal energy. A large exchange energy between nearest neighbor moments favors a uniform spin alignment and dominates the magnetic ordering at smaller length scales. However, the exchange interaction decreases rapidly with increasing distance, so other terms strongly affect the magnetic structure at longer length scales. The *magnetic dipolar energy*, which varies as the cube of the separation between dipole pairs, becomes significant over macroscopic samples, and therefore contributes to the formation of a number of different magnetic domains. The domain size for Fe_3O_4 is estimated to be on the order of 100 nm, with properties of single domain behavior persisting to some larger grain sizes.[77,78] Domain structures in magnetite are very sensitive to crystal imperfections, which play a major role in hysteresis and remanence by hindering the motion of domain walls. The shape of a hysteresis loop is determined partly by the domain state. Loops for single domain materials are typically wider than loops for multi domain materials, reflecting the higher coercivity and remanence in single domain systems. The ratio M_r/M_s provides a method for differentiating between single and non-single domain particles. For an assembly of single domain grains with randomly oriented easy axes, M_r/M_s depends on the type of anisotropy. For uniaxial anisotropy this ratio is typically 0.5. As will be discussed in the following, superparamagnetic particles exhibit no remanence or coercivity, and mixture of mostly superparamagnetic grains with some multidomain grains will yield values for $M_r/M_s \ll 0.01$.[79]

10.4 Magnetism in Iron Oxide Nanoparticles

Similar to bulk iron oxide, nanoparticles also develop ferrimagnetic order below the Neel temperature. The short-range exchange interactions produce the same ferrimagnetic structure for Fe_3O_4 and γ-Fe_2O_3 nanoparticles as are observed in bulk samples. There are, however, a number of very significant differences between the magnetic properties of bulk and nanoscale iron oxide samples, some of which make these materials particularly well suited for biomedical applications.

10.5   Reduced Saturation Magnetization

One important difference between bulk and nanoscale iron oxide samples is the sharp reduction in the saturation magnetization observed in nanoparticle samples. Experimentally, the saturation magnetization for Fe_3O_4 nanoparticles normally varies from approximately 50–60 emu/g,[80–83] as compared to nearly 90 emu/ g for bulk Fe_3O_4, although some synthesis procedures can result in magnetizations close to bulk values at low temperatures.[84] This behavior is normally attributed to the very large surface to volume ratio of the iron oxide nanoparticles. Recall that the local alignment of magnetic moments is driven by interactions between adjacent, or nearby, dipoles. This means that spins on or near the surface of the nanoparticle will experience much smaller effective internal fields, since there are not as many neighboring moments to contribute to the interactions. As a result, these moments may not order magnetically, and therefore will not contribute to the net magnetization.

The thickness of this magnetically disordered layer can be estimated by comparing the saturation magnetization of bulk and nanoscale Fe_3O_4. Assuming a reduction of the room temperature saturation magnetization from 90 emu/g for bulk Fe_3O_4 to 60 emu/g would imply that roughly one third of the moments in the nanoparticles are not magnetically ordered and do not contribute to the magnetization. For 12 nm nanoparticles, this corresponds to a magnetically disordered surface layer of roughly 1.2 nm. More careful studies on the magnetization curves of iron oxide nanoparticles suggest the presence of a non-magnetic surface layer having a thickness of 0.5–1.5 nm depending on the synthesis procedure and nanoparticle size.[85,86] One should, however, recognize that these nanoparticles can be, and are often observed to be, perfectly crystalline. In this context "disordered" refers only to the distribution of the magnetic dipole moments, and does not reflect on any structural amorphous properties of the nanoparticle. Additional evidence for the presence of these surface spins in iron oxide nanoparticles can be inferred from temperature-dependent magnetization studies, which show anomalies in the susceptibility at very low temperatures. These anomalies may be attributed to the freezing of these disordered surface spins,[87–90] although other mechanisms have also been proposed, including interactions among the nanoparticles at low temperatures.[91,92]

10.6   Magnetic Interactions

While the reduction in the saturation magnetization in iron oxide nanoparticles provides an example of the quantitative differences between the properties of bulk and nanoscale systems, the qualitative differences in the response of magnetic nanoparticles can often be more relevant for biomedical applications. Bulk ferromagnetic (or ferrimagnetic) materials have a remnant magnetization, and therefore show strong magnetic interactions even at zero external field. For many biomedical applications, any strong interactions between nanoparticles could be detrimental because these magnetic forces would lead to particle agglomeration. Significant agglomeration of the nanoparticles could in turn lead to blockage in arteries and other undesirable effects. In iron oxide nanoparticles, however, there is no magnetic hysteresis since the zero field magnetization is dominated by thermal fluctuation effects, resulting in qualitatively different behavior than bulk iron oxide.

10.7   Magnetic Hysteresis

As mentioned previously, the magnetic domain size for Fe_3O_4 is on the order of 100 nm. This is considerably larger than a typical size for ultrasmall iron oxide nanoparticles, which typically have diameters below 20 nm. Because of their small size, iron oxide nanoparticles cannot support the formation of domain walls, so they are single domain magnets. This means that the discussion of magnetic hysteresis in bulk magnets is not applicable to iron oxide nanoparticles.

10.8   Magnetic Moment of Fe_3O_4 and Dipolar Interaction

Because they are single domain particles, we can model the magnetization of each iron oxide nanoparticles as a colossal (classical) magnetic moment arising from the sum of the net moment on each unit cell. For Fe_3O_4, the moment on each unit cell arising from the antiferromagnetic ordering of the Fe spins is roughly $40/\mu_B$. This is due to the fact that the antiparallel spin structure means that the Fe^{3+} moments all cancel, leaving only the Fe^{2+} spins to provide a net moment, each of which is $5/\mu_B$. A spherical Fe_3O_4 nanoparticle having a diameter of 10 nm contains roughly 1,000 unit cells, leading to a net moment of approximately $40,000/\mu_B$ well below the Neel temperature, assuming full magnetic ordering. In practice, this simple picture is not completely correct because of surface spin effects; but modeling the net magnetization on each iron oxide nanoparticles as a single giant moment allows for a fairly robust description of their magnetic nanoparticles.

Although the spins inside each iron nanoparticles are assumed to be fully ordered, there is no ordering is expected among different nanoparticles, unless the particle separation is small. The short-range exchange interaction responsible for ferrimagnetic ordering drops very quickly with distance, and is negligible between spins on neighboring nanoparticles. The magnetic dipolar interaction between nearby iron oxide nanoparticles becomes significant only for very small separations. Assuming a magnetic dipole moment of 30,000 μ_B on each nanoparticle, the maximum dipolar energy, given by $E = (\mu_0/2\pi r^3)M^2$ for M the moment on each particle and r the separation,[67] becomes comparable to the room temperature thermal energy (25 meV) only for a center-to-center particle separation on the order of 30 nm. This means that the separation between nanoparticles is 20 nm, and the volume concentration approaches 2%. This is a nanoparticle concentration that is much larger than typically considered for biomedical applications, so for the purposes of the following discussion, the giant moments on each nanoparticle can be taken as non-interacting. The effects of nanoparticle interactions can, however, be seen for highly concentrated nanoparticle samples, where a crossover to a strong interacting glassy phase can be observed as the volume fraction of iron oxide nanoparticles in a suspension is increased.[93-95]

10.9 Superparamagnetism

The behavior of non-interacting, freely fluctuating moments in an external magnetic field as a function of temperature can be understood very well using statistical mechanics. At a fixed temperature, the magnetization of a collection of non-interacting moments as a function of applied magnetic field follows a Brillouin function.[67,68] The magnetization is zero at zero applied field and increases monotonically toward the saturation magnetization with increasing field. This behavior is called *paramagnetism*. At small fields, the magnetization is almost linear with magnetic field, with concave curvature developing at larger magnetic fields. In the classical limit, suitable for the giant moments on iron oxide nanoparticles, the magnetization curve can be approximated by the Langevin function, $L(x) = \coth(x) - 1/x$, where x is proportional to the size of the moment and the applied field, and inversely proportional to temperature and is given by MH/kT.[68] Experimentally, the room temperature magnetization curves for iron oxide nanoparticles can often be fit very well by the Langevin function. This is shown in Fig. 10.3a, which plots the magnetization as a function of applied field at 300 K for a nanoparticle sample consisting of Fe_3O_4 nanoparticles having a diameter of approximately 12 nm. At low fields the magnetization is almost linear with the applied field, but the magnetization saturates at a value near 50 emu/g for this particular sample.

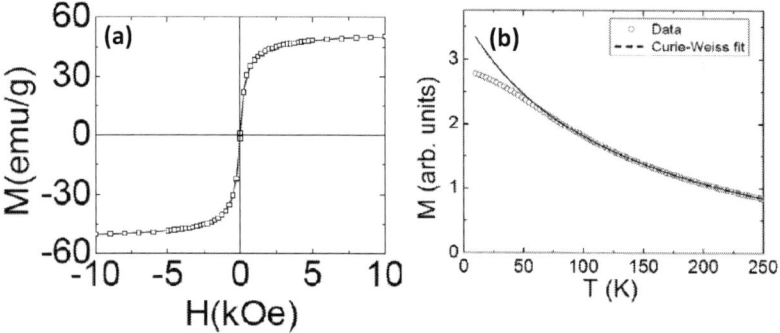

Figure 10.3 (a) Room temperature magnetization curve for a superparamagnetic iron oxide sample. (b) Field-cooled $M(T)$ curve for iron oxide nanoparticles. The dashed line is a fit to the Curie–Weiss law.

Paramagnetic systems also exhibit a characteristic temperature dependence. Expanding the Brillouin (or Langevin) function shows that the magnetization is inversely proportional to temperature at sufficiently small external fields, which is commonly known as Curie's law. In practice, Curie's law is often modified to account for weak interactions among the magnetic units in the system. This interaction energy is introduced through the Weiss temperature, T_θ, and the resultant equation, $M = C/(T - T_\theta)$, is called the *Curie–Weiss law*. In many systems, a small temperature independent background term is also seen. This temperature dependence is often approximately observed in iron oxide nanoparticles as well, as shown in Fig. 10.3b. While the field dependence and temperature dependence of the magnetization of iron oxide nanoparticles (Figs. 10.3a,b) are qualitatively similar to the properties of paramagnets, the magnetic susceptibility and magnetic moments are orders of magnitude larger for iron oxide nanoparticles than for atomic spins. Because of these properties, iron oxide nanoparticles are often referred to as *superparamagnetic* iron oxide.

10.10 Relaxation Mechanisms of Magnetic Nanoparticles

The similarities between the response of the giant moments on iron oxide nanoparticles and the properties of paramagnetic spins depend on the fact that the magnetization on each iron oxide nanoparticle is fluctuating, so that the equilibrium distribution is purely statistical and depends only on the temperature and applied field. This assumption is valid at high temperatures, but does not hold at low temperatures. In the simplest model for magnetic dynamics on an individual magnetic nanoparticle, first discussed by Neel[96] and Brown[97] with the properties of single domain particles elucidated previously by Stoner and Wohlfarth,[98] the individual spins on the nanoparticle undergo a collective, coherent change in direction remaining fully ordered. While recent studies point toward rather more complex spin reversal dynamics in some nanoparticle systems,[99] the simple model of Neel–Brown relaxation is sufficient to qualitatively and quantitatively understand many features of the magnetic dynamics of nanoparticles.

The Neel–Brown relaxation of magnetic iron oxide nanoparticle moments is driven by thermal fluctuations. While the magnetic energy would be degenerate for all directions in an isotropic system, magnetocrystalline anistropy selects low energy and high-energy directions in the iron oxide crystallite. This magnetocrystalline anistropy reflects the crystal structure of the nanoparticle, so the anistropy has cubic symmetry for cubic Fe_3O_4. The first order magnetocrystalline anisotropy energy density, K_1, is approximately -1.2×10^4 J/m^3 in bulk Fe_3O_4 at room temperature,[74] reflecting the minimum energy for the magnetization directed along a body diagonal. For simplicity, however, this anisotropy energy is often modeled as being uniaxial, with the lowest energy states being directed along a single axis of the nanoparticle. In order to switch between the two lowest energy states, with the nanoparticle moment directed along the two opposite directions of this axis, the system must overcome an energy barrier KV, with V the nanoparticle volume. This re-orientation energy can be provided by thermal fluctuations, leading to thermally activated behavior for magnetization reversal with an activation energy barrier, E_A, approximated by KV. The magnitude of this energy barrier for moment reversal then depends on the materials, through K, and also on the size of the nanoparticle, through V. This is schematically illustrated in Fig. 10.4, which plots the increase in energy for different directions of nanoparticle moments relative to a vertical easy axis.

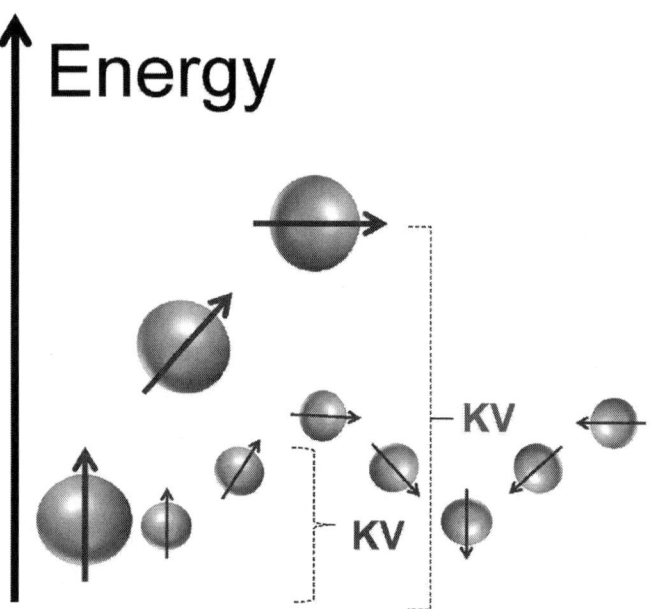

Figure 10.4 Variation of the energy barrier (KV) with the orientation of the nanoparticle moment. This energy barrier is higher for larger nanoparticles.

10.11   Role of Blocking Temperatures in Magnetic Properties

This thermally activated magnetization reversal produces a strong temperature dependence in the magnetic response of the iron oxide nanoparticles. At high temperatures, when the average thermal energy is comparable to this activation energy, the nanoparticle moments fluctuate freely. In this regime, the distribution of the nanoparticle moments can be modeled statistically and depends only on the temperature and applied field, in an identical manner to paramagnetic spins. In this high temperature regime, the nanoparticles are *superparamagnetic*. As the temperature is reduced, a much smaller fraction of the thermal fluctuations provide sufficient energy to overcome the anisotropy energy barrier, so the nanoparticle moments remain frozen. This change from a regime where the nanoparticles exhibit superparamagnetic behavior to the state where the nanoparticle moments are thermally blocked is often referred to *blocking* or *freezing*.

This blocking temperature does not represent a thermodynamic phase transition but a crossover from a fluctuating to kinetically trapped response for the nanoparticles.[100] The blocking temperature is not precisely defined for a collection of nanoparticles for two reasons. The first is that the thermal fluctuations are statistical, so even if the average thermal energy is much smaller than the activation energy required for the nanoparticle moment to overcome the anisotropy barrier, there remains a non-zero probability for sufficiently large fluctuations to allow for magnetization reversal. The crossover temperature between fluctuating and blocked behavior therefore depends on the measuring time, since very slow moment reversals will not be observed in faster measurements. Very generally, the relationship between the blocking temperature and anistropy energy barrier, KV, is taken as $KV = 25\ k_B\ T_B$. The numerical pre-factor of 25 comes from a typical measuring time of 60 s assuming a microscopic attempt frequency of 10^9 Hz. For the ultrasmall iron oxide nanoparticles typically considered for biomedical applications, the blocking temperature is generally below 100 K, or −173°C. This means that these nanoparticles will be firmly in the superparamagnetic regime for the physiological temperatures relevant for these applications.

The second challenge for precisely identifying the crossover temperature between the superparamagnetic and blocked regime for magnetic nanoparticles is that these studies are invariably done on a large number of particles. The anisotropy energy depends linearly on nanoparticle volume, and therefore on the cube of the nanoparticle diameter. As will be discussed briefly in the context of nanoparticle synthesis, it is extremely challenging to make highly monodisperse samples of iron oxide nanoparticles. While polydispersities on the order of 3%[101] have been obtained, more typical values for the polydispersity are 10% or larger.[102,103] A 10% polydispersity leads to variations of more than 30% in the anisotropy energy of a nanoparticle sample and a correspondingly large variation in blocking temperature. Experimentally, the magnetocrystalline anisotropy energy estimated for iron oxide nanoparticles can differ significantly from the values for bulk samples. Although the magnetocrystalline anisotropy for bulk Fe_3O_4 is approximately 1.2×10^4 J/m^3, estimates for K for nanoparticle samples, often inferred from measurements of the approximate blocking temperature, are typically an order of magnitude larger, depending on particle size.[104,105] A number of factors can contribute to this discrepancy, including the difficulty in accurately determining the blocking temperature, the presence of a non-magnetic surface layer that changes the effective magnetic volume of the nanoparticle, and modifications of the anisotropy due to finite size effects.

10.12   Zero Field–Cooled and Field-Cooled Behavior of Magnetic Nanoparticles

The crossover between superparamagnetic and blocked magnetic behavior in iron oxide nanoparticles can be readily observed through field-cooled (FC) zero field–cooled (ZFC) measurements. For FC magnetization measurements, a small magnetic field is applied to the nanoparticles in the superparamagnetic phase. This partially aligns the moments, as described by the Brillouin function, and this alignment persists as the sample is cooled, increasing as approximately described by Curie's law. Below the blocking temperature, the nanoparticle moments are frozen in a configuration with a finite magnetization. The magnetization is measured on warming from low temperatures with the same small magnetic field applied. In the blocked state, the moments do not fluctuate on the timescale of the magnetic measurements, so the magnetization is roughly constant. As the temperature increases above the blocking temperature, the now superparamagnetic particles begin to fluctuate and the magnetization follows the dependence described by Curie's law.

The ZFC magnetization data are qualitatively different. In zero applied field, the magnetic moments on the iron oxide nanoparticles are randomly oriented, so there is no net magnetization. After cooling the sample to base temperature at zero field, a small magnetic field is applied and the magnetization is measured on warming. At the lowest temperatures, the magnetization is still zero because the nanoparticle moments are blocked and unable to align with the external field. As the temperature is increased, more thermal energy is available to reverse these moments so the magnetization increases as well. This behavior continues until the nanoparticle moments are freely fluctuating at the blocking temperature, above which they follow the Curie law behavior and the magnetization decreases with a further increase in temperature.

Sample ZFC–FC data for an iron oxide nanoparticle sample are plotted in Fig. 10.5a. At high temperatures, the ZFC and FC curves coincide and the temperature dependence of the magnetization can be approximated by Curie law behavior. The ZFC and FC curves separate at lower temperatures, which indicates that some of the larger nanoparticles are thermally blocked and were not able to fully align with the applied field. At still lower temperatures, the ZFC curve shows a maximum before decreasing as the temperature is further reduced, while the FC curve becomes approximately temperature independent at low temperatures. The maximum in the ZFC curve is used to approximate the blocking temperature, although this peak can be very broad for polydisperse samples. In practice, the degree of polydispersity can be estimated by fitting the ZFC magnetization curve.[106–108]

The blocking temperature for a collection of nanoparticles depends on the magnitude of the applied magnetic field. An external magnetic field has the effect of increasing the energy for moments having an antiparallel orientation with respect to the field and lowering the energy for parallel moments. This changes the effective energy barrier, which, for non-interacting uniaxial iron oxide nanoparticles in a magnetic field becomes $E_A(H)$ =

$(KV - H/H_K)^\alpha$, where H_K is an internal magnetic field associated with the magnetocrystalline anisotropy and α is typically close to 1.5.[109] This gives an expected shift in the blocking temperature that varies with the square of the applied field. Interactions among the magnetic nanoparticles can alter this dependence to $T_B(H) \sim H^{2/3}$,[110,111] which is the same functional form expected for the de Almeida–Thouless line for spin glasses.[112]

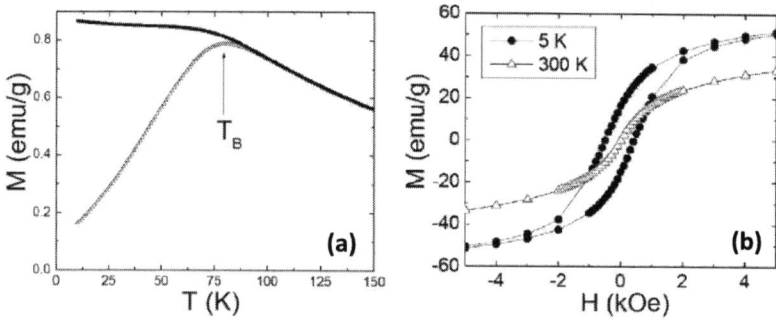

Figure 10.5 (a) ZFC–FC curves for iron oxide nanoparticles measured in a small magnetic field. (b) $M(H)$ curves for iron oxide nanoparticles measured at 5 and 300 K as indicated.

The crossover between superparamagnetic and blocked states for iron oxide nanoparticles is also reflected in their magnetization curves measured at fixed temperatures. At high temperatures, superparamagnetic iron oxide nanoparticles have zero coercivity. In the absence of an applied magnetic field, the nanoparticle moments flip freely and will not support any net magnetization. However, at low temperatures, below the blocking temperature, the nanoparticles can be aligned by very large magnetic fields, which can eliminate the anisotropy barrier to moment reversal if $H \gg H_K$. However, as the field is reduced at these low temperatures, the moments are blocked, leading to a remnant magnetization and a finite coercivity. Magnetic hysteresis curves at high temperature, in the superparamagnetic regime, and low temperature, in the thermally blocked regime, are shown in Fig. 10.5b. Unlike bulk iron oxide, the coercivity at low temperatures does not arise from domain wall motion, but rather reflects the fact that the nanoparticle moments are frozen below the blocking temperature. As such, the coercivity, H_c, for iron oxide nanoparticles shows a strong temperature dependence, which can generally be fit to $H_c = H_0$

$[1 - (T/T_B)^{,1/2}]$ with T_B the blocking temperature, although particle interactions can modify this behavior.[113]

10.13 AC Susceptibility Measurements

The frequency dependence of the crossover from fluctuating moments to blocked moments in a collection of iron oxide nanoparticles can be investigated by measurements of the ac magnetic susceptibility. As the temperature for a superparamagnetic sample is reduced, the frequency at which the nanoparticle moments can reverse is reduced. This means that at higher measuring frequencies, the nanoparticle moments are blocked at higher temperatures. This behavior can be seen in Fig. 10.6, which

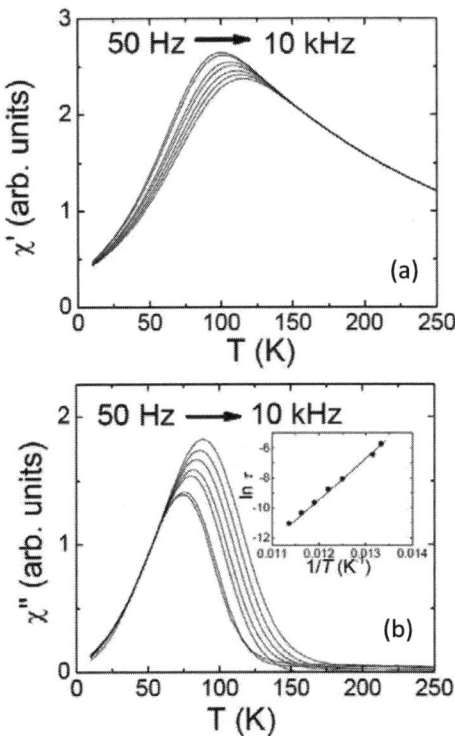

Figure 10.6 (a) Real and (b) imaginary components of the temperature-dependent ac magnetic susceptibility of iron oxide nanoparticles measured at different frequencies. The inset to panel (b) plots the temperature dependence of the loss peak.

plots the real (Fig. 10.6a) and imaginary (Fig. 10.6b) components of the ac susceptibility of an iron oxide nanoparticle sample as a function of temperature for a number of different measuring frequencies. The peak in susceptibility, which is approximately the blocking temperature, shifts to higher temperatures with increasing frequency. The frequency dependence of this peak temperature can be modeled by a Debye relaxation process, $\omega_p = \omega_0 \exp(E_A/k_B T_p)$, with ω_p the measuring frequency, ω_0 the microscopic attempt frequency, and T_p the peak temperature. A fit to this functional form, as log τ plotted against $1/T$, with $\tau = 1/\omega$, is shown in the inset to Fig. 10.6b. The attempt frequency is normally assumed to be in the range of 10^9 to 10^{12} Hz.[114-116] Significantly larger values are often found in fitting such relaxation curves, but values on the order of 10^{15} Hz or larger are generally unphysical[117] and may indicate that pure Debye relaxation is broken by interactions. For non-interacting moments, the relative shift in the peak temperature with measurement frequency, defined as $\phi = \Delta T_p/T_p \log(\Delta\omega)$, is generally found to be larger than 0.1, while for strongly interacting systems, such as spin glasses, this parameter is normally smaller than 0.01.[115,118]

10.14 Magnetohydrodynamics of Ferrofluids

Many of the potential biomedical applications for iron oxide nanoparticles involve suspensions of these nanoparticles in a solvent, generally referred to as *ferrofluids*. It is therefore important to understand the properties of iron oxide nanoparticles in solution and, more specifically, how these respond to external magnetic fields. Although the density of iron oxide is larger than that of water, or in fact of most commonly used solvents, isolated nanoparticles will typically remain well suspended in solution due to Brownian motion. There can be, however, a challenge in ensuring that the iron oxide nanoparticles remain isolated, since agglomeration will typically cause the nanoparticles to precipitate out of solution rendering the sample unsuitable for applications.

There are two primary mechanisms for preparing ferrofluid solutions of well-suspended iron oxide nanoparticles. *Steric stabilization* involves coating the nanoparticles with a surfactant, often an organic polymer, which will inhibit aggregation.[119] A layer of polymer, such as polyethylene glycol, polyacrylamide, or dextran, among many other possibilities, will provide a steric hindrance to keep the nanoparticles well separated and minimize agglomeration. Furthermore, these coatings can provide additional functionalities to the nanoparticles to promote certain biomedical applications, such as providing binding sites for drug loading and reducing the absorption of blood plasma proteins. *Electrostatic stabilization* offers another approach to maintain the stability of ferrofluids by suppressing agglomeration.[120,121] Introducing a surface charge to the iron oxide nanoparticles either through attaching charged ligands or the adsorption of charged ions can also reduce aggregation. Because of the Coulomb repulsion between two like-charged nanoparticles, an effective charge on the iron oxide particles will tend to keep these well separated and therefore stable in solution. Both steric and electrostatic stabilization depend very strongly on the specific solvent being used to prepare the ferrofluid. Whether the solution is polar or non-polar will change the effectiveness of certain steric stabilization agents, and the pH of aqueous solutions strongly affects the surface charge on nanoparticles.

10.14.1 Properties of Ferrofluids

A number of important properties of ferrofluids depend strongly on the dynamical response of the iron oxide nanoparticles suspended in solution. In particular, one of the physical properties that characterizes a ferrofluid is the hydrodynamic radius of the nanoparticles. Because the iron oxide nanoparticles are almost always coated with a surfactant to inhibit agglomeration, the hydrodynamic radius is typically larger than the hard sphere size of the iron oxide core. The hydrodynamic size for nanoparticles in the ferrofluid is normally measured experimentally rather than computed, often by dynamic light scattering, because this parameter depends strongly on both the nature of the surfactant and on the solvent.[48] This technique measures the autocorrelation function for scattering off of the nanoparticles, uses these data to calculate the diffusion constant, D_t, and then computes the hydrodynamic radius, r, using the Stokes–Einstein, $D_t = k_B T / 6\pi\eta r$, with η the solvent viscosity.[122] One other important property of the nanoparticles is the *zeta potential*, which provides a measure of the electrical potential near the particle surface.[123,124] Nanoparticle having a magnitude of the zeta potential greater than 30 mV, whether positive or negative, are presumed to have reasonable stability.

Iron oxide nanoparticles in solution are particularly interesting for biomedical applications because of their response to both ac and dc external magnetic fields. While ferrofluids can be considered as homogeneous magnetic liquids, and therefore treated by the theory of magnetohydrodynamics pioneered by Alfven,[125,126] over some range of parameter space, the behavior of the suspended iron oxide nanoparticles is particularly relevant for the applications we are considering, so we focus our discussion on these properties.

A magnetic field, $\mathbf{B}$, will exert a torque, $\boldsymbol{\tau}$, on the dipole moment of the nanoparticle, $\boldsymbol{\mu}$, given by $\boldsymbol{\tau} = \boldsymbol{\mu} \times \mathbf{b}$. This torque will tend to align the moment on the iron oxide nanoparticle with the applied field, which can involve either a physical rotation of the suspended particle, called Brownian relaxation, or a change in the direction of the magnetization without any motion of the nanoparticles, called Neel relaxation. Both Brownian and Neel relaxation will be discussed in detail in the following. In case of a static (dc) $\mathbf{B}$, the alignment of the magnetic dipole moments with this field has remarkable consequences for the distribution of iron oxide nanoparticles in the fluid. The development of a net moment on each nanoparticle aligned with $\mathbf{B}$ produces dipolar interactions among the particles. These dipolar interactions favor a highly anisotropic distribution of nanoparticles in the ferrofluid, with chains of nanoparticles forming parallel to the direction of $\mathbf{B}$.[127-129] Because of the relatively long timescales required for this anisotropic distribution to emerge from the homogeneous ferrofluid, discussed in the following, this effect is only observed in dc magnetic fields. This chain-like morphology minimizes the dipolar energy in the system, although it is believed that additional interactions, possibly van der Waals forces, are required to stabilize these nanoparticle chains against thermal fluctuations.[130,131]

10.14.2 Light Scattering Experiments with Ferrofluids

The dynamics of nanoparticle chain formation in iron oxide ferrofluids can be investigated using light scattering studies. Measurements of the forward scattering of laser light in iron oxide ferrofluids in a small applied magnetic field have found that the time constant for the development of chains of nanoparticles (which later coarsen into columns at much longer timescales) is inversely proportional to the magnitude of the field.[132,133] This time constant for the development of nanoparticle chains after the field is applied ranges from approximately 200 s at a field of 100 Oe, decreasing at higher fields. When the magnetic field is removed, the nanoparticle chains dissociate resulting in homogeneous scattering, with the time constant for this process being an order of magnitude smaller than for chain formation. These investigations provide experimental estimates for both the growth and decay times of non-equilibrium distributions of iron oxide nanoparticles in suspension.[132] Although these specific rates are expected to vary with the physical parameters characterizing the ferrofluid, including viscosity, the hydrodynamic size of the iron oxide nanoparticles, and the magnitude of the moment on each particle, these estimates remain relevant in developing biomedical applications for ferrofluids under static magnetic fields. Specifically, the time constant for chain formation in moderate magnetic fields is on the order of tens or hundreds of seconds, while the time constant for dissociation is on the order of seconds or tens of seconds.

While a static magnetic field can modulate the nanoparticle distribution by inducing dipolar interactions among the particles, a magnetic field gradient produces a force in individual iron oxide nanoparticles. A non-uniform magnetic field **B** will produce a force **F** on the magnetic dipole moment of a nanoparticle, **μ**, given by **F** = grad **μ·B**. This force will accelerate the nanoparticles toward larger magnetic fields, which can, in principle, be used to steer magnetic nanoparticles in vivo or for magnetic separation.[134,135]

10.15 Hyperthermia and Relaxation in AC Magnetic Fields

As mentioned previously, non-interacting magnetic moments of iron oxide nanoparticles in a ferrofluid exhibit two distinct relaxation mechanisms. In the Brownian relaxation mechanism, the nanoparticle physically rotates in the carrier liquid, so the time constant for relaxation is determined by the hydrodynamic size of the nanoparticle and the viscosity of the solvent along with the temperature. The Brownian relaxation time, τ_B, for a nanoparticle of hydrodynamic radius r in a solvent with viscosity

η at temperature T is given by $\tau_B = 4\pi r^3 \eta / k_B T$.[136] Brownian relaxation will be suppressed if the motion of the nanoparticle is constrained. Iron oxide nanoparticles fixed in polymer networks may exhibit negligible Brownian relaxation,[137] and this relaxation mechanism is completely suppressed for nanoparticles embedded in a solid matrix such as wax[138,139] or resin,[140,141] or below the freezing temperature of the solvent. Additionally, magnetic dipole interactions between nanoparticles can also modify the Brownian relaxation mechanism.[142–144]

The second mechanism for relaxation of the local moment on individual iron oxide nanoparticles is Neel relaxation, discussed previously in the context of superparamagnetic blocking. This is a thermally activated magnetization reversal over a magnetocrystalline anisotropy energy barrier. The Neel relaxation time, τ_N, can be approximately modeled in terms of the microscopic relaxation time scale and energy barrier E_A as $\tau_N = \tau_0 \exp(E_A / k_B T)$. In this expression the activation energy E_A is approximated by nanoparticle magnetocrystalline energy, KV, and the microscopic timescale τ_0 is generally in the range of 10^{-9} to 10^{-12} s.

Most generally, both Brownian and Neel relaxation mechanisms will be relevant for understanding the magnetic response of a system of iron oxide nanoparticles. Because these represent independent relaxation mechanisms, the relaxation times should be added in parallel to determine the net relaxation time τ_{net}. Thus, $\tau_{net} = \tau_N \tau_B / (\tau_N + \tau_B)$, and the shorter time constant will generally dominate the relaxation behavior. In real systems, particularly biological systems where the iron oxide nanoparticles may be localized by cell membranes or otherwise have their free rotation impeded, the effects of Brownian relaxation may be reduced. However, since the Neel relaxation mechanism is relatively insensitive to the external environment of the nanoparticle, beyond the temperature and magnetic field, τ_N is practically constant for all applications.

10.16 Heat Dissipation in Magnetic Nanoparticles

Much of the interest in understanding the magnetic relaxation mechanisms in iron oxide nanoparticles stems from the fact that heat is dissipated through these processes. Under the application of an alternating magnetic field at some angular frequency ω, the moments on individual iron oxide nanoparticles will oscillate with this excitation. Each oscillation of the nanoparticle moment results in a small amount of heat being produced, either because of frictional forces between the nanoparticle and liquid for Brownian relaxation or effects of magnetic viscosity for Neel relaxation. The magnetic dissipation for the nanoparticle system is parameterized in terms of χ'', the out-of-phase component of the magnetic susceptibility. In the case of a single relaxation time, τ, for the system, this can be approximated by $\chi'' = \chi_0(\omega\tau)/(1 + (\omega\tau)^2)$, where χ_0 is the static susceptibility. The resonance condition is defined by $\omega\tau = 1$ and will give the maximum value for χ''. The energy lost in each cycle is approximated by $\Delta E = 1/2\mu_0(\chi''H)H$, with H the magnitude of the ac field and μ_0 the permeability of free space.[145] With this expression, the power dissipated by a system of nanoparticles to an applied field with angular frequency ω is $P = 1/2\mu_0\omega\chi''H^2$. This energy should be understood as being extracted from the applied magnetic field.

The power dissipated by this magnetic relaxation will result in local heating, referred to as *(magnetic) hyperthermia*. In the simplest model presented above, the power dissipated increases as the square of the magnitude of the applied field. In practice, it can be very challenging to design systems to apply very large magnetic fields at high frequencies because of the self-inductance in the magnetic coils. The magnetic power dissipated depends on the frequency through two different contributions. The explicit dependence of power on frequency yields a linear increase in heating with the frequency of the applied magnetic field. However, the power also has an implicit dependence on frequency through the variation in χ''. Since χ'' will be maximized at resonance, exciting the nanoparticles near their resonant frequency can produce larger powers. Furthermore, the amplitude of the peak in χ'' at resonance in a real ferrofluid system depends on the quality factor of the system, which will be larger for more monodisperse nanoparticles having a narrower distribution of resonant frequencies.

10.17 Specific Absorption Rate

The heating produced by this magnetic dissipation is often expressed as the *specific absorption rate*, or SAR (also called the *specific loss power* [SLP]) defined as the power produced per unit mass of nanoparticle in solution.[43,146,147] This definition is motivated by the observation that so long as nanoparticle interactions can be completely neglected, χ_0 and hence both χ'' and P, should vary linearly with the mass of iron oxide nanoparticles present in the system. This simple relationship is not perfectly realized experimentally—due to the presence of the non-magnetic surface layers discussed previously, the static susceptibility depends on both nanoparticle mass and morphology—but this remains an important metric for parameterizing nanoparticle heating.

The SAR for ferrofluid samples can be estimated by measuring the temperature change on exposure to an ac magnetic field. For these measurements to be accurate, the thermometer used should be insensitive to alternating magnetic fields, which precludes metallic thermometers in which eddy currents could produce local heating, and the ferrofluid sample should be thermally isolated. This thermal isolation should minimize heat transfer from the sample to the environment although, in practice, this can be difficult to control. Several examples of experimental setups for measuring SAR are available in the literature,[148,149] with commercial systems now beginning to appear on the market.[150] Representative heating curves for different concentrations of iron oxide nanoparticles in an aqueous solution are plotted in Fig. 10.7. The heat produced in the ferrofluids by the applied magnetic field varies monotonically with the nanoparticle concentration.

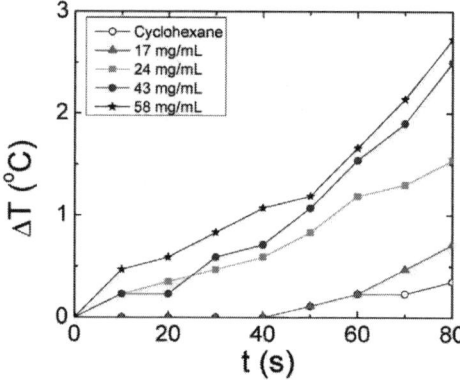

Figure 10.7 Heating curves for different concentrations of iron oxide nanoparticles suspended in cyclohexane under an applied ac magnetic field.

The SAR is typically computed from the initial slope of the heating curve. This protocol is followed for two reasons. At high temperatures, far above room temperature, the ambient heat loss from the ferrofluid to the environment will be larger, even with good thermal isolation, eventually becoming as large as the heat input into the ferrofluid through magnetic dissipation. This ambient heat loss qualitatively explains the saturation of the heating curves at high temperatures. A second reason to define SAR in terms of the initial slope is that the magnetic dissipation for iron oxide nanoparticles in solution shows strong temperature dependence. Since SAR is normally understood to be determined at ambient conditions, the initial slope of the heating curve should be used for this calculation. The initial slope, in units of "$°C\ s^{-1}$" can be converted into a power by approximating the heat capacity of the ferrofluid by the heat capacity of the solvent alone, which is a reasonable assumption for dilute aqueous solution.

The SAR for iron oxide–based ferrofluids shows considerable variation in the literature, with values ranging from 10 to 400 W/g for iron oxide nanoparticles,[147,151,152] with some reports of even higher values. Because the magnetic power, hence SAR, depends strongly on both frequency and the magnitude of the applied field, the discrepancies can, in part, be understood by recognizing that the measurements were done on different instruments at different frequencies and with different magnitudes of the applied fields. With this in mind, it is often useful to define the *intrinsic loss power* (ILP), which scales the SAR value by the ac frequency and square of the applied magnetic field.[153] This provides a system-independent way to define the power dissipated by magnetic relaxation as an intrinsic property of the ferrofluid.

For practical applications, the temperature change produced by magnetic relaxation of nanoparticles depends not only on the heating produced by the magnetic dissipation, but also on the thermal properties of the sample containing the iron oxide and the surrounding environment. The heat produced by the nanoparticles will diffuse from the sample into the surroundings, with this heat flow depending on the thermal conductivity. For biomedical applications, the diffusion of heat from locations containing iron oxide nanoparticles into the surrounding tissue can be large, as these regions may be in good thermal contact. Many of the issues involved with maintaining large temperature gradients in the body have been discussed in the literature.[154–156]

One of the advantages of using iron oxide nanoparticles to produce hyperthermia is that in this system the heating is produced on extremely small length scales, with individual nanoparticles having volumes many orders of magnitude smaller than a cell nucleus. This could potentially allow for local heating at length scale of nanometers or tens of nanometers, although this remains a topic for debate in the literature. While experimental studies have demonstrated that temperature-sensitive cation channels in cells can be activated by heating from magnetic nanoparticles,[157] theoretical models for nanoparticle heating in solution suggest that rapid thermal processes should preclude any significant temperature gradient near the surface of the nanoparticle.[158]

One possible resolution to this apparent discrepancy may arise from a separation of timescales in the problem. While the theoretical model assumes that the power produced by magnetic dissipation is continuous, and computes thermal diffusion times on the order of tens of nanoseconds, the magnetic relaxation processes in typical nanoparticle systems can have time constants of microseconds for reasonable choices of physical parameters. If we suppose that the average power per nanoparticle is 10^{-14} W, then magnetization reversals with a timescale of 10 μs would deposit 10^{-19} J of energy with each reversal, corresponding to an energy density of approximately 1 J/cm^3, assuming a volume of 100 nm^3 for the nanoparticle. This large energy density could plausibly produce some transient local heating, although because of the large surface to volume ratio of nanoparticles, this heat is likely to diffuse very quickly into the surrounding solvent. Beyond modeling the heat production in iron oxide nanoparticles arising from magnetic relaxation, it will be crucial to develop a better understanding of thermal conduction mechanisms in ferrofluids and other systems containing nanoparticles at length scales from nanometers to microns to centimeters in order to realize the full potential of magnetic hyperthermia for developing biomedical applications.

10.18 Magnetic Resonance Imaging

Superparamagnetic iron oxide nanoparticles are also of great interest for biomedical applications because of their utility as contrast agents for magnetic resonance imaging.[159] The signal produced in magnetic resonance imaging (MRI) depends on the precession frequency of protons, ω, which in turn depends on the amplitude of the magnetic field as $\omega = \gamma \mathbf{B}$. In this expression γ is the gyromagnetic ratio for protons, 42.56 MHz/T, and $\mathbf{B}$ is the magnetic field. Applying a field gradient across the sample produces a systematic shift in the resonant frequency with position, which allows the MR signal to be associated with a specific location in the sample.

Very roughly, and considered classically, the magnetic resonance signal arises as follows. A dc magnetic field, B, exerts a torque, τ, on the magnetic moments associated with proton moments, $\boldsymbol{\mu}$, that are not fully aligned with the field, with the torque given by $\boldsymbol{\tau} = \boldsymbol{\mu} \times \mathbf{B}$, so $\tau = \mu \mathbf{B} \sin \theta$ with θ the angle between $\boldsymbol{\mu}$ and $\mathbf{B}$. By Newton's second law, this torque is proportional to the time rate of change of the angular momentum, which is, in turn, proportional to the moment, so the effect of $\boldsymbol{\tau}$ is to cause $\boldsymbol{\mu}$ to precess around the $\mathbf{B}$ field axis. The torque, hence precession frequency, is proportional to the magnitude of the $\mathbf{B}$ field, as described above.

There are two separate relaxation mechanisms that are relevant in determining the MR signal. The first, the spin-lattice relaxation, involves moments at some finite angle θ to the applied magnetic field changing direction to align with $\mathbf{B}$. After the magnetization of a population of spins is rotated to align at some angle to $\mathbf{B}$, with the spins rotated to be perpendicular to $\mathbf{B}$ for many MR sequences, the magnetization will recover along $\mathbf{B}$ with an exponential time constant given by T_1.[160] The internal energy lost by the spins as they align with $\mathbf{B}$ is distributed to the lattice, hence the term "spin-lattice" relaxation. The second relaxation mechanism, spin–spin relaxation, arises from transient interactions that modify the precession frequency during the scattering. These fluctuations in the precession frequency cause a collection of spins, initially precessing about the $\mathbf{B}$ field in unison, to dephase. This dephasing follows an exponential relaxation with time constant T_2.[161] This dephasing by spin–spin interactions is different from the dephasing produced by non-uniformities in the applied magnetic field, which is characterized by a time T_2^*, which is smaller than T_2.

A standard approach for measuring the T_2 signal, and distinguish this from effects arising from inhomogeneities in the static field, is called the Hahn spin-echo method.[159,162] This involves applying a large dc magnetic field to partially align the proton moments, as described by the Brillouin function presented previously. A radiofrequency electromagnetic pulse is applied to system, selected to tip the moments to align perpendicularly to the **B** field, referred to as a $\pi/2$-pulse to reflect the 90° in the direction of the magnetization. This causes the moments to precess about the **B** field. This precession depends on both the local field, assumed to be static, if inhomogeneous, and spin–spin scattering. This means that the precession for individual moments will be different, causing a dephasing of the initial alignment of the magnetic moments. After some well-defined delay time, t, a second rf pulse, called a π-pulse is applied to the spins to flip them by 180°. This has the effect of leaving the moments in the plane perpendicular to the **B** field, but reverse the direction of their precession, since $\sin(\theta + 180°) = -\sin\theta$. The precession frequency of individual moments is still determined by the local field, so after the elapse of this same waiting time t, the dephasing that arose from an inhomogeneous static field will be corrected, so the only residual phase differences arise from spin–spin scattering. The magnetic signal measured as a function of this waiting time t will have a characteristic exponential decay, with a time constant proportional to T_2.[159]

In tissues, T_1 relaxation times vary from roughly hundreds of milliseconds to thousands of milliseconds, while T_2 relaxation times vary from tens of milliseconds to hundreds of milliseconds.[159] Because these relaxation times vary with the type of tissue, selecting MR scans to selectively weight the T_1 or T_2 relaxation times can provide higher quality images with better resolution. These MR images can be improved by the use of contrast agents, which can vary the T_1 or T_2 relaxation times. The presence of magnetic iron oxide nanoparticles can dramatically shorten the T_2 relaxation time by providing fluctuating, dynamic local magnetic interactions that result in transient variations in the local spin precession rate for protons. Note that iron oxide nanoparticles affect the T_2 relaxation time, rather than the T_2^* relaxation time, because the nanoparticle moments fluctuate very rapidly on the timescale set by T_2. In one study the spin–spin relaxation rate R_2, with R_2 defined as $1/T_2$, increased dramatically from 6 mM^{-1}s^{-1} to 80 mM^{-1}s^{-1} in a given sample, varying monotonically with nanoparticle size.[163] Iron oxide nanoparticles have also been found to increase to R_2/R_1 ratio from anywhere from approximately 10[164] to nearly 100.[163] There are also suggests that iron oxide nanoparticles could be considered for use as a T_1 contrast agent, replacing gadolinium chelates for certain applications.[165]

One of the particular advantages of using iron oxide nanoparticles as a contrast agent for MR imaging is that for certain well-defined geometries, it is possible to estimate the local susceptibility of the sample from the MR signal, and hence determine the concentration of iron oxide nanoparticles. This potential for iron oxide nanoparticles to allow for a non-invasive measurement of the local concentration using MR is particularly attractive when combined with other applications, such as targeted drug delivery. Since iron oxide nanoparticles are a T_2 contrast agent, and so reduce the MR signal by increasing the amount of dephasing, portions of the sample containing iron oxide nanoparticles will appear dark in T_2-weighted images. By comparing the amplitude of this signal relative to portions of the sample without iron oxide nanoparticles, it is possible to estimate the magnetic susceptibility of the nanoparticle-rich sample. The susceptibility estimated using MR differs by the value measured directly by the magnetometer by less than 10% for a carefully controlled sample geometry.[166]

10.19 Synthesis

Having discussed the physical properties of iron oxide nanoparticles in some detail, we now make a brief digression to touch on some of the common techniques used to prepare these materials. There are a number of outstanding review articles that provide a broad introduction to many of the standard approaches for synthesizing iron oxide nanoparticles, and functionalizing these nanoparticles for biomedical applications, and provide a comprehensive set of references to the literature in this area. Therefore, this discussion of synthesis is structured with a very limited scope, with the intent to highlight some of the most widely used techniques for preparing iron oxide nanoparticles and touch briefly on functionalization. A more complete overview of this important aspect of incorporating magnetic iron oxide nanoparticles into biomedical applications can be found in the literature,[66,167] which touches on a number of other synthesis techniques as well.[168]

As often arises when synthesizing nanomaterials, there are a number of viable techniques for preparing iron oxide nanoparticles, each of which has certain advantages and disadvantages. Ideally, the synthesis technique should be inexpensive, readily scalable to yield commercially viable quantities of material, produce monodisperse nanoparticles having a well-controlled size and high magnetizations, and be readily compatible with approaches for adding surfactants to the nanoparticles, or otherwise functionalizing the material. Unfortunately, none of the synthesis techniques meets all of these ideal target goals, so in practice, the specific synthesis approach used will be dictated by the experimental considerations for each particular study. For the vast majority of biomedical applications, in addition to preparing the core iron oxide nanoparticles it is also necessary to coat them, either to introduce new functionalities, or simply to ensure that the nanoparticles are stable in solution. These coatings can consist of a wide range of materials, including simple molecules, polymers, or inorganic shells. Some of the different possible approaches for coating the iron oxide nanoparticles will also be presented, although this is also covered in much greater detail in review articles in the literature.

10.19.1   Co-Precipitation

One of the most straightforward, inexpensive, and accessible techniques for preparing iron oxide nanoparticles is co-precipitation.[169,170] In this approach, Fe^{2+} and Fe^{3+} ions are mixed in an aqueous solution at an appropriate stoichiometric ratio, for example, 1:2 to prepare Fe_3O_4. In basic solutions, the surplus of highly oxidizing OH^- ions, will promote the formation of iron oxide nanoparticles. For example, the formation of Fe_3O_4 proceeds as Fe^{2+} + $2Fe^{3+}$ + $8OH^- \rightarrow Fe_3O_4 + 4H_2O$. Since Fe_3O_4 is highly susceptible to oxidation to produce Fe_2O_3, this reaction must proceed in the absence of oxygen. In practice, black Fe_3O_4 can readily be distinguished from brownish γ-Fe_2O_3 although, as discussed previously, the two iron oxide materials share similar magnetic characteristics. The details of the nanoparticle structure depend on the pH of the solution, reaction temperature, and other properties.

One specific recipe for preparing Fe_3O_4 nanoparticles using co-precipitation that we have used extensively is as follows. In order to obtain the aqueous solution of Fe ions, 4.0 g of $FeCl_2 \cdot 4H_2O$ is dissolved in 10 mL of 2 M HCl and 10.8 g of $FeCl_3 \cdot 6H_2O$ is dissolved separately in 40 mL of 2 M HCl before the two solutions are mixed together. To this solution, 500 mL of 1 M NH_4OH is added drop wise while stirring the mixture, with the stirring continuing for 10 min after the last of the basic solution is added. This will produce a brown precipitate that turns black. The supernatant should be removed from these black magnetite nanoparticles by rinsing several times with distilled water to a final pH of 7. This process can be accelerated by using a strong permanent magnet to attract the nanoparticles to the bottom of the beaker to allow the supernatant to be removed more easily. This process will produce 4–5 mg of Fe_3O_4 nanoparticles having an average diameter of roughly 10–12 nm and a relatively broad size distribution. In practice, it is necessary to coat these nanoparticles with some surfactant before they can be incorporated for use in most applications.[48]

The co-precipitation method for preparing iron oxide nano-particles offers a number of advantages. Preparing nanoparticles by co-precipitation requires minimal infrastructure and uses only readily available inexpensive precursors. The synthesis is very rapid, typically taking less than an hour, and the approach is highly scalable, allowing for the production of large amounts of nanoparticles. Most significantly, because the synthesis can be done at ambient temperatures, it is possible to combine the preparation of iron oxide nanoparticles with surface functionalization, although the weakly basic conditions required for precipitation can prove detrimental to some surfactants. There are also disadvantages to using co-precipitation to prepare iron oxide nanoparticles. Nanoparticle samples synthesized by co-precipitation often show a relatively broad size distribution. This polydispersity can limit certain applications for the iron oxide nanoparticles. Additionally, as noted above, there can be some incidental oxidation of the nanoparticles during synthesis (particularly Fe_3O_4), leading to the formation of secondary oxide phases and reducing the magnetization.

10.19.2 Hydrothermal Synthesis

Hydrothermal methods can also be used to prepare relatively higher quality iron oxide nanoparticles. In this approach, solutions containing Fe ions are oxidized and heated, often under pressure. The reaction temperature typically varies from 80 to 200°C. In some cases, the iron oxide nanoparticles can be precipitated in a basic solution, similar to the approach used in co-precipitation, with the reaction product heated and pressurized to prepare higher quality nanoparticles. The geometry, including shape and size, of the nanoparticles formed by hydrothermal synthesis is found to depend on the specific solvent used to prepare the sample.[171] One particular advantage of this technique is that at the elevated temperatures and pressures used in this process, polar organic reagents can become miscible with water. This allows the synthesis of iron oxide nanoparticles to proceed in a solution containing both water and surfactants, leading to simultaneous synthesis and surface modification.[172]

10.19.3 Microemulsion

Highly uniform iron oxide nanoparticles having a widely tunable size distribution can be prepared using a microemulsion technique. This approach uses an oil phase, a surfactant phase, and an aqueous phase to produce well-defined segregated phases. The reverse microemulsion system consists of small bubbles of the aqueous solution surrounded by the oil matrix, with the size of the bubbles depending on the ratio of the water to surfactant concentrations. The nanoparticle co-precipitation occurs in aqueous droplets containing Fe ions, so the resulting size of the iron oxide particles is highly constrained by the size of the water droplets. Koutzarova et al. have used this approach, with n-hexanol as the oil phase and CTAB and n-butanol acting as surfactants, to prepare Fe_3O_4 nanoparticles having a narrow size distribution, with the mean diameter from 14 nm to 36 nm varying with the concentration of CTAB introduced.[173] Another approach to synthesizing iron oxide nanoparticles involves using two separate microemulsions, one containing the Fe ions in the aqueous solution and the other containing the alkaline medium required to precipitate the magnetite.[174] In this method, the iron oxide nanoparticles form when the two separate microemulsions mixed. Lopez Perez el al. have modified this approach to work at higher temperatures to produce iron oxide nanoparticles having a characteristic size of 2–8 nm that can be tuned by varying the synthesis procedure.[174]

While microemulsion offers the important benefits of a highly uniform size distribution for the iron oxide nanoparticles, and the ability to readily tune the average size by controlling the surfactant concentration in the microemulsion, this particular synthesis technique has a number of disadvantages as well. One of the most significant of these for practical applications is the difficulty in producing large quantities of nanoparticles by microemulsion. This makes microemulsion a desirable technique to use to investigate the fundamental properties of iron oxide nanoparticles, where a narrow size distribution is particularly important, but this approach may not scale satisfactorily for eventual biomedical applications. Together with the relatively low sample yield, some of the chemicals used in the synthesis can be expensive, particularly when used in large quantities, again making microemulsion a more suitable technique for preparing smaller batches of nanoparticles. Finally, the surfactants used in the synthesis may be difficult to remove from the nanoparticles, which could modify their cytotoxic properties.

10.19.4   Coating

In addition to the three techniques discussed above, there are a number of other approaches used for preparing iron oxide nanoparticles. Because each of these techniques has distinct advantages and disadvantages, the method for synthesis should be selected based on the specific needs of the study and available resources. In many cases, beyond simply synthesizing bare iron oxide nanoparticles, it is necessary to coat these particles with some surfactant or coating, whether to simply prevent agglomeration or to provide new functionalities. Iron oxide nanoparticles can also be embedded in larger structures to introduce a magnetic response into these other nanomaterials. However, in the following, we focus simply on core–shell structures with the core consisting of an iron oxide nanostructure and an organic or inorganic shell.

10.19.4.1   Organic coatings

Iron oxide nanoparticles coated with organic molecules or polymers are particularly attractive for many biomedical applications. These surfactants should be selected to be biocompatible and, in some cases, can introduce new functionalities. A very large number of different organic surfactants have been used to coat iron oxide nanoparticles, with each providing certain advantages. Fatty acids, in particular oleic acid, have been very widely used to coat iron oxide nanoparticles. These are short molecules, so increase the hydrodynamic diameter of the nanoparticle by a relatively modest amount, and are useful for making oil-soluble composites. Studies on iron oxide nanoparticles coated with lauric, myristic, and oleic acids find a systematic change in the hydrodynamic response of the nanoparticles with the surfactant chain length, suggesting that the physical properties of the nanoparticles can be selectively tuned by carefully varying the surfactant.[48]

A number of larger molecules or polymers are also widely used in coating iron oxide nanoparticles. Dextran[175,176] and chitosan[177-179] are naturally occurring polymers that have previously been used for other biomedical applications and have proven to be attractive candidates for applications involving iron oxide nanoparticles. Both of these materials are biocompatible and serve to stabilize the nanoparticles in solution. Furthermore, they provide binding sides for other molecules to functionalize the nanoparticle composite. Polyethylene glycol and PEO have been shown to mitigate the binding of blood plasma proteins to nanoparticles, which can increase circulation times, improving the potential of the nanoparticles for drug delivery.[180,181] Coating iron oxide nanoparticles with hyaluronic acid can improve their stability in solution and biocompatibility, while at the same time, potentially serving as a targeting agent.[182,183]

10.19.4.2 Inorganic coatings

Besides organic materials, there are a number of exciting possibilities that arise when coating iron oxide nanoparticles with inorganic materials. Gold[184-186] and silica[187,188] have been particularly widely studied. Synthesizing a gold surface layer over the iron oxide introduces a number of new functionalities. Gold chemistry is particularly well understood, having a great potential for attaching a range of different molecules and for self-assembly based on Au–thiol bonding. Furthermore, gold is a relatively inert material (although less so in nanoparticle form) and gold nanoparticles are believed to be non-cytotoxic.[189] Finally, coating the iron oxide nanoparticles with gold produces a strong absorption of infrared radiation, which, when combined with the magnetic properties arising from the iron oxide core, produces an extremely responsive composite with properties that can be tuned in vivo using distinctly different probes.

Coating iron oxide nanoparticles with silica also provides a number of benefits. Iron oxide/silica nanoparticles tend to exhibit excellent chemical stability and biocompatibility, with the silica shell serving to minimize nanoparticle agglomeration. The silica shell can also provide a better substrate for additional functionalization than the bare iron oxide nanoparticle. Beyond adding a single layer of an inorganic material on the iron oxide nanoparticles, more complicated geometries have also been investigated, including structures integrating CdSe quantum dots[190] and exchange-biased nanoparticles.[191,192]

10.20   Functional Groups

New functionalities can be introduced to the coated iron oxide nanoparticles by attaching specific targeting groups. For example, conjoining a fluorescent dye to the nanoparticle can allow it to be imaged using optical microscopy, which can, in turn, be used to investigate the distribution of nanoparticles inside a cell using confocal microscopy.[193,194] A variety of different targeting agents can also be attached to the nanoparticles. Tat peptide functionalized nanoparticles show a propensity to target the cell nucleus, while antibodies, or antibody fragments, can potentially target specific binding sites.[195–197]

The synthesis techniques highlighted in this section do not provide an exhaustive review of the methods used to prepare iron oxide nanoparticles. Since the size, morphology, and crystallinity of the nanoparticles is often highly sensitive to the synthesis details, it can be necessary to tune the process parameters to prepare nanoparticles to optimally match the required properties. When considering the design of iron oxide nanoparticles, it can be instructive to consider the interplay between the structural properties of the particles and their magnetic nanoparticles, as discussed previously. Furthermore, for biomedical applications, the response of the particles in vivo will be strongly affected by the surfactants used to coat the nanoparticles, and not just the iron oxide core itself. Preparing viable nanoparticle composites therefore requires a careful evaluation of the biochemical response of the particles beyond simply optimizing the magnetic properties.

10.21 Biomedical Applications

The physical properties of magnetic nanoparticles make them attractive candidates for inclusion in a wide range of biomedical applications. They integrate imaging functionality (through their characteristics as a contrast agent for magnetic resonance imaging) with targeting (through their response to external magnetic fields) and responsive thermal behavior (through magnetic hyperthermia) all combined with the innate properties of nanoparticles to penetrate cell membranes. This had led to a number of significant studies concerning potential applications for iron oxide nanoparticles, both in vivo and in vitro. A limited selection of these studies is presented in the following, with the objective to illustrate the breadth of applications of these materials.

10.21.1 Magnetic Hyperthermia

Although hyperthermia has a long history for treating cancer,[198] challenges associated with heating tumors to clinically relevant temperatures, typically taken as ~42°C, without damaging surrounding tissue have led to this treatment method being bypassed in favor of other approaches having less onerous side effects. Magnetic nanoparticles, however, offer a new approach to designing hyperthermia treatments. Alternating magnetic field themselves are generally believed to be non-harmful. These magnetic fields only deposit significant energy in regions containing magnetic nanoparticles, which means that the heating produced by these magnetic fields is highly localized to these regions, circumventing one of the major obstacles to developing hyperthermia as a viable treatment option.

One particularly exciting series of studies concerning the potential applications of iron oxide magnetic nanoparticles has been conducted by Jordan and co-workers.[46,199,200] In these clinical studies, patients with recurrent prostrate cancers were treated using magnetic hyperthermia, in conjunction with other treatment modalities. Ferrofluid samples containing well-characterized iron oxide nanoparticles were implanted in the tumors using different techniques, and heated using a custom-built system providing magnetic fields with an amplitude of up to 10 kA/m at frequencies of 100 kHz. The authors were able to achieve magnetic hyperthermia heating rates in the range 60–380 W/kg, producing a temperature of 40°C over a majority of the tumor. However, the 42°C coverage included only a small fraction of the tumor volume, leading to the conclusion that at least a modest increase in the magnetic heating would be required to improve the applicability of this technique to make this hyperthermia a viable approach to cancer treatment.[47]

These experiments have also provided some important background information on the feasibility of incorporating iron oxide nanoparticles into clinical treatments. Although nanoparticles were detected in the prostate 12 months after the hyperthermia treatment, there was no evidence for systemic toxicity in a time frame of up to two years.[201] Furthermore, long-term monitoring of the patients suggested that there were no continuing impairments to the quality of life. These studies offer promising, if somewhat sparse, evidence that there may be no fundamental limitations to using iron oxide magnetic nanoparticles for at least certain biomedical applications with regards to their long-term impact on patient health.

Moving beyond simple iron oxide nanoparticles, there have been some promising studies using a magnetic exchange bias, which arises at the interface between two magnetically dissimilar materials, in core–shell ferrite nanoparticles to dramatically enhance their hyperthermia response. Nanoparticles consisting of a Fe core surrounded by a Fe_3O_4 shell proved to be non-toxic, but produced a strong regression in tumors in mice when heated.[202] In another such study, magnetic nanoparticles having a $CoFe_2O_4$ core and $MnFe_2O_4$ shell were injected into tumors formed by grafting human brain cells onto mice, which were then exposed to an alternating magnetic field of 37 kA/m and 500 kHz to produce local heating.[203] Over the course of treatment, which ran for 18–26 days, the tumors treated with hyperthermia from these core–shell nanoparticles exhibited a more dramatic reduction than similar tumors treated with comparable concentrations of doxorubicin, confirming that magnetic nanoparticle hyperthermia may provide a viable approach to developing new techniques for treating cancer. While this particular study veered away from iron oxide nanoparticles, it is important to note that similar exchange bias effects have been recently observed in Fe_3O_4 core/ γ-Fe_2O_3 shell nanoparticles.[204] This suggests that core–shell iron oxide nanoparticles may also exhibit the enhanced hyperthermia desirable for clinical applications.

10.21.2   Controlled Drug Release

A particular advantage of iron oxide nanoparticles is the potential to use different characteristics of these materials to design multimodal imaging and treatment techniques. One promising example of this approach is to use the high surface area of the nanoparticles, typically in conjunction with a thermosensitive polymer, to obtain good drug loading, with the drug release triggered by local hyperthermia produced in the nanoparticles by an alternating magnetic field.[35,51,53] Such controlled release offers important practical advantages, including the ability to use magnetic resonance imaging to localize the nanoparticles in vivo before triggering the release. Furthermore, since magnetic fields can readily penetrate the entire body, this release can be triggered with the nanoparticles localized in any tissue.

The materials used for these studies often consist of iron oxide nanoparticles coated with a thermosensitive polymer, with poly(*N*-isopropylacrylamide) (PNIPAA) and poly-*n*-isopropylacrylamide (PNIPAM) being particularly widely used.[205-207] These polymers undergo conformal changes in solution near physiological temperatures, with the precise miscibility temperature often depending on the exact composition, which can then be tuned to fall slightly above body temperature. Some suitable compound, with anticancer drugs being widely investigated, is then loaded into the nanoparticle/polymer composite. The system is stable at low temperatures, but as the temperature is increased above the lower critical solution temperature of the composite, which is triggered by applying an alternating magnetic field, the conformal change in the polymer causes the drug to be ejected.

Investigations of PNIPAM/iron oxide nanoparticle composites have demonstrated the efficacy of this approach through in vitro studies. Kee et al. have shown a 41% release of doxorubicin after 101 h at 42°C.[208] Significantly, they also demonstrated a 14.7% release in vitro after treating the sample to magnetic hyperthermia treatment for 47 min. However, the release rate showed only a weak dependence on the lower critical solution temperature associated with the conformal change of the polymer. This observation is similar to results obtained by Regmi et al., who found a dramatically increased release rate of mitoxantrone, reaching 1%/min, for a PNIPAM/iron oxide composite heated by alternating magnetic fields as compared to heating in a water bath.[35] For both of these heating protocols, the release rate did not show any significant change at the lower critical solution temperature for PNIPAM, implying that other mechanisms beyond the polymer conformation must control the release dynamics.

Other studies have explored alternate methods for using iron oxide nanoparticles to trigger drug release with approaches that might prove compatible with certain clinical restrictions. Tai et al. have used iron oxide nanoparticles inserted in a thermosensitive liposome structure to allow drug release controlled by an external magnetic field without significantly increasing the ambient temperature.[53] This approach was pursued to allow for magnetically controlled release of the drug into tissue that may be adversely affected by even a relatively small amount of heating. The authors used carboxylfluorescein as a model drug and demonstrated the controlled release from the liposome-iron oxide composite by an alternating magnetic field with the sample injected into a rat forearm muscle. Liu et al. have demonstrated a "burst" response for magnetically controlled drug release based on irreversible disruption of polymer nanocapsules triggered by extreme heating of iron oxide nanoparticles.[209] In this study, the release rate for a model drug from the polymer composite produced by magnetic hyperthermia, with roughly 40% cumulative release in the first 5 min, was a factor of 20 larger than the release produced by heating in a water bath. The authors attributed this dramatic increase in the release rate to very high temperatures very near the iron oxide nanoparticles during magnetic heating leading to the disruption of the polymer nanocapsules. This versatility in using the novel properties of iron oxide nanoparticles to remotely trigger drug release in vivo is expected to allow for new clinical applications requiring more precise control over the local dosage.

10.21.3 Magnetic Resonance Imaging

Iron oxide nanoparticles have also been studied extensively as contrast agents for magnetic resonance imaging applications. Targeting iron oxide nanoparticles to tumor sites, for example, can allow for diagnostic imaging. One particularly fruitful approach has been to exploit the activity of the RES cells to iron oxide nanoparticles.[64] Because healthy Kupffer cells in the liver will readily uptake iron oxide nanoparticles while some cancerous liver cells do not, MR images taken after the injection of iron oxide nanoparticles can be used to distinguish tumors in the liver. This follows because healthy cells are darker in T_2-weighted images. Similar effects hold true for the spleen as well, leading to high-resolution images that allow for increased identification of lesions compared to other imaging techniques. Furthermore, since iron oxide nanoparticles typically accumulate in the liver and spleen, passive targeting is sufficient to allow for imaging of tumors in these organs.

Using iron oxide nanoparticles to enhance MR images in other organs typically requires some additional targeting agent. Since folate receptors are overexpressed on the surface of many types of cancer cells, functionalizing iron oxide nanoparticles with folic acid can serve to selectively target cancerous tumors. This approach was used by Choi et al. to target nasopharyngeal epidermal carcinoma cells in a mouse model.[210] They demonstrated that the tumor signal decreased by 38% in T_2-weighted images, which was significantly larger than the decrease in surrounding muscle tissue. Transferrin receptors can also be overexpressed on the surface of cancer cells, leading to the possibility of targeting iron oxide nanoparticles on functionalizing with transferrin. In vivo studies on rats using transferrin functionalized iron oxide nanoparticle to target SMT/ SA mammary gland carcinoma cells showed a correlation between the reduction in the tumors T_2 relaxation time and the tissues transferrin levels. Monoclonal antibodies also have the capability to specifically target cancerous tumors, and iron oxide nanoparticles conjugate with these targeting ligands have been shown to offer contrast enhancement in MR imaging on mouse and rat models.[211,212] However, these antibodies are generally rather large, which can limit the number of antibodies that can be attached to each nanoparticle and can dramatically increase the hydrodynamic size of the composite nanomaterial.

More recently, iron oxide nanoparticles have been proposed for use in dual-mode imaging techniques, combining terahertz molecular imaging (TMI) with conventional MRI.[213] Because the terahartz excitations interact strongly with vibrational modes in water, TMI can be used to distinguish cancer cells by their higher water content. Part et al. have used human ovarian cancer cells (SKOV3) both in vitro and in mouse models to characterize the response of the iron oxide nanoparticles to both terahertz and magnetic resonance imaging techniques. The two techniques both allowed for good imaging of the cancer cells containing magnetic nanoparticles in vitro and in vivo, suggesting the possibility that using iron oxide nanoparticles to facilitate such multimodal imaging could lead to improved diagnostic approaches for the early detection of cancer.

10.21.4 Gene Delivery

Nanoparticles are considered very valuable for facilitating targeted delivery of genes into tissues and cells. In gene delivery technique a gene of interest is introduced to express its encoded protein in a suitable host or a cell. The gene delivery systems that are currently employed are viral vectors such as retroviruses and adenoviruses,[214-217] nucleic acid electroporation and nucleic acid transfection.[33,218,219] The efficacy of the three techniques varies from 80–90% in the case of viral vector method to 20–30% for transfection methods. The transfection method, while not highly efficient, maintains cell viability for medical applications as compared to the viral vector method, which may insert viral vectors in nucleic acid sequence in the host genome causing potentially unwanted side effects. The process of electroporation is also highly efficient (50–70%); however, half of the recipient cells die due to the electrical stimulation. Magnetic nanoparticles have been used in a number studies to increase the transfection efficiencies of cultured cells. Magnetic nanoparticles and nucleic acid complexes are added to the cell culture media and then onto the cell surface by the application of an external magnetic field. Magnetic oxide nanoparticles such as MFe_2O_4, where M = Co, Ni, Mn exhibit better properties as compared to other magnetic materials for this application but these nanoparticles are highly toxic for the cells and expected to have only limited use in vivo applications.[220,221]

10.21.5 Nanoparticle Endocytosis

In order for iron oxide nanoparticles to serve as drug delivery platforms or function for many of the other applications discussed previously, they must pass through the cell membrane by endocytosis. There have been a large number of studies on the specific mechanisms for endocytosis of nanoparticles. It has been found that magnetic nanoparticles, when driven by an external magnetic field, can alter the functions of a cell promoting transport.[22,222] Several investigations have been made to understand the role of size, geometry of the magnetic nanoparticles, and membrane elasticity on the endocytosis rate[223-225] However, the direct effects of a magnetic field on the endocytosis of magnetic nanoparticles on the organization of cytoskeletal network and the membrane elasticity remains unclear. Two recent studies have reported a new aspect of the magnetomechanical response of cell membrane.[226,227] It was found that a low frequency magnetic field applied to a biofunctionalized magnetic microdisk attached to a cell membrane will produce oscillations in the disk. These vibrations transmit a mechanical force, roughly 20–40 pN, to the cell membrane causing cell death.

Another investigation by Furlani and Ng also serves to clarify the transport mechanisms. The authors propose a model to predict transport of magnetic nanoparticles in a passive magnetophoretic system to obtain an insight into the physics of the particle transport at the nanoscale.[228] Experimentally, McBain et al. developed an oscillating magnetic array system that provided lateral motion to the gene/magnetic nanoparticle complex to promote transfection.[229] Battarai developed a system consisting of hexanoyal chloride modified chitosan stabilized iron oxide nanoparticles for viral gene delivery for high efficiency delivery in both in vivo and in vitro systems.[230] In another important investigation, Kamaei et al. studied transfection using gold/iron oxide magnetic (GoldMAN) nanoparticles. It was shown that the adenovirus gene delivery vector (Ad) was clearly immobilized on GoldMAN and the Ad/ GoldMAN complex was introduced into the cell by an external magnetic field which increased gene expression by a factor of 1000 as compared that of Ad alone.[231] While there remains considerable work to be done to understand the details of endocytosis of iron oxide nanoparticles, these studies suggest that transport through the cell membrane may be strongly affected or influenced by the magnetic properties of the particles.

10.22 Conclusions

In summary, the size-dependent physical properties of magnetic nanoparticles allow a number of biomedical applications. These fundamental magnetic properties, including the saturation magnetization, magnetic interaction, superparamagnetism, and relaxation mechanisms dominate the magnetic behavior in ultrasmall iron oxide particles, and having a good understanding of these properties is important for the rational design of biomedical applications. A number of other concepts relevant for magnetic nanoparticles, including ac susceptibility, magnetohydrodynamics of ferrofluids including the light scattering experiments, role of relaxation mechanisms in hyperthermia and heat dissipation, specific absorption rate and magnetic resonance imaging, are also important for translating between the properties of nanoparticles and their potential applications. Different nanoparticle properties, including shape and size, and therefore the magnetic properties, of the nanoparticles are sensitive to the specific technique used to prepare the samples. Depending on the potential application, different techniques such as co-precipitation, hydrothermal process, and micro emulsion may be suitable for preparing the magnetic nanoparticles. The functionality also plays an important role in determining the interaction of the nanoparticle in a given environment; these properties are strongly affected by different coatings of the magnetic nanoparticles, including organic, dextran, chitosan, and polyethylene glycol, and inorganic such as gold and silica coatings. Iron oxide nanoparticles have already been explored in detail for a number of biomedical applications, including magnetic hyperthermia, controlled drug delivery, magnetic resonance imaging, and gene delivery. The possibility of understanding the size dependency, synthesis and consequently the physical properties of the magnetic nanoparticles, opens a new era in the frontiers of biomedical sciences.

Acknowledgements

We acknowledge many helpful and insightful conversations and contributions from colleagues and collaborators through the years, including Elizabeth Buc, Sudakar Chandran, Norman Cheng, Ambesh Dixit, Gali Hillman, Bhanu Jena, Suvra Laha, Vaman Naik, Humeshkar Nemala, David Oupicky, Jayanth Panyam, Maheshika Perera, Rajesh Regmi, Ronald Tackett, and Cornel Rablau. Support for this work has been provided by the donors of the American Chemical Society (PRF No. 46160-G10), by the Wayne State University OVPR, and by the Richard Barber Foundation.

References

1. Molenbroek, A. M., Nrskov, J. K., and Clausen, B. S. (2001). Structure and reactivity of Ni-Au nanoparticle catalysts. *J Phys Chem B*, **105**, 5450–5458.

2. Wang, S., Zhang, M., and Zhang, W. (2011). Yolk-shell catalyst of single Au nanoparticle encapsulated within hollow mesoporous silica microspheres. *ACS Catal*, **1**, 207–211.

3. Jin-nyeong, J., Hong-Gi, L., and Yeon-Tae, Y. (2011). Size effect of Au nanoparticle on electrocatalytic activity of Pt-Au/C composite catalysts for methanol oxidation. *Electrochem Solid-State Lett*, **14**, 89–91.

4. Chen, H.-T., Chang, J.-G., Ju, S.-P., and Chen, H.-L. (2010). Ethylene epoxidation on a Au nanoparticle versus a Au(111) surface: A DFT study. *J Phys Chem Lett*, **1**, 739–742.

5. Aragon, F. H., De Souza, P. E. N., Coaquira, J. A. H., Hidalgo, P., and Gouvea, D. (2012). Spin-glass-like behavior of uncompensated surface spins in NiO nanoparticulated powder. *Phys B: Condensed Matter*, **407**, 2601–2605.

6. Coey, J. M. D., Stamenov, P., Gunning, R. D., Venkatesan, M., and Paul, K. (2010). Ferromagnetism in defect-ridden oxides and related materials. *New J Phys*, **12**, 053025.

7. Coey, J. M. D., Wongsaprom, K., Alaria, J., and Venkatesan, M. (2008). Charge-transfer ferromagnetism in oxide nanoparticles. *J Phys D: Appl Phys*, **41**, 134012.

8. Sundaresan, A., and Rao, C. N. R. (2007). Ferromagnetism as a universal feature of inorganic nanoparticles. *Nano Today*, **4**, 96–106.

9. Gellman, A. J., and Shukla, N. (2009). Nanocatalysis: More than speed. *Nat Mater*, **8**, 87–88.

10. Lee, I., Delbecq, F., Morales, R., Albiter, M. A., and Zaera, F. (2009). Tuning selectivity in catalysis by controlling particle shape. *Nat Mater*, **8**, 132–138.

11. Serpell, C. J., Cookson, J., Ozkaya, D., and Beer, P. D. (2011). Core@ shell bimetallic nanoparticle synthesis via anion coordination. *Nat Chem*, **3**, 478–483.

12. Bowker, M. (2009). A prospective: Surface science and catalysis at the nanoscale. *Surf Sci*, **603**, 2359–2362.

13. Ranganath, K. V. S., and Glorius, F. (2011). Superparamagnetic nanoparticles for asymmetric catalysis—A perfect match. *Catal Sci Technol*, **1**, 13–22.

14. Roca, A. G., et al. (2009). Magnetite nanoparticles with no surface spin canting. *J Appl Phys*, **105**, 114309.

15. Peddis, D., Cannas, C., Piccaluga, G., Agostinelli, E., and Fiorani, D. (2010). Spin-glass-like freezing and enhanced magnetization in ultra-small $CoFe_2O_4$ nanoparticles. *Nanotechnology*, **21**, 125705–125710.

16. Obaidat, I. M., Mohite, V., Issa, B., Tit, N., and Haik, Y. (2009). Predicting a major role of surface spins in the magnetic properties of ferrite nanoparticles. *Crystal Res Technol*, **44**, 489–494.

17. Eun Jung, C., et al. (2003). Superparamagnetic relaxation in $CoFe_2O_4$ nanoparticles. *J Magn Magn Mater*, **262**, 198–202.

18. Jain, T. K., et al. (2008). Magnetic nanoparticles with dual functional properties: drug delivery and magnetic resonance imaging. *Biomaterials*, **29**, 4012–4021.

19. Laconte, L., Nitin, N., and Bao, G. (2005). Magnetic nanoparticle probes. *Mater Today*, **8**, 32–38.

20. Leslie-Pelecky, D. L., and Rieke, R. D. (1996). Magnetic properties of nanostructured materials. *Chem Mater*, **8**, 1770–1770.

21. O'Grady, K. (2009). Editorial: Progress in applications of magnetic nanoparticles in biomedicine. *J Phys D: Appl Phys*, **42**, 220301.

22. Pankhurst, Q. A., Thanh, N. K. T., Jones, S. K., and Dobson, J. (2009). Progress in applications of magnetic nanoparticles in biomedicine. *J Phys D: Appl Phys*, **42**, 224001–224015.

23. Wu, A., Ou, P., and Zeng, L. (2010). Biomedical applications of magnetic nanoparticles. *Nano*, **5**, 245–270.

24. Pankhurst, Q. A., Connolly, J., Jones, S. K., and Dobson, J. (2003). Applications of magnetic nanoparticles in biomedicine. *J Phys D (Appl Phys)*, **36**, 167–181.

25. Chertok, B., et al. (2008). Iron oxide nanoparticles as a drug delivery vehicle for MRI monitored magnetic targeting of brain tumors. *Biomaterials*, **29**, 487–496.

26. Hu, J., Qian, Y., Wang, X., Liu, T., and Liu, S. (2012). Drug-loaded and superparamagnetic iron oxide nanoparticle surface-embedded amphiphilic block copolymer micelles for integrated chemotherapeutic drug delivery and MR imaging. *Langmuir*, **28**, 2073–2082.

27. Kayal, S., and Ramanujan, R. V. (2010). Doxorubicin loaded PVA coated iron oxide nanoparticles for targeted drug delivery. *Mater Sci Eng C*, **30**, 484–490.

28. Mahmoudi, M., Simchi, A., Imani, M., and Hafeli, U. O. (2009). Superparamagnetic iron oxide nanoparticles with rigid cross-linked polyethylene glycol fumarate coating for application in imaging and drug delivery. *J Phys Chem C*, **113**, 8124–8131.

29. Ruiz-Hernandez, E., Baeza, A., and Vallet-Regi, M. (2011). Smart drug delivery through DNA/magnetic nanoparticle gates. *ACS Nano*, **5**, 1259–1266.

30. Xu, H., et al. (2011). Polymer encapsulated upconversion nanoparticle/iron oxide nanocomposites for multimodal imaging and magnetic targeted drug delivery. *Biomaterials*, **32**, 9364–9373.

31. Chao-Bin, C., Ji-Yao, C., and Wen-Chien, L. (2009). Fast transfection of mammalian cells using superparamagnetic nanoparticles under strong magnetic field. *J Nanosci Nanotechnol*, **9**, 2651–2659.

32. McBain, S. C., Yiu, H. H. P., Haj, A. E., and Dobson, J. (2007). Polyethyleneimine functionalized iron oxide nanoparticles as agents for DNA delivery and transfection. *J Mater Chem*, **17**, 2561–2565.

33. Ran, N., et al. (2010). Hybrid superparamagnetic iron oxide nanoparticle-branched polyethylenimine magnetoplexes for gene transfection of vascular endothelial cells. *Biomaterials*, **31**, 4204–4213.

34. Sudakar, C., et al. (2008). Fe_3O_4 incorporated AOT-alginate nanoparticles for drug delivery. *IEEE Trans Magn*, **44**, 2800–2803.

35. Regmi, R., et al. (2010). Hyperthermia controlled rapid drug release from thermosensitive magnetic microgels. *J Mater Chem*, **20**, 6158–6163.

36. Shen, C.-R., et al. (2011). Characterization of quaternized chitosan-stabilized iron oxide nanoparticles as a novel potential magnetic resonance imaging contrast agent for cell tracking. *Polymer Int*, **60**, 945–950.

37. Tromsdorf, U. I., Bruns, O. T., Salmen, S. C., Beisiegel, U., and Weller, H. (2009). A highly effective, nontoxic T_1 MR contrast agent based on ultrasmall PEGylated iron oxide nanoparticles. *Nano Lett*, **9**, 4434–4440.

38. Yong Il, P., et al. (2011). Transformation of hydrophobic iron oxide nanoparticles to hydrophilic and biocompatible maghemite nanocrystals for use as highly efficient MRI contrast agent. *J Mater Chem*, **21**, 11472–11477.

39. Tsai, Z.-T., et al. (2010). In situ preparation of high relaxivity iron oxide nanoparticles by coating with chitosan: a potential MRI contrast agent useful for cell tracking. *J Magn Magn Mater*, **322**, 208–213.

40. Baker, I., Qi, Z., Weidong, L., and Sullivan, C. R. (2006). Heat deposition in iron oxide and iron nanoparticles for localized hyperthermia. *J Appl Phys*, **99**, 8–106.

41. Guandong, Z., Yifeng, L., and Baker, I. (2010). Surface engineering of core/shell iron/iron oxide nanoparticles from microemulsions for hyperthermia. *Mater Sci Eng C (Mater Biol Appl)*, **30**, 92–97.

42. Hayashi, K., Shimizu, T., Asano, H., Sakamoto, W., and Yogo, T. (2008). Synthesis of spinel iron oxide nanoparticle/organic hybrid for hyperthermia. *J Mater Res*, **23**, 3415–3424.

43. Hergt, R., Dutz, S., Muller, R., and Zeisberger, M. (2006). Magnetic particle hyperthermia: nanoparticle magnetism and materials development for cancer therapy. *J Phys Condensed Matter*, **18**, 2919–2934.

44. Kita, E., et al. (2010). Heating characteristics of ferromagnetic iron oxide nanoparticles for magnetic hyperthermia. *J Appl Phys*, **107**, 09–321.

45. Laurent, S., Dutz, S., Hafeli, U. O., and Mahmoudi, M. (2011). Magnetic fluid hyperthermia: focus on superparamagnetic iron oxide nanoparticles. *Adv Colloid Interface Sci*, **166**, 8–23.

46. Jordan, A., Scholz, R., Wust, P., Faehling, H., and Felix, R. (1999). Magnetic fluid hyperthermia (MFH): cancer treatment with AC magnetic field induced excitation of biocompatible superparamagnetic nanoparticles. *J Magn Magn Mater*, **201**, 413–419.

47. Wust, P., et al. (2006). Magnetic nanoparticles for interstitial thermotherapy-feasibility, tolerance and achieved temperatures. *Int J Hyperthermia*, **22**, 673–685.

48. Regmi, R., et al. (2009). Effects of fatty acid surfactants on the magnetic and magnetohydrodynamic properties of ferrofluids. *J Appl Phys*, **106**, 119302.

49. Hergt, R., et al. (2004). Enhancement of AC-losses of magnetic nanoparticles for heating applications. *J Magn Magn Mater*, **280**, 358–368.

50. Hayashi, K., et al. (2010). High-frequency, magnetic-field-responsive drug release from magnetic nanoparticle/organic hybrid based on hyperthermic effect. *ACS Appl Mater Interfaces*, **2**, 1903–1911.

51. Lien, Y.-H., Wu, T.-M., Wu, J.-H., and Liao, J.-W. (2011). Cytotoxicity and drug release behavior of PNIPAM grafted on silica-coated iron oxide nanoparticles. *J Nanoparticle Res*, **13**, 5065–5075.

52. Yue, Z.-G., et al. (2011). Iron oxide nanotubes for magnetically guided delivery and pH-activated release of insoluble anticancer drugs. *Adv Funct Mater*, **21**, 3446–3453.

53. Lin-Ai, T., et al. (2009). Thermosensitive liposomes entrapping iron oxide nanoparticles for controllable drug release. *Nanotechnology*, **20**, 135101–135109.

54. Lin, M. M., Kim, D. K., El Haj, A. J., and Dobson, J. (2008). Development of superparamagnetic iron oxide nanoparticles (SPIONS) for translation to clinical applications. *IEEE Trans Nanobiosci*, **7**, 298–305.

55. Tong, S., Hou, S., Zheng, Z., Zhou, J., and Bao, G. (2010). Coating optimization of superparamagnetic iron oxide nanoparticles for high T_2 relaxivity. *Nano Lett*, **10**, 4607–4613.

56. Majewski, P., and Thierry, B. (2007). Functionalized magnetite nanoparticles: synthesis, properties, and bio-applications. *Crit Rev Solid State Mater Sci*, **32**, 203.

57. Koo, O. M., Rubinstein, I., and Onyuksel, H. (2005). Role of nanotechnology in targeted drug delivery and imaging: a concise review. *Nanomed Nanotechnol, Biol Med*, **1**, 193–212.

58. Arruebo, M., Fernandez-Pacheco, R., Ibarra, M. R., and Santamaria, J. (2007). Magnetic nanoparticles for drug delivery. *Nano Today*, **2**, 22–32.

59. Namdeo, M., et al. (2008). Magnetic nanoparticles for drug delivery applications. *J Nanosci Nanotechnol*, **8**, 3247–3271.

60. Farokhzad, O. C., and Langer, R. (2009). Impact of nanotechnology on drug delivery. *ACS Nano*, **3**, 16–20.

61. Huang, S.-H., and Juang, R.-S. (2011). Biochemical and biomedical applications of multifunctional magnetic nanoparticles: a review. *J Nanoparticle Res*, **13**, 4411–4430.

62. Kobayashi, T. (2011). Cancer hyperthermia using magnetic nano-particles. *Biotechnol J*, **6**, 1342–1347.

63. Thiesen, B., and Jordan, A. (2008). Clinical applications of magnetic nanoparticles for hyperthermia. *Int J Hyperthermia*, **24**, 467–474.

64. Rosen, J. E., Chan, L., Shieh, D.-B., and Gu, F. X. (2012). Iron oxide nanoparticles for targeted cancer imaging and diagnostics. *Nanomed Nanotechnol, Biol, Med*, **8**, 275–290.

65. Corr, S. A., Rakovich, Y. P., and Gun'Ko, Y. K. (2008). Multifunctional magnetic-fluorescent nanocomposites for biomedical applications. *Nanoscale Res Lett*, **3**, 87–104.

66. Gupta, A. K., and Gupta, M. (2005). Synthesis and surface engineering of iron oxide nanoparticles for biomedical applications. *Biomaterials*, **26**, 3995–4021.

67. Ashcroft, N., and Mermin, D. (1976), *Solid State Physics*, Thomson Learning, New York.

68. Blundell, S. (2001). *Magnetism in Condensed Matter* (Oxford Master Series in Physics), Oxford University Press, USA.

69. Geertsma, W., and Khomskii, D. (1996). Influence of side groups on 90 superexchange: a modification of the Goodenough–Kanamori–Anderson rules. *Phys Rev B*, **54**, 3011–3014.

70. Weihe, H., and Guedel, H. U. (1997). Quantitative interpretation of the Goodenough-Kanamori rules: a critical analysis. *Inorg Chem*, **36**, 3632–3632.

71. Danan, H., Herr, A., and Meyer, A. J. P. (1968). New determinations of saturation magnetization of nickel and iron. *J Appl Phys*, **39**, 669–670.

72. Liu, Q., Barron, V., Torrent, J., Qin, H., and Yu, Y. (2010). The magnetism of micro-sized hematite explained. *Phys Earth Planetary Interiors*, **183**, 3–4, 387–397.

73. Calhoun, B. A. (1954). Magnetic and electric properties of magnetite at low temperatures. *Phys Rev*, **94**, 1577–1585.

74. Kakol, Z., and Honig, J. M. (1989). Influence of deviations from ideal stoichiometry on the anisotropy parameters of magnetite $Fe_{3(1-\delta)}O_4$. *Phys Rev B*, **40**, 9090–9097.

75. Xu, S., and Merrill, R. T. (1992). Stress, grain size and magnetic stability of magnetite. *J Geophys Res*, **97**, 4321–4329.

76. Hodych, J. P. (1982). Magnetostrictive control of coercive force in multidomain magnetite. *Nature*, **298**, 542–544.

77. Dunlop, D. J., Song, X., and Heider, F. (2004). Alternating field demagnetization, single-domain-like memory, and the Lowrie–Fuller test of multidomain magnetite grains (0.6–356 m). *J Geophys Res*, **109**, 10, Doi: 10.1029/2004JB003006.

78. Worm, H. U., Ryan, P. J., and Banerjee, S. K. (1991). Domain size, closure domains, and the importance of magnetostriction in magnetite. *Earth Planetary Sci Lett*, **102**, 71–78.

79. Banerjee, S. K., and Moskowitz, B. M. (1985) in *Magnetite Biomineralization and Magnetoreception in Organisms: A New Magnetism* (ed J. L. Kirschvink), Plenum Publishing, New York.

80. Wensheng, L., Yuhua, S., Anjian, X., and Weiqiang, Z. (2010). Green synthesis and characterization of superparamagnetic Fe_3O_4 nanoparticles. *J Magn Magn Mater*, **322**, 1828–1833.

81. Chao, H., et al. (2008). Large-scale Fe_3O_4 nanoparticles soluble in water synthesized by a facile method. *J Phys Chem C*, **112**, 11336–11339.

82. Zhao, S., and Asuha, S. (2010). One-pot synthesis of magnetite nanopowder and their magnetic properties. *Powder Technol*, **197**, 295–297.

83. Pei, W., Kumada, H., Natsume, T., Saito, H., and Ishio, S. (2007). Study on magnetite nanoparticles synthesized by chemical method. *J Magn Magn Mater*, **310**, 2375–2377.

84. Parekh, K., Upadhyay, R. V., and Aswal, V. K. (2009). Monodispersed superparamagnetic Fe_3O_4 nanoparticles: synthesis and characterization. *J Nanosci Nanotechnol*, **9**, 2104–2110.

85. Ali-zade, R. A. (2006). Magnetite nanoparticles with an antiferromagnetic surface layer. *Inorg Mater*, **42**, 1215–1221.

86. Ozkaya, T., et al. (2009). Synthesis of Fe_3O_4 nanoparticles at 100°C and its magnetic characterization. *J Alloys Compounds*, **472**, 18–23.

87. Aquino, R., et al. (2005). Magnetization temperature dependence and freezing of surface spins in magnetic fluids based on ferrite nanoparticles. *Phys Rev B*, **72**, 184435–184431.

88. Nadeem, K., Krenn, H., Traussing, T., and Letofsky-Papst, I. (2011). Distinguishing magnetic blocking and surface spin-glass freezing in nickel ferrite nanoparticl. *J Appl Phys*, **109**, 013912.

89. Alves, C. R., et al. (2006). Surface spin freezing of ferrite nanoparticles evidenced by magnetization measurements. *J Appl Phys*, **99**, 8–905.

90. Tackett, R. J., Bhuiya, A. W., and Botez, C. E. (2009). Dynamic susceptibility evidence of surface spin freezing in ultrafine $NiFe_2O_4$ nanoparticles. *Nanotechnology*, **20**, 445705 (445707 pp.).

91. Cador, O., Grasset, F., Haneda, H., and Etourneau, J. (2004). Memory effect and super-spin-glass ordering in an aggregated nanoparticle sample. *J Magn Magn Mater*, **268**, 232–236.

92. Tamura, I., and Mizushima, T. (2002). Explanation for magnetic properties of interacting iron oxide nanocrystals. *J Magn Magn Mater*, **250**, 241–248.

93. Jonsson, T., et al. (1995). Aging in a magnetic particle system. *Phys Rev Lett*, **75**, 4138–4141.

94. Jonsson, T., Svedlindh, P., and Hansen, M. F. (1998). Static scaling on an interacting magnetic nanoparticle system. *Phys Rev Lett*, **81**, 3976–3979.

95. Zabenkin, V. N., Axelrod, L. A., Gordeev, G. P., Kraan, W. H., Lazebnik, I. M., Vorobiev, A. A. (2004). Effect of thermomagnetic treatment on the magnetic state of a ferrofluid: A polarised neutron study, *Phys B: Condensed Matter*, **350**, e211–e215.

96. Neel, L. (1949). Théorie du traînage magnétique des ferromagnétiques en grains fins avec application aux terres cuites. *Ann de Géophysique*, **5**, 99–136.

97. Brown, W. F., Jr. (1963). Thermal fluctuations of a single-domain particle. *Phys Rev*, **130**, 1677–1686.

98. Stoner, E. C., and Wohlfarth, E. P. (1991). A mechanism of magnetic hysteresis in heterogeneous alloys. *IEEE Trans Magn*, **27**, 3475–3518.

99. Feygenson, M., et al. (2011). Low-energy magnetic excitations in Co/CoO core/shell nanoparticles. *Phys Rev B*, **83**, 174414.

100. De Biasi, E., Zysler, R. D., Ramos, C. A., and Knobel, M. (2008). A new model to describe the crossover from superparamagnetic to blocked magnetic nanoparticles. *J Magn Magn Mater*, **320**, 312–315.

101. Bronstein, L. M., et al. (2007). Influence of iron oleate complex structure on iron oxide nanoparticle formation. *Chem Mater*, **19**, 3624–3632.

102. Forge, D., et al. (2010). Development of magnetic chromatography to sort polydisperse nanoparticles in ferrofluids. *Contrast Media Mol Imaging*, **5**, 126–132.

103. Gonzales-Weimuller, M., Zeisberger, M., and Krishnan, K. M. (2009). Size-dependant heating rates of iron oxide nanoparticles for magnetic fluid hyperthermia. *J Magn Magn Mater*, **321**, 1947–1950.

104. Caizer, C., Savii, C., and Popovici, M. (2003). Magnetic behaviour of iron oxide nanoparticles dispersed in a silica matrix. *Mater Sci Eng B*, **B97**, 129–134.

105. Demortiere, A., et al. (2011). Size-dependent properties of magnetic iron oxide nanocrystals. *Nanoscale*, **3**, 225–232.

106. Bradbury, A., Menear, S., and Chantrell, R. W. (1985) in *Proceedings of the International Conference on Magnetism: ICM '85, 26–30 August 1985*. 745–746.

107. Azeggagh, M., and Kachkachi, H. (2007). Effects of dipolar interactions on the zero-field-cooled magnetization of a nanoparticle assembly. *Phys Rev B*, **75**, 174410–174411.

108. Chakraverty, S., et al. (2005). Memory in a magnetic nanoparticle system: polydispersity and interaction effects. *Phys Rev B*, **71**, 54401–54401.

109. Wernsdorfer, W., et al. (1997). Experimental Evidence of the Neel–Brown Model of Magnetization Reversal. *Phys Rev Lett*, **78**, 1791–1794.

110. Mamiya, H., and Nakatani, I. (1997) (Elsevier) in *International Conference on Magnetism 1997, 27 July–1 Aug. 1997*. 966–967.

111. Dormann, J. L., Fiorani, D., and El Yamani, M. (1987). Field dependence of the blocking temperature in the superparamagnetic model: H2/3 coincidence. *Phys Lett A*, **120**, 95–99.

112. de Almeida, J. R. L., and Thouless, D. J. (1978). Stability of the Sherrington–Kirkpatrick solution of a spin glass model. *J Phys A*, **11**, 983–990.

113. Kneller, E. F., and Luborsky, F. E. (1963). Particle size dependence of coercivity and remanence of single-domain particles. *J Appl Phys*, **34**, 656–658.

114. Singh, V., Seehra, M. S., and Bonevich, J. (2009). AC susceptibility studies of magnetic relaxation in nanoparticles of Ni dispersed in silica. *J Appl Phys*, **105**, 07B518.

115. Tackett, R. J., et al. (2010). Evidence of low-temperature superparamagnetism in Mn_3O_4 nanoparticle ensembles. *Nanotechnology*, **21**, 365703–365708.

116. Vouille, C., Thiaville, A., Miltat, J. (2003). Thermally activated switching of nanoparticles: A numerical study", *J Magn Magn Mater*, **272–276**, e1237–e1238.

117. Sharma, S. K., et al. (2007). Magnetic study of $Mg_{0.95}Mn_{0.05}Fe_2O_4$ ferrite nanoparticles. *Solid State Commun*, **141**, 203–208.

118. Dormann, J. L., Bessais, L., and Fiorani, D. (1988). A dynamic study of small interacting particles: superparamagnetic model and spin-glass laws. *J Phys C*, **21**, 2015–2034.

119. Williams, D. N., Gold, K. A., Holoman, T. R. P., Ehrman, S. H., and Wilson, O. C., Jr. (2006). Surface modification of magnetic nanoparticles using gum arabic. *J Nanoparticle Res*, **8**, 749–753.

120. Jianzhong, H., Liyi, S., Xin, F., and Lin, X. (2009). Electrostatic and electrosteric stabilization of aqueous suspensions of barite nanoparticles. *Powder Technol*, **192**, 166–170.

121. Jingying, W., et al. (2009). Solvothermal synthesis and magnetic properties of size-controlled nickel ferrite nanoparticles. *J Alloys Compounds*, **479**, 791–796.

122. Pecora, R. (2000). Dynamic light scattering measurement of nanometer particles in liquids. *J Nanoparticle Res*, **2**, 123–131.

123. Chen, Z. P., et al. (2008). Preparation and characterization of water-soluble monodisperse magnetic iron oxide nanoparticles via surface double-exchange with DMSA. *Colloids Surf A*, **316**, 210–216.

124. Lopez-Cruz, A., Barrera, C., Calero-DdelC, V. L., and Rinaldi, C. (2009). Water dispersible iron oxide nanoparticles coated with covalently linked chitosan. *J Mater Chem*, **19**, 6870–6876.

125. Alfven, H. (1942). Existence of electromagnetic-hydrodynamic waves. *Nature*, **150**, 405–406.

126. Chandrasekhar, S. (1961). *Hydrodynamic and Hydromagnetic Stability*, Oxford University Press, Oxford.

127. Ivanov, A. O., Kantorovich, S. S., Mendelev, V. S., and Pyanzina, E. S. (2006). Ferrofluid aggregation in chains under the influence of a magnetic field. *J Magn Magn Mater*, **300**, 206–209.

128. Mendelev, V. S., and Ivanov, A. O. (2004). Ferrofluid aggregation in chains under the influence of a magnetic field. *Phys Rev E*, **70**, 51502–51501.

129. Patel, R., and Chudasama, B. (2009). Hydrodynamics of chains in ferrofluid-based magnetorheological fluids under rotating magnetic field. *Phys Rev E*, **80**, 012401 (012404 pp.).

130. Kruse, T., Krauthauser, H. G., Spanoudaki, A., and Pelster, R. (2003). Agglomeration and chain formation in ferrofluids: two-dimensional x-ray scattering. *Phys Rev B*, **67**, 94206–94201.

131. Zhang, Q., Wang, J., and Zhu, H. (1996). Three-dimensional dynamic configuration simulation of ferrofluid based on a model of interaction among pre-existing aggregates. *J Mater Sci Lett*, **15**, 2101–2102.

132. Rablau, C., et al. (2008). Magnetic-field-induced optical anisotropy in ferrofluids: a time-dependent light-scattering investigation. *Phys Rev E*, **78**, 051502 (051504 pp.).

133. Rablau, C., Vaishnava, P., Sudakar, C., Tackett, R., Lawes, G., Naik, R., (2008). Nanoparticle aggregation and relaxation effects in ferrofluids:

studied through anisotropic light scattering, *Proc SPIE—Int Soc Opt Eng*, **7032**, 70320Z.

134. Kinoshita, T., Seino, S., Mizukoshi, Y., Nakagawa, T., and Yamamoto, T. A. (2007). Functionalization of magnetic gold/iron-oxide composite nanoparticles with oligonucleotides and magnetic separation of specific target. *J Magn Magn Mater*, **311**, 255–258.

135. Min-Cheol, K., et al. (2006). Dynamic characteristics of super-paramagnetic iron oxide nanoparticles in a viscous fluid under an external magnetic field. *IEEE Trans Magn*, **42**, 979–982.

136. Liang, T., Ying, J., Yuanpeng, L., and Jian-Ping, W. (2011). Real-time measurement of Brownian relaxation of magnetic nanoparticles by a mixing-frequency method. *Appl Phys Lett*, **98**, 213702–213703.

137. Vaishnava, P. P., et al. (2007). Magnetic relaxation and dissipative heating in ferrofluids. *J Appl Phys*, **102**, 063914–063911.

138. Spinu, L., et al. (Elsevier) in *Joint European Magnetic Symposium JEMS'01, 28 August–1 September 2001*. 604–607.

139. Noginova, N., Weaver, T., Andreyev, A., Radocea, A., and Atsarkin, V. A. (2009). NMR and spin relaxation in systems with magnetic nanoparticles: effects of size and molecular motion. *J Phys Condensed Matter*, **21**, 255301 (255307 pp.).

140. Kortaberria, G., Arruti, P., Jimeno, A., Mondragon, I., and Sangermano, M. (2008). Local dynamics in epoxy coatings containing iron oxide nanoparticles by dielectric relaxation spectroscopy. *J Appl Polym Sci*, **109**, 3224–3229.

141. Mira, J., et al. (IEEE) in *1997 IEEE International Magnetics Conference (INTERMAG '97), 1–4 April 1997*, 5th ed., 3724–3726.

142. Chamberlin, R. V., et al. (2002). Percolation, relaxation halt, and retarded van der Waals interaction in dilute systems of iron nanoparticles. *Phys Rev B*, **66**, 172403–172401.

143. Denisov, S. I., and Trohidou, K. N. (2001). Fluctuation theory of magnetic relaxation for two-dimensional ensembles of dipolar interacting nanoparticles. *Phys Rev B*, **64**, 184433–184431.

144. Iglesias, O., and Labarta, A. (2004). Magnetic relaxation in terms of microscopic energy barriers in a model of dipolar interacting nanoparticles. *Phys Rev B*, **70**, 144401–144401.

145. Rosensweig, R. E. (Elsevier) in *9th International Conference on Magnetic Fluids, 23–27 July 2001*, 1–3 ed. 370–374.

146. Hiergeist, R., et al. (Elsevier) in *Eighth International Conference on Magnetic Fluids, 29 June–3 July 1998*. 420–422.

147. Kita, E., et al. (2010). Ferromagnetic Nanoparticles for Magnetic Hyperthermia and Thermoablation Therapy. *J Phys D*, **43**, 474011–474019.

148. Natividad, E., Castro, M., and Mediano, A. (2009). Adiabatic vs. non-adiabatic determination of specific absorption rate of ferrofluids. *J Magn Magn Mater*, **321**, 1497–1500.

149. Natividad, E., Castro, M., and Mediano, A. (2011). Adiabatic magnetothermia makes possible the study of the temperature dependence of the heat dissipated by magnetic nanoparticles under alternating magnetic fields. *Appl Phys Lett*, **98**, 243119.

150. http://www.nanotherics.com/magnetherm.htm.

151. Minhong, J., et al. (2012). Physical limits of pure superparamagnetic Fe_3O_4 nanoparticles for a local hyperthermia agent in nanomedicine. *Appl Phys Lett*, **100**, 092406 (092404 pp.).

152. Hergt, R., and Dutz, S. (2007). Magnetic particle hyperthermia-biophysical limitations of a visionary tumour therapy. *J Magn Magn Mater*, **311**, 187–192.

153. Kallumadil, M., et al. (2009). Suitability of commercial colloids for magnetic hyperthermia. *J Magn Magn Mater*, **321**, 1509–1513.

154. Haemmerich, D., dos Santos, I., Schutt, D. J., Webster, J. G., and Mahvi, D. M. (2006). In vitro measurements of temperature-dependent specific heat of liver tissue. *Med Eng Phys*, **28**, 194–197.

155. Liangruksa, M., Ganguly, R., and Puri, I. K. (2011). Parametric investigation of heating due to magnetic fluid hyperthermia in a tumor with blood perfusion. *J Magn Magn Mater*, **323**, 708–716.

156. Craciunescu, O. I., Das, S. K., McCauley, R. L., Macfall, J. R., and Samulski, T. V. (2001). 3D numerical reconstruction of the hyperthermia induced temperature distribution in human sarcomas using DE-MRI measured tissue perfusion: validation against non-invasive MR temperature measurements. *Int J Hyperthermia*, **17**, 221–239.

157. Heng, H., Delikanli, S., Hao, Z., Ferkey, D. M., and Pralle, A. (2010). Remote control of ion channels and neurons through magnetic-field heating of nanoparticles. *Nat Nanotechnol*, **5**, 602–606.

158. Keblinski, P., Cahill, D. G., Bodapati, A., Sullivan, C. R., and Taton, T. A. (2006). Limits of localized heating by electromagnetically excited nanoparticles. *J Appl Phys*, **100**, 54305–54301.

159. Haacke, E. M., Brown, R. W., Thompson, M. R., and Venkatesan, R. (1999). *Magnetic Resonance Imaging: Physical Principles and Sequence Design*, Wiley, New York.

160. Finegan, K. (1988). MRI-T_1 and T_2 weighting "for the uninitiated". *Radiographer*, **35**, 77–80.

161. Yilmaz, A., et al. (2002). Paramagnetic contribution of serum iron to the spin–spin relaxation rate ($1/T_2$) measured by MRI. *Appl Magn Reson*, **22**, 11–22.

162. Watanabe, Y., Perera, G. M., and Mooij, R. B. (2002). Image distortion in MRI-based polymer gel dosimetry of Gamma Knife stereotactic radiosurgery systems. *Med Phys*, **29**, 797–802.

163. Morales, M. P., Bomati-Miguel, O., Perez de Alejo, R., Ruiz-Cabello, J., Veintemillas-Verdaguer, S., O'Grady, K. (2003). Contrast agents for MRI based on iron oxide nanoparticles prepared by laser pyrolysis, *J Magn Magn Mater*, **266**, 102–109.

164. Pouliquen, D., et al. (1989). Superparamagnetic iron oxide nanoparticles as a liver MRI contrast agent: contribution of microencapsulation to improve biodistribution. *Magn Reson Imaging*, **7**, 619–627.

165. Miguel, O. B., et al. (2007). Comparative analysis of the 1H NMR relaxation enhancement produced by iron oxide and core–shell iron-iron oxide nanoparticles. *Magn Reson Imaging*, **25**, 1437–1441.

166. Shen, Y., et al. (2008). Quantifying magnetic nanoparticles in non-steady flow by MRI. *Magn Reson Mater Phys, Biol, Med*, **21**, 345.

167. Wei, W., Quanguo, H., and Changzhong, J. (2008). Magnetic iron oxide nanoparticles: synthesis and surface functionalization strategies. *Nano Scale Res Lett*, **3**, 397–415.

168. Zagorodni, A. A., Salazar-Alvarez, G., and Muhammed, M. (2006). Novel flow injection synthesis of iron oxide nanoparticles with narrow size distribution. *Chem Eng Sci*, **61**, 4625–4633.

169. Keshavarz, S., et al. (2010). Relaxation of Polymer Coated Fe_3O_4 Magnetic nanoparticles in aqueous solution. *IEEE Trans Magn*, **46**, 1541–1543.

170. Petcharoen, K., and Sirivat, A. (2012). Synthesis and characterization of magnetite nanoparticles via the chemical co-precipitation method. *Mater Sci Eng B*, **177**, 421–427.

171. Song, G., et al. (2009). Facile hydrothermal synthesis of iron oxide nanoparticles with tunable magnetic properties. *J Phys Chem C*, **113**, 13593–13599.

172. Takami, S., et al. (2007). Hydrothermal synthesis of surface-modified iron oxide nanoparticles. *Mater Lett*, **61**, 4769–4772.

173. Koutzarova, T., Kolev, S., Ghelev, C., Paneva, D., and Nedkov, I. (2006). Microstructural study and size control of iron oxide nanoparticles

produced by microemulsion technique. *Phys Status Solidi (c)*, **3**, 1302–1307.

174. Perez, J. A. L., Quintela, M. A. L., Mira, J., Rivas, J., and Charles, S. W. (1997). Advances in the preparation of magnetic nanoparticles by the microemulsion method. *J Phys Chem B*, **101**, 8045–8047.

175. Herrera, A. P., Rodriguez, H. L., Torres-Lugo, M., and Rinaldi, C. (American Institute of Chemical Engineers) in *05AIChE: 2005 AIChE Annual Meeting and Fall Showcase, October 30–November 4, 2005.* 3044.

176. Pardoe, H., Chua-Anusorn, W., St. Pierre, T. G., and Dobsonb, J. (Elsevier) in *Third International Conference on Scientific and Clinical Applications of Magnetic Carriers, 3–6 May 2000*, 1–2 ed., 1–46.

177. Gui-yin, L., Yu-ren, J., Ke-long, H., Ping, D., and Jie, C. (2008). Preparation and properties of magnetic Fe_3O_4-chitosan nanoparticles. *J Alloys Compounds*, **466**, 451–456.

178. Linlin, L., et al. (2007). Magnetic and fluorescent multifunctional chitosan nanoparticles as a smart drug delivery system. *Nanotechnology*, **18**, 405102–405101.

179. Salehizadeh, H., Hekmatian, E., Sadeghi, M., and Kennedy, K. (2012). Synthesis and characterization of core–shell Fe_3O_4-gold-chitosan nanostructure. *J Nanobiotechnol*, **10**.

180. Aggarwal, P., Hall, J. B., McLeland, C. B., Dobrovolskaia, M. A., and McNeil, S. E. (2009). Nanoparticle interaction with plasma proteins as it relates to particle biodistribution, biocompatibility and therapeutic efficacy. *Adv Drug Deliv Rev*, **61**, 428–437.

181. Chen, H., et al. (2010). Reducing non-specific binding and uptake of nanoparticles and improving cell targeting with an antifouling PEO-b-PMPS copolymer coating. *Biomaterials*, **31**, 5397–5407.

182. Dong-Eun, L., et al. (2011). Hyaluronidase-sensitive SPIONs for MR/optical dual imaging nanoprobes. *Macromol Res*, **19**, 861–867.

183. Kumar, A., et al. (2007). Development of hyaluronic acid-Fe_2O_3 hybrid magnetic nanoparticles for targeted delivery of peptides. *Nanomed Nanotechnol, Biol, Med*, **3**, 132–137.

184. Iglesias-Silva, E., et al. (2010). Synthesis of gold-coated iron oxide nanoparticles. *J Non-Crystalline Solids*, **356**, 1233–1235.

185. Pal, S., Morales, M., Mukherjee, P., and Srikanth, H. (2009). Synthesis and magnetic properties of gold coated iron oxide nanoparticles. *J Appl Phys*, **105**, 07–504.

186. Sun, Q., et al., (2007). Theoretical study on gold-coated iron oxide nanostructure: magnetism and bioselectivity for amino acids. *J Phys Chem C*, **111**, 4159–4163.

187. Bumb, A., et al. (2008). Synthesis and characterization of ultra-small superparamagnetic iron oxide nanoparticles thinly coated with silica. *Nanotechnology*, **19**, 335601 (335606 pp.).

188. Souza, D. M., et al. (2008). Synthesis and in vitro evaluation of toxicity of silica-coated magnetite nanoparticles. *J Non-Crystalline Solids*, **354**, 4894–4897.

189. Shukla, R., et al. (2005). Biocompatibility of gold nanoparticles and their endocytotic fate inside the cellular compartment: a microscopic overview. *Langmuir*, **21**, 10644–10654.

190. Insin, N., et al. (2008). Incorporation of iron oxide nanoparticles and quantum dots into silica microspheres. *ACS Nano*, **2**, 197–202.

191. Baker, C., Hasanain, S. K., and Shah, S. I. (2004). The magnetic behavior of iron oxide passivated iron nanoparticles. *J Appl Phys*, **96**, 6657–6662.

192. Xiaowei, T., and Hong, Y. (2007). Iron oxide shell as the oxidation-resistant layer in $SmCo_5@Fe_2O_3$ core–shell magnetic nanoparticles. *J Nanosci Nanotechnol*, **7**, 356–361.

193. Heitsch, A. T., Smith, D. K., Patel, R. N., Ress, D., and Korgel, B. A. (2008). Multifunctional particles: magnetic nanocrystals and gold nanorods coated with fluorescent dye-doped silica shells. *J Solid State Chem*, **181**, 1593–1602.

194. Mahmoudi, M., Serpooshan, V., and Laurent, S. (2011). Engineered nanoparticles for biomolecular imaging. *Nanoscale*, **3**, 3007–3026.

195. Ivkov, R., et al. (SPIE—The International Society for Optical Engineering) in *Thermal Treatment of Tissue: Energy Delivery and Assessment IV, 20 Jan. 2007.* 64400I (64408 pp.).

196. Oghabian, M. A., et al. (2011). Detectability of Her2 positive tumors using monoclonal antibody conjugated iron oxide nanoparticles in MRI. *J Nanosci Nanotechnol*, **11**, 5340–5344.

197. Vigor, K. L., et al. (2010). Nanoparticles functionalized with recombinant single chain Fv antibody fragments (scFv) for the magnetic resonance imaging of cancer cells. *Biomaterials*, **31**, 1307–1315.

198. Bicher, H. I. (1980). The physiological effects of hyperthermia. *Radiology*, **137**, 511–513.

199. Jordan, A., et al. (Elsevier) in *Third International Conference on Scientific and Clinical Applications of Magnetic Carriers, 3–6 May 2000*, 1–2 ed., 118–126.

200. Johannsen, M., et al. (2005). Clinical hyperthermia of prostate cancer using magnetic nanoparticles: presentation of a new interstitial technique. *Int J Hyperthermia*, **21**, 637–647.

201. Johannsen, M., et al. (2007). Morbidity and quality of life during thermotherapy using magnetic nanoparticles in locally recurrent prostate cancer: results of a prospective phase I trial. *Int J Hyperthermia*, **23**, 315–323.

202. Rachakatla, R. S., et al. (2010). Attenuation of mouse melanoma by A/C magnetic field after delivery of Bi-magnetic nanoparticles by neural progenitor cells. *ACS Nano*, **4**, 7093–7104.

203. Jae-Hyun, L., et al. (2011). Exchange-coupled magnetic nanoparticles for efficient heat induction. *Nat Nanotechnol*, **6**, 418–422.

204. Hwang, Y., et al. (2012). Exchange bias behavior of monodisperse $Fe_3O_4/g\text{-}Fe_2O_3$ core/shell nanoparticles. *Curr Appl Phys*, **12**, 808–811.

205. Lien, Y.-H., and Wu, T.-M. (2008). Preparation and characterization of thermosensitive polymers grafted onto silica-coated iron oxide nanoparticles. *J Colloid Interface Sci*, **326**, 517–521.

206. Rubio-Retama, J., et al. (2007). Synthesis and characterization of thermosensitive PNIPAM microgels covered with superparamagnetic $\gamma\text{-}Fe_2O_3$ nanoparticles. *Langmuir*, **23**, 10280–10285.

207. Shengmao, Z., Linna, Z., Benfang, H., and Zhishen, W. (2008). Preparation and characterization of thermosensitive PNIPAA-coated iron oxide nanoparticles. *Nanotechnology*, **19**, 325608.

208. Kee, I. H. C., et al. (2009). Thermoresponsive core–shell magnetic nanoparticles for combined modalities of cancer therapy. *Nanotechnology*, **20**, 305101–305111.

209. Ting-Yu, L., Kun-Ho, L., Dean-Mo, L., San-Yuan, C., and Chen, I. W. (2009). Temperature-sensitive nanocapsules for controlled drug release caused by magnetically triggered structural disruption. *Adv Funct Mater*, **19**, 616–623.

210. Choi, H., Choi, S. R., Zhou, R., Kung, H. F., and Chen, I. W. (2004). Iron oxide nanoparticles as magnetic resonance contrast agent for tumor imaging via folate receptor-targeted delivery. *Academic Radiol*, **11**, 996–1004.

211. Weissleder, R., Lee, A. S., Khaw, B. A., Shen, T., and Brady, T. J. (1992). Antimyosin-labeled monocrystalline iron oxide allows detection of myocardial infarct: MR antibody imaging. *Radiology*, **182**, 381–385.

212. Remsen, L. G., et al. (1996). MR of carcinoma-specific monoclonal antibody conjugated to monocrystalline iron oxide nanoparticles: the potential for noninvasive diagnosis. *Am J Neuroradiol*, **17**, 411–418.

213. Park, J.-Y, Choi, H. J., Nam, G.-E., Cho, K.-S., Son, J.-H. (2011). In vivo dual-modality terahertz/magnetic resonance imaging using superparamagnetic iron oxide nanoparticles as a dual contrast agent, *IEEE Trans Terahertz Sci Technol*, **2**, 93–98.

214. Scherer, F., et al. (2002). Magnetofection: enhancing and targeting gene delivery by magnetic force in vitro and in vivo. *Gene Ther*, **9**, 102–109.

215. Tresilwised, N., et al. (2010). Boosting oncolytic adenovirus potency with magnetic nanoparticles and magnetic force. *Mol Pharm*, **7**, 1069–1089.

216. Hashimoto, M., and Hisano, Y. (2010). Directional gene-transfer into the brain by an adenoviral vector tagged with magnetic nanoparticles. *J. Neurosci Methods*, **194**, 316–320.

217. Mah, C., et al. (2002). Improved method of recombinant AAV2 delivery for systemic targeted gene therapy. *Mol Ther*, **6**, 106–112.

218. Ang, D., et al. (2011). Insights into the mechanism of magnetic particle assisted gene delivery. *Acta Biomaterialia*, **7**, 1319–1326.

219. Yunfeng, S., et al. (2010). In situ preparation of magnetic nonviral gene vectors and magnetofection in vitro. *Nanotechnology*, **21**, 115103 (115108 pp.).

220. Sun, X., Gutierrez, A., Yacaman, M. J., Dong, X., and Jin, S. (Elsevier) in *Symposium A-Nanostructured Materials, The Fifth IUMRS International Conference on Advanced Materials (ICAM '99), 13–18 June 1999*, 1 ed., 157–160.

221. Tomitaka, A., Kobayashi, H., Yamada, T., Jeun, M., Bae, S. Takemura, Y. (2010). Magnetization and self-heating temperature of $NiFe_2O_4$ nanoparticles measured by applying ac magnetic field, *J Phys: Conf Ser*, **200**, 122010.

222. Del Pino, P. et al. (2010). Gene silencing mediated by magnetic lipospheres tagged with small interfering RNA. *Nano Lett*, **10**, 3914–3921.

223. Nel, A. E., et al. (2009). Understanding biophysicochemical interactions at the nano-bio interface. *Nat Mater*, **8**, 543–557.

224. Huajian, G., Wendong, S., and Freund, L. B. (2005). Mechanics of receptor-mediated endocytosis. *Proc Natl Acad Sci USA*, **102**, 9469–9474.

225. Lunov, O., et al. (2011). Modeling receptor-mediated endocytosis of polymer-functionalized iron oxide nanoparticles by human macrophages. *Biomaterials*, **32**, 547–555.

226. Dong-Hyun, K., et al. (2010). Biofunctionalized magnetic-vortex microdiscs for targeted cancer-cell destruction. *Nat Mater*, **9**, 165–171.

227. Kim, D.-H., et al. (2011). Mechanoresponsive system based on submicron chitosan-functionalized ferromagnetic disks. *J Mater Chem*, **21**, 8422–8426.

228. Furlani, E. P., and Ng, K. C. (2008). Nanoscale magnetic biotransport with application to magnetofection. *Phys Rev E*, **77**, 061914 (061918 pp.).

229. McBain, S. C., et al. (2008). Magnetic nanoparticles as gene delivery agents: enhanced transfection in the presence of oscillating magnet arrays. *Nanotechnology*, **19**, 405102 (405105 pp.).

230. Bhattarai, S. R., et al. (2008). Hydrophobically modified chitosan/gold nanoparticles for DNA delivery. *J Nanoparticle Res*, **10**, 151–162.

231. Kamei, K., et al. (2009). Direct cell entry of gold/iron-oxide magnetic nanoparticles in adenovirus mediated gene delivery. *Biomaterials*, **30**, 1809–1814.

Chapter 11

Atomic Force Microscopy Imaging of DNA Delivery Nanosystems

Yi Zou,[a] David Oupicky,[b] and Guangzhao Mao[a]

[a]*Department of Chemical Engineering and Materials Science,
Wayne State University, Detroit MI 48202*
[b]*Department of Pharmaceutical Sciences, Wayne State University, Detroit MI 48202*

gzmao@eng.wayne.edu

Polymeric gene delivery nanosystems are attractive alternatives to virus-based systems due to safety and other concerns. The release dynamics of DNA encapsulated in the nanosystems is key to the understanding and tuning of gene delivery efficacy. In this chapter, we describe the use of atomic force microscope (AFM) to the understanding and tuning of DNA release dynamics from bioreducible polymer-based nanosystems, including polyplex nanoparticles and layer-by-layer (LbL) thin films.

11.1 Introduction

AFM has become the microscope of choice for the investigation of biological, biophysical, biochemical, and biomimetic processes. AFM is non-destructive and is capable of imaging both dry and wet samples. Furthermore, its application has been extended to

NanoCellBiology: Multimodal Imaging in Biology and Medicine
Edited by Bhanu P. Jena and Douglas J. Taatjes
Copyright © 2014 Pan Stanford Publishing Pte. Ltd.
ISBN 978-981-4411-79-0 (Hardcover), 978-981-4411-80-6 (eBook)
www.panstanford.com

investigating physiochemical properties of thin films,[1,2] lipid bilayers,[3,4] live cells,[5,6] and other biological samples.[7,8] In contrast to the electron microscopes such as cryo-TEM,[9] which is another widely used tool to obtain structural understanding at the nanometer scale, biological sample preparation for AFM is relatively straightforward permitting intact nanostructure to be imaged in biologically compatible solution environment. AFM has been used extensively to study morphology, adsorption, and condensation of nucleic acids for more than two decades.[10–13] In situ real-time AFM has also been reported in the studies of DNA condensation[14] and release[15] dynamics. AFM has contributed significant knowledge to DNA condensation and release dynamics in nanosystems.

There are two basic AFM operational methods: the contact mode and tapping mode (also called the AC mode). In the contact mode, AFM operates at a constant deflection so that the interaction force between the sample and probe is constant. In the tapping mode, the AFM probe oscillates near its resonant frequency and scans the surface with constant damped amplitude. Due to its oscillatory nature, the tapping mode usually provides clearer images for soft biological samples,[16] especially in solution.[17] By imaging continuously in solution, it becomes possible to monitor the disassembly of DNA/polymer assemblies in real time and at the molecular scale. Figure 11.1 shows the in situ AFM setup for DNA release studies.

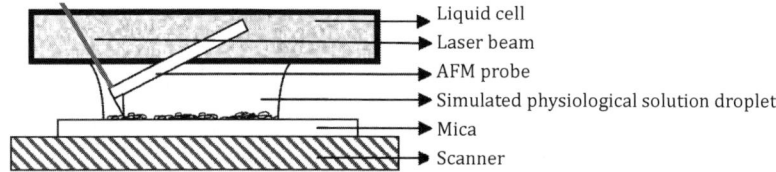

    Liquid cell
    Laser beam
    AFM probe
    Simulated physiological solution droplet
    Mica
    Scanner

Figure 11.1 A typical setup for in situ AFM operation. The AFM probe scans over immobilized polyplexes or LbL films deposited on a 2D surface at a minimal contact force in simulated physiologic solution.

The promise of gene therapy to treat a variety of genetic and acquired diseases has fueled research on gene delivery nanosystems. Virus-based gene delivery nanosystems have shortcomings, including delivery capacity, toxicity, immune response, residual pathogenicity, and cause of secondary carcinogenesis.[18–21] Nonviral gene carriers, especially positively charged synthetic polymers, such as poly(ethylenimine) (PEI)[22] and poly(L-lysine) (PLL),[23] are attractive alternatives because they show lower safety risks and can be tailored to specific therapeutic needs. But a major challenge for nonviral gene carriers is their low transfection efficiency. Currently the field is dominated by "black-box" strategies that test reporter gene expression levels of various formulations. A critical question to the overall gene delivery efficiency is how and when plasmid DNA is dissociated from its complexes with polycations.[24–29] Tools such as AFM that allow correlating physiochemical properties of gene delivery carriers with in vitro and in vivo gene delivery results will provide understanding that leads to fine tuning of the gene delivery efficiency.

One strategy to enhance gene delivery efficiency while maintain cytotoxicity is to use stimuli-responsive polymers, i.e., bioreducible polymers.[30,31] Scheme 11.1 lists examples of bioreducible polymers that have been used in our studies. These polymers contain disulfide bonds that are bioreducible by redox agents such as glutathione inside the nucleus or thiol-containing membrane proteins. The

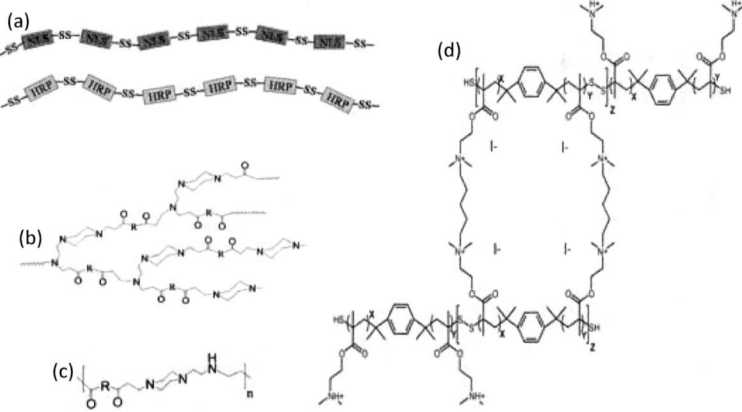

Scheme 11.1   Bioreducible polymers investigated. (a) Polypeptides of his-tidine-rich peptide (CKHHHKHHHKC) and nuclear localiza-tion signal (CGAGPKKKRKVC) peptide. (b) Reducible hyper-branched (RHB) PAA. The R groups represent two different amide monomers. *N,N′*-cystaminebisacrylamide (CBA) con-tains the disulfide bond while *N,N′*-methylenebisacrylamide (MBA) does not. (c) Linear poly(amido amine) (PAA). (d) Cross-linked poly(2-dimethylaminoethyl methacrylate) (rP-DMAEMA).

disulfide linkers in the bioreducible polymers are cleaved during the thiol-disulfide exchange reaction. High-molecular-weight polycations are degraded into low-molecular-weight oligomers with lower binding affinity to the DNA thus allowing it to be released from the polyplexes. This chapter focuses on recent advances in AFM imaging of the DNA release processes from bioreducible nanosystems, including (1) polyplex nanoparticles and (2) DNA/polycation LbL films in simulated physiologic conditions. While we focus primarily on work conducted in our labs, DNA release dynamics have been studied by other groups using AFM,[32,33] flow cytometry,[34,35] and Förster resonance energy transfer (FRET).[35,36] The polyplexes are suitable for systemic gene delivery while the LbL platform is a promising localized gene delivery strategy (Scheme 11.2). 1,4-dithiothreitol (DTT) is used to simulate the reducing environment in vivo.[37]

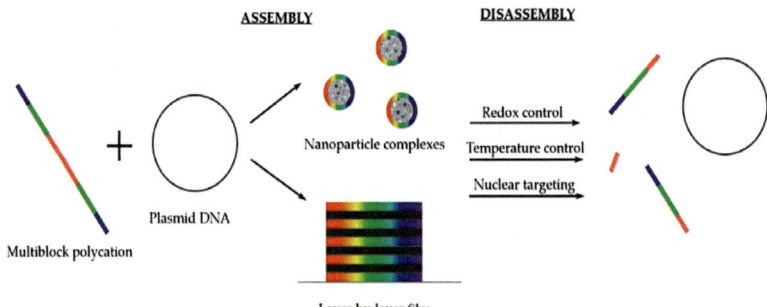

Scheme 11.2 Multicomponent polycations with redox-responsive (in red), temperature-responsive (in green), and nuclear targeting (in blue) blocks for systemic and localized delivery of plasmid DNA.

11.2 Materials and Methods

11.2.1 Materials

Plasmid DNA vectors, gWiz High-Expression GFP plasmid (6.7 kb) and gWiz High-Expression Secreted Alkaline Phosphatase (SEAP) plasmid (5.8 kb), are purchased from Aldevron. Dithiothreitol (DTT, Sigma), 1-(2-aminoethyl)piperazine (AEPZ, Aldrich), 1-methylpiperazine (Aldrich), *N,N'*-methylenebisacrylamide (MBA, Aldrich), and *N,N'*-cystaminebisacrylamide (CBA, Polysciences) are purchased in the highest commercially available purity and used without further purification. Hyperbranched and linear bioreducible PAAs are synthesized via Michael addition copolymerization.[38] Bioreducible poly(2-dimethylaminoethyl methacrylate)s are synthesized via reversible addition–fragmentation chain transfer polymerization.[39] 1,5-diiodopentane (DIP) is purchased from Acros Organics and used without further purification. Water is deionized to 18 MΩ × cm resistivity using the Nanopure system from Barnstead. Grade V5 muscovite mica is purchased from Ted Pella. Polished n-type silicon wafers (resistivity 50–75 Ω cm) are purchased from Wafer World.

11.2.2 Polyplex Preparation

Concentrated DNA stock (1 g/L) is diluted in 30 mM acetate buffer (pH 4.5 or 5.0). DNA solution containing 20 mg/L DNA is used to prepare all polyplexes at various N/P ratio (amine-to-DNA phosphate molar ratio). The polymer solution is added to the DNA solution and mixed by vortexing at 3200 rpm (Fisher Scientific Vortex Mixer) for 10 s, then the solution is incubated at room temperature for 30 min following previously developed procedures.[28,40]

11.2.3 LbL Preparation

In order to characterize the nanostructure of LbL films, before film deposition, substrates are treated to ensure surface cleanness and homogeneity. For instance, silicon wafer is cleaned in piranha solution (volume ratio 3:1 of concentrated sulfuric acid to 30% aqueous hydrogen peroxide solution). Then, 1 × 1 cm^2 substrate piece is immersed into polycation solution for 15 min, followed by rinsing with deionized water three times for 2 min each. Deposition operation is repeated alternatively in DNA solution and polycation solution (DNA concentration 0.25 g/L, polycation concentration 1–2 g/L) until desired number of layers is obtained.

11.2.4 AFM Imaging

AFM imaging is conducted using Multimode IIIa from digital instrument and Dimension 3100 AFM from VEECO. Polyplexes are immobilized on freshly cleaved mica and extra solute is removed by rinsing with deionized water for three times. Tapping mode in liquid is performed using silicon nitride probes (NP type, Veeco) with a nominal radius of curvature of 20 nm and cantilever spring constant of 0.38 N/m as provided by the manufacturer. Usually, the polyplexes are imaged in 50 µL simulated physiological solution. The surface is imaged continuously at an average rate of 1–2 Hz on a 2 × 2 or 5 × 5 µm^2 area until no significant changes are observed. The ranges of frequency, amplitude, integral, and proportional gains used are 7.5–8.5 kHz, 0.5–1 V, 0.5–2, and 0.75–3, respectively. For LbL films, disassembly is conducted in DTT solution (pH 5–7, salt concentration 0–0.2M) prior to imaging, and the samples are imaged in air. All AFM images are analyzed using Nanoscope software version 5.12b by Veeco.

11.3 Results and Discussion

11.3.1 Polyplex Self-Assembly

As shown in Fig. 11.2, AFM images illustrate the morphology evolution with reaction time (incubation time). The result suggests that the self-assembly of polyplexes is kinetics dominated.[28,40] Rods are favorable at the beginning of the reaction while toroids emerge later. This agrees with a previous study, which reported the kinetically dominated polyplex formation phenomenon using transmission electron microscope (TEM).[41] In order to maximize

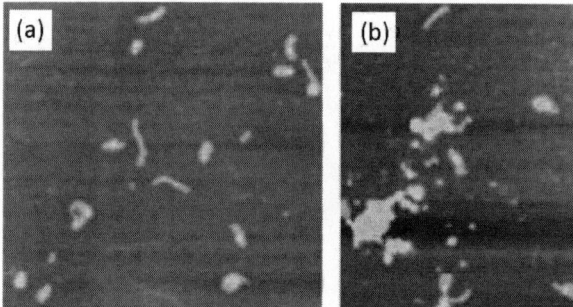

Figure 11.2 AFM height images of time dependence in linear PAA polyplex formation (disulfide content 15%, Mw 54,500 g/mol, N/P 4). (a) 10 min and (b) 60 min. The images are captured in tapping mode in air. The scan size is 2 µm, and the z-range is 10 nm. Reprinted from Wan, L., et al. *J. Phys. Chem. B*, 2009, **113**(42), 13735–13741.

characteristic toroid population, 30 min incubation is used for most of the study. For a 6732 bp plasmid DNA, monomolecular polyplex theoretical volume is calculated to be 1.4 × 10^4 nm^3 by assuming interhexagonal separation between a neighboring polycation/DNA chain of 2.7 nm.[42] Compared with the histogram of bearing volume analysis for RHB/DNA polyplex, it can be concluded that most of the polyplexes only contain a single DNA chain with three characteristic morphologies, which are toroid, rod, and spheroid (Fig. 11.3).

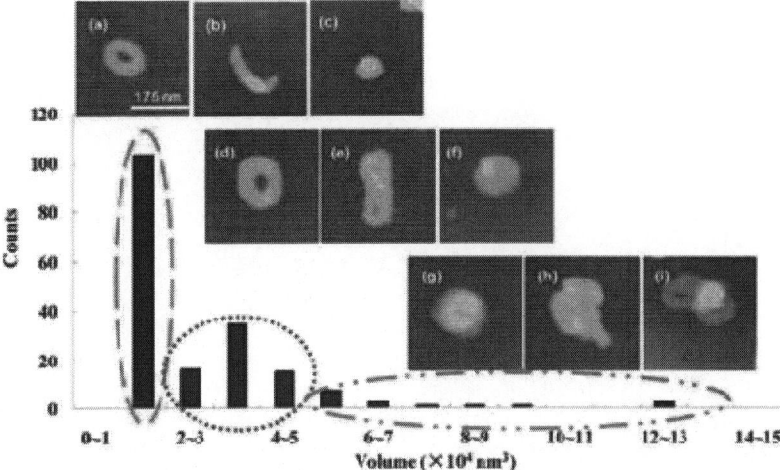

Figure 11.3 Size and morphology distribution. The scan size is 350 nm. The z-range is 15 nm for (g–i) and 8 nm for all others. Polyplexes whose volume is in the range of 1–2 × 10^4 nm^3 are represented by (a–c). Polyplexes whose volume is in the range of 2–5 × 10^4 nm^3 are represented by (d–f). Reprinted from Wan, L., et al. *J. Phys. Chem. B*, 2009, **113**, 13735–13741.

This kinetically dominated process is also indicated by invariant size and morphology distribution of polyplex regardless of polycation molecular weight (25,000–130,000 g/mol), chain architecture (linear vs. hyperbranched), and disulfide content (0–100% CBA). At the same incubation time, the average outer diameter of toroids remains to be 100 nm, which is also supported by light scattering data. In addition, colloidal stability of polyplexes benefits from their highly charged surface, which is measured to be larger than +40 mV.

11.3.2 Cooperative DNA Release vs. Non-Cooperative DNA Release

The DNA release dynamics is studied using reducible polypeptide—poly(HRP) and poly(NLS). There are two bioreducible polypeptides with two peptide monomers: (1) histidine-rich peptide (HRP), which is capable of buffering endosomal pH; (2) nuclear localization signal (NLS), which provides polyplexes with potential nuclear localization capability. Their release behavior appears to be similar in gel electrophoresis and light scattering data.[28,43] However, at the molecular level, the real-time AFM results show that there are two significantly different release mechanisms. Figure 11.4 shows the

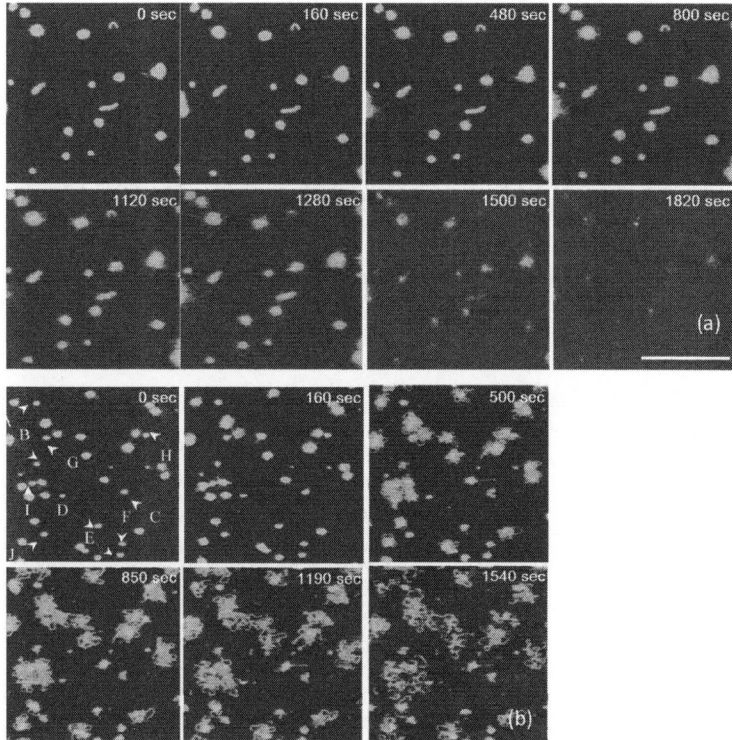

Figure 11.4 Real time AFM sequence of DNA release from (a) NLS[6+] polyplexes and (b) HRP[10+] polyplexes in 20 mM DTT and 0.4 M NaCl solution. Time zero corresponds to the addition of DTT. The z-range is 10 nm. Scan size is 2.6 × 2.6 μm^2 for (a) and 2.0 × 2.0 μm^2 for (b). Reprinted from Wan, L., et al. *Langmuir*, 2008, **24**, 12474–12482.

time-lapse DNA release sequences from the two polyplexes under identical release condition (0.2M NaCl and 20mM DTT). The NLS polyplexes release DNA abruptly regardless of their initial size and morphology. Meanwhile, the HRP polyplexes release DNA gradually and partially. In other words, DNA release from the NLS polyplexes displays an abrupt and size-independent release dynamics while the HRP polyplexes display a cooperative and size dependent disassembly mechanism. The difference is attributed to the charge valence of the reduced oligocation fragments (NLS^{6+} vs. HRP^{10+}). The abrupt non-cooperative release behavior for the NLS polyplexes may correlate with their low in vitro transfection efficiency. It is possible that the disulfide bonds in poly(NLS) are cleaved in the reducing environment of the cytosol resulting in premature release of the DNA molecules before reaching the nucleus. This result also suggests a potential strategy on the control of DNA release timing by subtly varying the charge valence of the staring oligocation monomers during polymer synthesis.

11.3.3 A Proposed Disassembly Route for Bioreducible Polyplexes

Another in situ AFM study on DNA release triggered by polycation degradation has been conducted with bioreducible PAA polyplexes.[40] Recently, bioreducible PAAs have become promising DNA delivery vectors because of their low cytotoxicity and high transfection efficiency.[44–46]

The PAAs are synthesized via Michael addition copolymerization reaction.[47,48] In addition to their potential pharmaceutical application, by varying the feed ratio of three monomers, a series of bioreducible PAAs with several variables are obtained providing a good model to study the correlation between their physiochemical properties and biological activity. Despite the variation in disulfide content, molecular weight, and polymer chain architecture, a common morphological route of DNA release has been observed.

The morphological pathway of polyplex disassembly is visualized by in situ AFM (Fig. 11.5). Because of depolymerization induced by thiol-disulfide exchange reaction, high-molecular-weight polycation is converted to low-molecular-weight oligomers. Then, the transition energy is lowered to allow various forms to converge into the lowest energy form, which is toroid structure in this case. As shown in

Figs. 11.5a,b, DNA release begins with morphological transition from metastable rod and spherical particles to the toroid form. Afterward, at the intermediate stage as shown by Fig. 11.5c, the toroids interact with each other by aggregation and fusion. Depolymerization also weakens the electrostatic interaction thus enables DNA strands to rearrange from kinetically constrained binding sites. In the last stage (Fig. 11.5d), DNA worm-like chains gradually unravel from the polyplex resulting in loose loops/tails that are held by a central compact core.

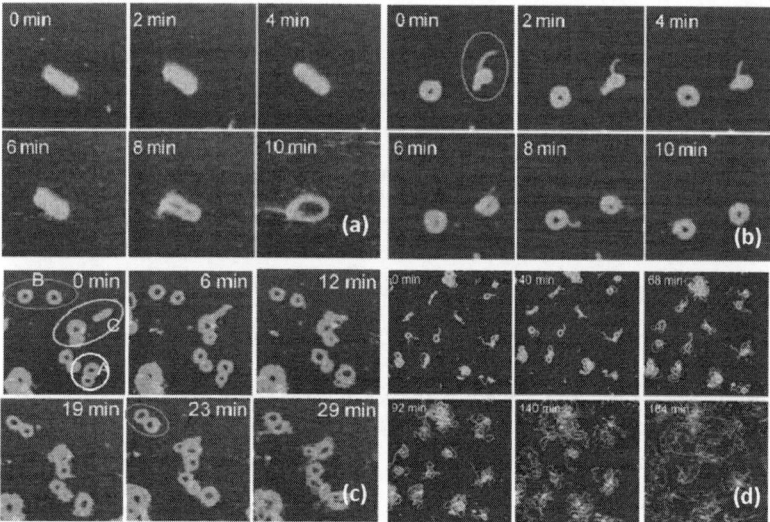

Figure 11.5 Real time AFM images showing (a) morphology transition from rod to toroid, (b) morphology transition from sphere to toroid, (c) particle-particle interaction and aggregation (A, B, and C are three polyplexes groups containing two individual ones at the beginning and fused into each other in the end), and (d) a typical DNA release sequence including all three stages. Time zero corresponds to the injection of the DTT solution. The scan size is 500 nm for (a), 600 nm for (b), 1 μm for (c), and 2 μm for (d). The z-range is 7 nm for (a), 8 nm for (b), 10 nm for (c), and 6 nm for (d). Reprinted from Wan, L., et al. *J. Phys. Chem. B*, 2009, **113**(42), 13735–13741

11.3.4 LbL Film Assembly

The first AFM study of bioreducible polycation/DNA LbL films is reported by the Mao group in 2007 using disulfide-containing reducible poly(trans-activating transcriptional activator) (PTAT) and nonreducible PLL as control. As reported previously with non-bioreducible polycation/DNA LbL films,[49] surface characteristics are crucial for cell growth and adhesion. Prior to film disassembly study, the change of physical properties with layer growth is monitored up to 17 layers. According to AFM topographic images, root-mean-square roughness for 5 × 5 μm^2 area increases from 0.5 nm at $(PLL/DNA)_1$ to 6.0 nm at $(PLL/DNA)_{8.5}$ (films are denoted as $(polycation/DNA)_{n/2}$ with n being the number of layers). Film thickness measured by ellipsometer shows an exponential growth mode of film thickness. Such growth mode may be attributed to the polycation chain mobility and diffusion out of multilayers. This is also consistent with film morphology change with increasing layers observed by AFM. It is shown that smooth, particle-free surface with strand-like texture dominates the initial layers (Fig. 11.6a), while large particles begin to appear at higher layers (Fig. 11.6b). Similar growth behavior has also been observed on PAA LbL films.[50]

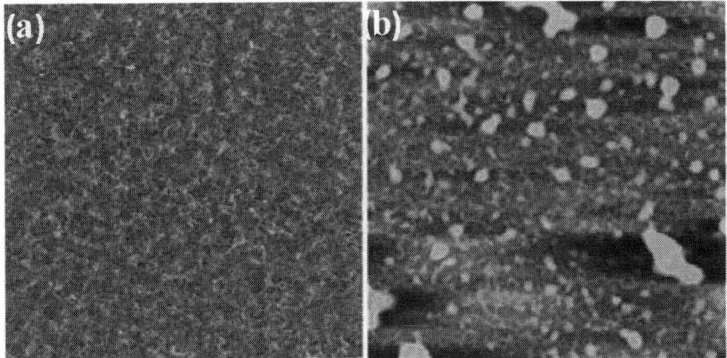

Figure 11.6 AFM height images of the PTAT/DNA films with different number of layers. (a) $(PTAT/DNA)_3$ 3 bilayer film; (b) $(PTAT/DNA)_{8.5}$ 8.5 bilayer film. The scan size is 5 × 5 μm^2. The z-range is 20 nm for (a) and 50 nm for (b). Reprinted from Blacklock, J., et al. *Biomaterials,* 2007, **28**, 117–124.

11.3.5 LbL Film Disassembly

For disassembly study, the $(PTAT/DNA)_{8.5}$ film mentioned above is treated by simulated physiological solution containing DTT as reducing agent. Film morphology change up to 7 h is captured by

AFM. Some key frames are shown in Fig. 11.7. From morphological aspect, DNA release is observed after 10 min of reaction as indicated by gradual disappearance of large particles as well as small strands. With increase of reaction time, network texture continues to break apart and roughness also decreases. After 60 min, the surface is devoid of the LbL characteristic textures.

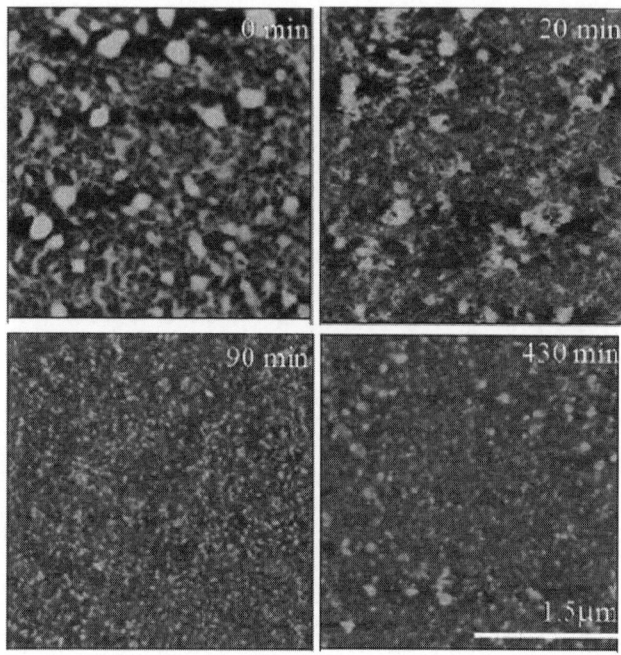

Figure 11.7 Time-lapse AFM height images of (PTAT/DNA)$_{8.5}$ film disassembly in 10 mM DTT and 0.5 M NaCl. The scan size is 3 μm. The z-range is 35 nm. Reprinted from Blacklock, J., et al. *Biomaterials*, 2007, **28**, 117–124.

Disassembly study of hyperbranched PAA (RHB)/DNA LbL films provides a more detailed view of DNA release dynamics, highlighting nanostructures, toroids, and bundles as dominant intermediate structures. Polyplex providing DNA in pre-condensed form is also incorporated in LbL films in order to modulate the timing of DNA release. LbL films with up to three layers with different terminating layers (RHB/DNA, RHB/DNA/RHB, RHB/DNA/Polyplex) are studied by AFM as well as X-ray reflectivity, ellipsometry, and Fourier transform infrared spectroscopy. Similar to DNA release dynamics from reducible polypeptide, salt is found to accelerate the overall rate of film disassembly. Under identical reducing condition, polyplex layer disassembles faster than the DNA layer. Side-by-side comparison between three kinds of films is presented in Fig. 11.8. Predominant intermediate structure for the DNA layer during film disassembly is found to be fiber bundle structure. For RHB/DNA and RHB/DNA/RHB films, after erosion, DNA strands start to emerge. The transition from a homogeneous film to a porous film occurs in the first 2 h of degradation. For RHB/DNA film, the DNA layer coverage is decreased and broken up after 4 h. The degradation is found to be consistent with a gradual loss of film material rather than a sudden loss or peel-off of the whole layer. After 6 h, most DNA (>90%) is released from the film. RHB/DNA/RHB film, which shares similar intermediate structure with RHB/DNA film, degrades slower. It takes about two more hours to finish degradation. For RHB/DNA/polyplex film, toroid structure, which has the lowest energy as intermediate structure, is also observed for polyplex layer (Fig. 11.8c). It takes about 10 h to achieve complete release.

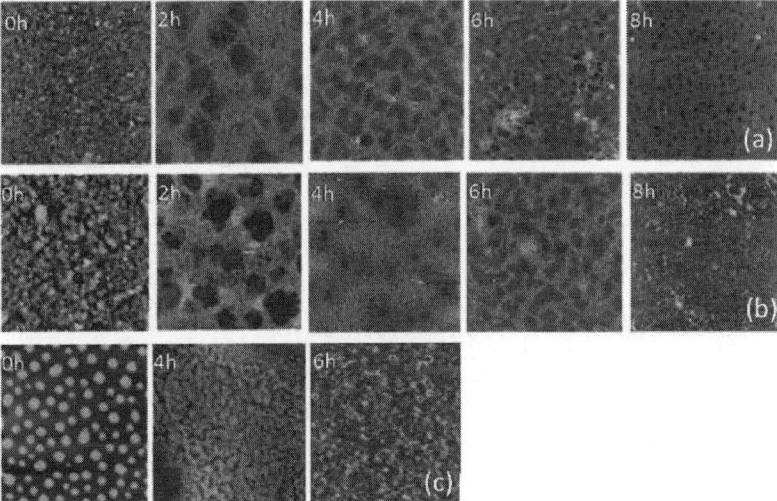

Figure 11.8 Time-lapse AFM height images of (a) RHB/DNA film, (b) RHB/DNA/RHB film, and (c) RHB/DNA/polyplex film disassembly in 20 mM DTT and 0.1M NaCl. Reproduced with permission from Blacklock, J., Mao, G., Oupickyì, D., Möhwald, H., DNA release dynamics from bioreducible layer-by-layer films. *Langmuir* 2010. Copyright 2010 American Chemical Society.

11.3.6 Cross-Linking of LbL Films

Surface topography, microstructure, and functionality influence cell-surface interactions and cell proliferation.[51-53] One approach in tuning surface characteristics is chemical cross-linking of polymer chains in the thin films.[54,55] In our AFM study, the effect of cross-linking is studied using 1,5-diiodopentane (DIP) as cross-linker.[1] DIP chemically reacts with rPDMAEMA in the LbL film consisting 15 bilayers of rPDMAEMA and DNA. AFM as well as ellipsometry results show the changes in film morphology, film swelling behavior, and film rigidity. Besides the morphological characterization that has been mentioned, nano-indentation, which measures the mechanical property of thin film, is another important application of AFM. Figure 11.9a shows the comparison of film rigidity before and after

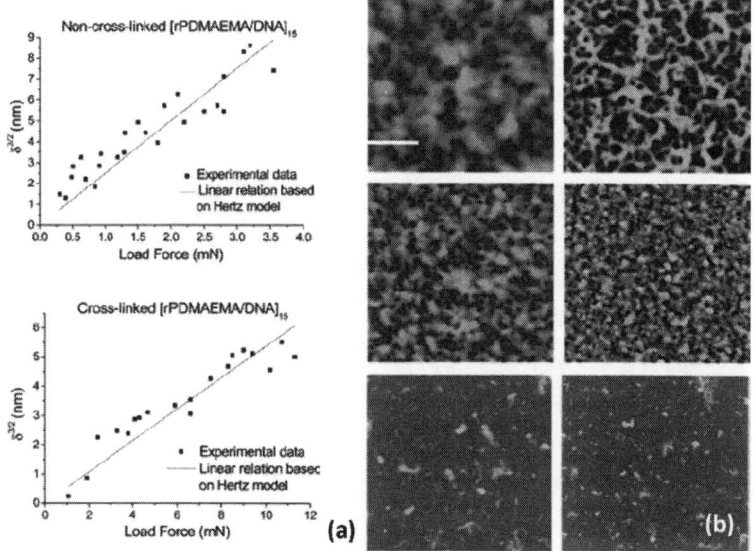

Figure 11.9 (a) Indentation vs. load force curve obtained by AFM and the fit to the Hertz model for the same [rPDMAEMA/DNA]$_{15}$ film before cross-linking (top) and after cross-linking (bottom). (b) Side-by-side comparison of DNA release sequence before and after cross-linking for (rPDMAEMA/DNA)$_{15}$ film. The left column is time-elapse AFM height images for film before cross-linking while right column is after cross-linking. The scan size is 5 μm. The z-range is 190 nm, 100 nm, and 40 nm, respectively. (b) Reprinted from Blacklock, J., et al. *J. Phys. Chem. B*, 2010.

cross-linking. Force curves are fitted with Hertz model: $\delta = AF^b$, where δ is indentation, F is load force, and b is assumed to be $2/3$. It is shown that the elastic modulus increases from 1.8 ± 0.8 MPa to 7.9 ± 1.1 MPa after cross-linking for the same film using the same AFM probe. The effect of cross-linking on film disassembly by DTT is found to be insignificant even though elastic modulus has increased by about fourfold (Fig. 11.9b). Moreover, biological experiments show that film cross-linking could enhance cell adhesion and prolong the duration of cellular transfection. These results contribute to the development of bioreducible polymer coatings for localized gene delivery.

11.4 Summary

Gene delivery nanosystems must overcome multiple extracellular and intracellular barriers to achieve their desired biological efficacy. It is crucial to understand DNA release dynamics at nanoscale and correlate the knowledge with in vitro and in vivo results. This chapter covers recent AFM studies of DNA release dynamics from polymeric gene delivery nanosystems under simulated physiologic conditions. The understanding of DNA release mechanism will guide future research to formulate efficient and safe gene delivery vectors.

References

1. Blacklock, J., Sievers, T. K., Handa, H., You, Y.-Z., Oupickyì, D., Mao, G., and Molhwald, H. (2010). Cross-linked bioreducible layer-by-layer films for increased cell adhesion and transgene expression, *J Phys Chem B*, **114**(16), 5283–5291.

2. Schneider, A., Richert, L., Francius, G., Voegel, J. C., and Picart, C. (2007). Elasticity, biodegradability and cell adhesive properties of chitosan/hyaluronan multilayer films, *Biomed Mater*, **2**(1), S45–S51.

3. Steltenkamp, S., Muller, M. M., Deserno, M., Hennesthal, C., Steinem, C., and Janshoff, A. (2006). Mechanical properties of pore-spanning lipid bilayers probed by atomic force microscopy, *Biophys J*, **91**(1), 217–226.

4. Dufrene, Y. F., Boland, T., Schneider, J. W., Barger, W. R., and Lee, G. U. (1998). Characterization of the physical properties of model

biomembranes at the nanometer scale with the atomic force microscope, *Faraday Discuss*, **111**, 79–94.

5. Cogollo, J. F. S., Tedesco, M., Martinoia, S., and Raiteri, R. (2011). A new integrated system combining atomic force microscopy and micro-electrode array for measuring the mechanical properties of living cardiac myocytes, *Biomed Microdevices*, **13**(4), 613–621.

6. A-Hassan, E., Heinz, W. F., Antonik, M. D., D'Costa, N. P., Nageswaran, S., Schoenenberger, C. A., and Hoh, J. H. (1998). Relative microelastic mapping of living cells by atomic force microscopy, *Biophys J*, **74**(3), 1564–1578.

7. Butt, H. J., Cappella, B., and Kappl, M. (2005). Force measurements with the atomic force microscope: technique, interpretation and applications, *Surf Sci Rep*, **59**(1–6), 1–152.

8. Alessandrini, A., and Facci, P. (2005). AFM: a versatile tool in biophysics, *Meas Sci Technol*, **16**(6), R65–R92.

9. Carnerup, A. M., Ainalem, M.-L., Alfredsson, V., and Nylander, T. (2009). Watching DNA condensation induced by poly(amido amine) dendrimers with time-resolved cryo-TEM, *Langmuir*, **25**(21), 12466–12470.

10. Lindsay, S. M., Nagahara, L. A., Thundat, T., Knipping, U., Rill, R. L., Drake, B., Prater, C. B., Weisenhorn, A. L., Gould, S. A. C., and Hansma, P. K. (1989). STM and AFM images of nucleosome DNA under water, *J Biomol Struct Dyn*, **7**(2), 279–287.

11. Weisenhorn, A. L., Gaub, H. E., Hansma, H. G., Sinsheimer, R. L., Kelderman, G. L., and Hansma, P. K. (1990). Imaging single-stranded-DNA, antigen-antibody reaction and polymerized langmuir-blodgett-films with an atomic force microscope, *Scanning Microsc*, **4**(3), 511–516.

12. Hansma, H. G., Weisenhorn, A. L., Gould, S. A. C., Sinsheimer, R. L., Gaub, H. E., Stucky, G. D., Zaremba, C. M., and Hansma, P. K. (1991). Progress in sequencing deoxyribonucleic-acid with an atomic force microscope, *J Vac Sci Technol B*, **9**(2), 1282–1284.

13. Bustamante, C., Vesenka, J., Tang, C. L., Rees, W., Guthold, M., and Keller, R. (1992). Circular DNA molecules imaged in air by scanning force microscopy, *Biochemistry*, **31**(1), 22–26.

14. Martin, A. L., Davies, M. C., Rackstraw, B. J., Roberts, C. J., Stolnik, S., Tendler, S. J. B., and Williams, P. M. (2000). Observation of DNA-

polymer condensate formation in real time at a molecular level, *FEBS Lett*, **480**(2–3), 106–112.

15. Li, B. S., Sattin, B. D., and Goh, M. C. (2006). Direct and real-time visualization of the disassembly of a single RecA-DNA-ATPγS complex using AFM imaging in fluid, *Nano Lett*, **6**(7), 1474–1478.

16. Putman, C. A. J., Vanderwerf, K. O., Degrooth, B. G., Vanhulst, N. F., and Greve, J. (1994). tapping mode atomic-force microscopy in liquid, *Appl Phys Lett*, **64**(18), 2454–2456.

17. Schabert, F. A., and Engel, A. (1994). Reproducible acquisition of Escherichia coli porin surface topographs by atomic force microscopy, *Biophys. J.*, **67**(6), 2394–2403.

18. Friedmann, T., and Roblin, R. (1972). Gene therapy for human genetic disease?, *Science*, **175**(4025), 949–955.

19. Crystal, R. G. (1995). Transfer of genes to humans: early lessons and obstacles to success, *Science*, **270**(5235), 404–410.

20. Tripathy, S. K., Black, H. B., Goldwasser, E., and Leiden, J. M. (1996). Immune responses to transgene-encoded proteins limit the stability of gene expression after injection of replication-defective adenovirus vectors, *Nat Med*, **2**(5), 545–550.

21. Tiyaboonchai, W., Woiszwillo, J., and Middaugh, C. R. (2003). Formulation and characterization of DNA-polyethylenimine-dextran sulfate nanoparticles, *Eur J Pharm Sci*, **19**(4), 191–202.

22. Godbey, W. T., Ku, K. K., Hirasaki, G. J., and Mikos, A. G. (1999). Improved packing of poly(ethylenimine)/DNA complexes increases transfection efficiency, *Gene Ther*, **6**(8), 1380–1388.

23. Zauner, W., Ogris, M., and Wagner, E. (1998). Polylysine-based transfection systems utilizing receptor-mediated delivery, *Adv Drug Deliv Rev*, **30**(1–3), 97–113.

24. Chen, J., Wu, C., and Oupicky, D. (2009). Bioreducible Hyperbranched Poly(amido amine)s for Gene Delivery, *Biomacromolecules*, **10**(10), 2921–2927.

25. Honore, I., Grosse, S., Frison, N., Favatier, F., Monsigny, M., and Fajac, I. (2005). Transcription of plasmid DNA: influence of plasmid DNA/polyethylenimine complex formation, *J Control Release*, **107**(3), 537–546.

26. Pollard, H., Remy, J. S., Loussouarn, G., Demolombe, S., Behr, J. P., and Escande, D. (1998). Polyethylenimine but not cationic lipids promotes

transgene delivery to the nucleus in mammalian cells, *J Biol Chem*, **273**(13), 7507–7511.

27. Schaffer, D. V., Fidelman, N. A., Dan, N., and Lauffenburger, D. A. (2000). Vector unpacking as a potential barrier for receptor-mediated polyplex gene delivery, *Biotechnol Bioeng*, **67**(5), 598–606.

28. Wan, L., Manickam, D. S., Oupicky, D., and Mao, G. Z. (2008). DNA release dynamics from reducible polyplexes by atomic force microscopy, *Langmuir*, **24**(21), 12474–12482.

29. Zabner, J., Fasbender, A. J., Moninger, T., Poellinger, K. A., and Welsh, M. J. (1995). Cellular and molecular barriers to gene-transfer by a cationic lipid, *J Biol Chem*, **270**(32), 18997–19007.

30. Oupicky, D., Parker, A. L., and Seymour, L. W. (2002). Laterally stabilized complexes of DNA with linear reducible polycations: strategy for triggered intracellular activation of DNA delivery vectors, *J Am Chem Soc*, **124**(1), 8–9.

31. Oupicky, D., Ogris, M., Howard, K. A., Dash, P. R., Ulbrich, K., and Seymour, L. W. (2002). Importance of lateral and steric stabilization of polyelectrolyte gene delivery vectors for extended systemic circulation, *Mol Ther*, **5**(4), 463–472.

32. Shim, M. S., Wang, X., Ragan, R., and Kwon, Y. J. (2010). Dynamics of nucleic acid/cationic polymer complexation and disassembly under biologically simulated conditions using in situ atomic force microscopy, *Microsc Res Tech*, **73**(9), 845–856.

33. Li, C., Tian, H., Duan, S., Liu, X. N., Xu, P. F., Qiao, R. Z., and Zhao, Y. F. (2011). Controllable DNA condensation-release induced by simple azaheterocyclic-based metal complexes, *J Phys Chem B*, **115**(45), 13350–13354.

34. Bechler, S. L., and Lynn, D. M. (2012). Characterization of degradable polyelectrolyte multilayers fabricated using DNA and a fluorescently-labeled poly(beta-amino ester): shedding light on the role of the cationic polymer in promoting surface-mediated gene delivery, *Biomacromolecules*, **13**(2), 542–552.

35. Schneider, S., Lenz, D., Holzer, M., Palme, K., and Suss, R. (2010). Intracellular FRET analysis of lipid/DNA complexes using flow cytometry and fluorescence imaging techniques, *J Control Release*, **145**(3), 289–296.

36. Chen, H. H., Ho, Y. P., Jiang, X., Mao, H. Q., Wang, T. H., and Leong, K. W. (2009). Simultaneous non-invasive analysis of DNA condensation and stability by two-step QD-FRET, *Nano Today*, **4**(2), 125–134.

37. Oupicky, D., Bisht, H. S., Manickam, D. S., and Zhou, Q.-H. (2005). Stimulus-controlled delivery of drugs and genes, *Expert Opin Drug Deliv*, **2**(4), 653–665.

38. Blacklock, J., You, Y.-Z., Zhou, Q.-H., Mao, G., and Oupick, D. (2009). Gene delivery in vitro and in vivo from bioreducible multilayered polyelectrolyte films of plasmid DNA, *Biomaterials*, **30**(5), 939–950.

39. You, Y. Z., Manickam, D. S., Zhou, Q. H., and Oupicky, D. (2007). A versatile approach to reducible vinyl polymers via oxidation of telechelic polymers prepared by reversible addition fragmentation chain transfer polymerization, *Biomacromolecules*, **8**(6), 2038–2044.

40. Wan, L., You, Y., Zou, Y., Oupicky, D., and Mao, G. (2009). DNA release dynamics from bioreducible poly(amido amine) polyplexes, *J Phys Chem B*, **113**(42), 13735–13741.

41. Vilfan, I. D., Conwell, C. C., Sarkar, T., and Hud, N. V. (2006). Time study of DNA condensate morphology: implications regarding the nucleation, growth, and equilibrium populations of toroids and rods, *Biochemistry*, **45**, 8174–8183.

42. Dauty, E., Remy, J.-S., Blessing, T., and Behr, J.-P. (2001). Dimerizable cationic detergents with a low cmc condense plasmid DNA into nanometric particles and transfect cells in culture, *J Am Chem Soc*, **123**(38), 9227–9234.

43. Manickam, D. S., and Oupicky, D. (2006). Multiblock reducible copolypeptides containing histidine-rich and nuclear localization sequences for gene delivery, *Bioconjugate Chem*, **17**(6), 1395–1403.

44. Christensen, L. V., Chang, C.-W., Kim, W. J., Kim, S. W., Zhong, Z., Lin, C., Engbersen, J. F. J., and Feijen, J. (2006). Reducible poly(amido ethylenimine)s designed for triggered intracellular gene delivery, *Bioconjugate Chem*, **17**(5), 1233–1240.

45. Lin, C., Zhong, Z., Lok, M. C., Jiang, X., Hennink, W. E., Feijen, J., and Engbersen, J. F. J. (2006). Linear poly(amido amine)s with secondary and tertiary amino groups and variable amounts of disulfide linkages: synthesis and in vitro gene transfer properties, *J Control Release*, **116**(2), 130–137.

46. Hoon Jeong, J., Christensen, L. V., Yockman, J. W., Zhong, Z., Engbersen, J. F. J., Jong Kim, W., Feijen, J., and Wan Kim, S. (2007). Reducible poly(amido ethylenimine) directed to enhance RNA interference, *Biomaterials*, **28**(10), 1912–1917.

47. Wu, D., Liu, Y., Chen, L., He, C., Chung, T. S., and Goh, S. H. (2005). 2A$_2$ + BB′B″ approach to hyperbranched poly(amino ester)s, *Macromolecules*, **38**(13), 5519–5525.

48. Blacklock, J., You, Y.-Z., Zhou, Q.-H., Mao, G., and Oupický, D. (2009). Gene delivery in vitro and in vivo from bioreducible multilayered polyelectrolyte films of plasmid DNA, *Biomaterials*, **30**(5), 939–950.

49. Wittmer, C. R., Phelps, J. A., Saltzman, W. M., and Van Tassel, P. R. (2007). Fibronectin terminated multilayer films: protein adsorption and cell attachment studies, *Biomaterials*, **28**(5), 851–860.

50. Blacklock, J., Mao, G., Oupickyì, D., and Möhwald, H. (2010). DNA release dynamics from bioreducible layer-by-layer films, *Langmuir*, **26**(11), 8597–8605.

51. Pelham Jr, R. J., and Wang, Y. L. (1997). Cell locomotion and focal adhesions are regulated by substrate flexibility, *Proc Natl Acad Sci USA*, **94**(25), 13661–13665.

52. Discher, D. E., Janmey, P., and Wang, Y. L. (2005). Tissue cells feel and respond to the stiffness of their substrate, *Science*, **310**(5751), 1139–1143.

53. Blacklock, J., Vetter, A., Lankenau, A., Oupicky, D., and Mohwald, H. (2010) Tuning the mechanical properties of bioreducible multilayer films for improved cell adhesion and transfection activity, *Biomaterials*, **31**(27), 7167–7174.

54. Moussallem, M. D., Olenych, S. G., Scott, S. L., Keller, T. C. S., and Schlenoff, J. B. (2009). smooth muscle cell phenotype modulation and contraction on native and cross-linked polyelectrolyte multilayers, *Biomacromolecules*, **10**(11), 3062–3068.

55. Richert, L., Schneider, A., Vautier, D., Vodouhe, C., Jessel, N., Payan, E., Schaaf, P., Voegel, J. C., and Picart, C. (2006). Imaging cell interactions with native and crosslinked polyelectrolyte multilayers, *Cell Biochem Biophys*, **44**(2), 273–285.

Chapter 12

Impedance Spectroscopy for Characterization of Biological Function

Brian C. Riggs,[a] Janet L. Paluh,[b] G. E. Plopper,[c] and Douglas B. Chrisey[d]

[a]Department of Physics and Engineering Physics, Tulane University,
6400 Freret Street, 2001 Percival Stern Hall, New Orleans, LA 70118, USA
[b]Nanodevices and Stem Cell Engineering,
College of Nanoscale Science and Engineering,
University at Albany-SUNY,
NanoFab East, 4424, 257 Fuller Road, Albany, NY 12203, USA
[c]Department of Biology, Rensselaer Polytechnic Institute,
110 8th St Bio Tech 2nd Fl, Troy NY 12180, USA
[d]Department of Physics and Engineering Physics,
Tulane University, 6400 Freret Street, 2001 Percival Stern Hall,
New Orleans, LA 70118, USA

riggsb2@gmail.com

The application of electrical impedance spectroscopy (EIS) is a long-standing method for the electrical characterization of conducting systems, such as metallurgical material systems, and more recently has been used to reveal exciting mechanistic discoveries in biology. Equivalent circuit models, in combination with physical models, provide insight into the various mechanisms of charge transfer throughout the material systems being probed often indicating deeper mechanistic insights. In this chapter, fundamental impedance theory is introduced, and we present

NanoCellBiology: Multimodal Imaging in Biology and Medicine
Edited by Bhanu P. Jena and Douglas J. Taatjes
Copyright © 2014 Pan Stanford Publishing Pte. Ltd.
ISBN 978-981-4411-79-0 (Hardcover), 978-981-4411-80-6 (eBook)
www.panstanford.com

the construction of equivalent circuit models and mathematical models using epithelial cells grown on interdigitated electrodes as an example to guide the reader in performing EIS experiments. Electrical impedance spectroscopy has been applied to biological systems in the past two decades allowing for extensive studies of biological phenomena in a real-time and non-invasive manner. This chapter concludes with a discussion on the using EIS to advance the understanding of fundamental biological processes, application in point-of-care treatments in the healthcare industry, and the development of "in the field" environmental monitoring.

12.1   Introduction

The observation of biological samples with electrical impedance sensing (EIS) requires addressing several challenges that include the relative time scale being observed, the effects of the measurement on the sample, and the sensitivity of the measurement. Most conventional procedures for biological sample investigation make observations at individual time points that are initially selected based on an educated guess of the significant processes involved. Although this has been used effectively when information on the kinetics of the system is available, this may not always be the case, leaving it up to the investigator to muse on the rate of the reactions in play or when a single reaction is to occur. The effect of any characterization method on behavior is also an issue that spans across all fields of science from biological samples, inorganic material samples, or chemical analyses. This latter point is generally addressed by assuming that the effect of the measurement is negligible with respect to the behavior of interest (e.g., staining assays require a fixing step that prevents further alteration from the staining procedure or radiation from x-ray imaging does not change the presence or integrity of bones). When concerning oneself with sensitive changes in cell behavior or chemical reactions, the observational method should be scalable to be considered to have negligible effects. Lastly, the sensitivity of the measurement must be fine enough for practical application; atomic resolution is not necessary when looking for a broken bone. However, it is highly desirable when looking at the surface roughness of implants. When faced with characterizing biological samples such as tissue layers, individual cell behavior, antibody–antigen reactions, and enzyme reactions researchers often turn to a method that conventionally is used for studying electrochemical systems: electrical impedance spectroscopy. Impedance spectroscopy is founded on measuring the resistance in movement of charge throughout the sample of interest. By applying an alternating voltage to a sample over a range of frequencies, the electrical response, which is directly related to physical behavior, can be characterized. This chapter will describe the theory and application of impedance spectroscopy for biological systems and provide a guide for using and understanding impedance measurements.

12.1.1 History

Over the past two centuries, several key contributions to impedance spectroscopy have been realized, outlined in Fig. 12.1. Most developments in characterization are founded in mathematic principles that allow for more exact measurements or simpler data analysis. The foundation of impedance spectroscopy was established in the late 19th century by Oliver Heaviside with the discovery of two operator transformations that allowed the conversion between Fourier and Laplacian space.[1] This allowed for the conversion of complicated integral equations describing circuit elements into much simpler algebraic expression that be solved without the aid of computational software. Warburg published a paper in 1899 applying these transforms to solve the impedance reaction for the diffusion of species to a surface.[2] This evolved into the Warburg element that is used to describe diffusion to a semi-infinite planar source. Electrical analogs used to describe the mechanism of charge transport came into fruition in the 1940s with the works of Dolin, Ershler, and Randles.[3,4] Using electrical circuit components, a material system could be described, although incompletely, by the arrangement of several ideal elements. As mathematical techniques developed to benefit analysis, hardware improvements increased the range and resolution at which measurements could be made. The advent of the electrical potentiostat (1940) and frequency response analyzer[5] (1970) allowed for the control of sub-millihertz frequencies, which is necessary to measure electrochemical reactions at electrode interfaces and corrosion mechanisms. These developments, in tandem with the analytical mathematics, allowed for the maturation of EIS into the field that is exploited for a wide range of applications. More recent advances include mathematical solutions for different physical systems, innovations in methods of mathematical analysis,[6,7] or improvements in hardware due primarily to the progress made in the electronics industry enabling better measurement. Today impedance spectroscopy is a well-established method for chemical characterization that allows for detailed examination into the mechanisms of electrochemical systems.

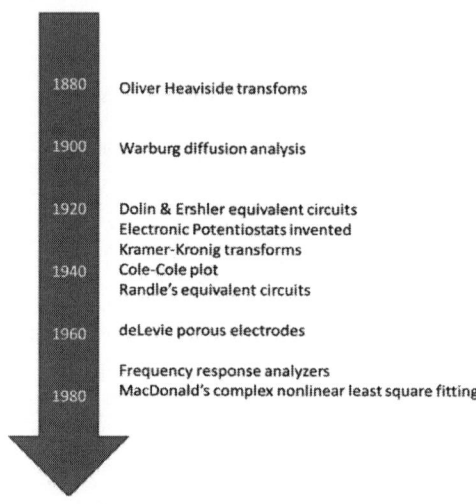

1880	Oliver Heaviside transfoms
1900	Warburg diffusion analysis
1920	Dolin & Ershler equivalent circuits
	Electronic Potentiostats invented
	Kramer-Kronig transforms
1940	Cole-Cole plot
	Randle's equivalent circuits
1960	deLevie porous electrodes
	Frequency response analyzers
1980	MacDonald's complex nonlinear least square fitting

Figure 12.1 Timeline of advancements for impedance spectroscopy.

12.1.2   General Information

Impedance spectroscopy measurements quantify the resistance to the movement of charge throughout the sample of interest. If a constant potential is applied to a segment of copper wire, it is found that electrons will move freely through the material toward the positive potential with *relatively* minimal resistance. This material system contains only one component and, as such, is relatively simple. If a constant field is applied to a glass of water, the water molecules will align their intrinsic polarity along the applied field. If an alternating field is applied, so that the polarization of the field changes with time, the polar molecules will rotate back and forth following the field movement. This rotation takes a certain amount of time and will be unable to keep pace with the changes of the field at high frequencies. At a molecule's resonant frequency for rotation, the amount of time it takes to rotate is equal to the inverse of the frequency and the molecule absorbs the most energy from the field. The most common product that applies this concept of resonant electromagnetic absorption to a material is a microwave oven. With more complex mixtures, containing many different charge carriers, interfaces, and complex geometries, a wide range of frequencies measurements, in combination with analogous models, help characterize the different components and reactions in the solution by the appearance of different resonant frequencies.

12.1.3 Experimental Setup

An EIS experiment utilizes an applied alternating voltage (V) with constant amplitude applied to a sample of known dimensions via conducting electrodes while the current is measured. Using Ohm's law ($V = I \times Z$), the impedance (Z) can be calculated. Impedance measurements can be done over a wide range of frequencies from <1 mHz to 100 GHz and there are several commercial impedance spectroscopy systems available.[8-10] In order to accurately characterize the impedance of a material the contribution of the electrode to the sample must be subtracted away. To do this, the geometry and contact area of the electrode must be known. There are different configurations for attaching the electrodes depending on the material to be measured, which will be discussed in Section 2.2.

12.1.4 More Than Just Equipment

The majority of the work involved in EIS is in the analysis and modeling of the physical system. Figure 12.2 shows a schematic flow chart of the process to obtain applicable, real, useful information. EIS is an iterative analysis process comparing several models against one another until a satisfactory relation is found. The first step is to develop a physical model that describes the material system of interest. The physical model should address possible kinetic mechanisms of the system such as a charge transfer at the electrode interface or the diffusion of ions within the material. From a general physical model a mathematical model can be created that fully describes the system that relies on fundamental physics and kinetics, including the application of Fick's laws and basic electronic behavior. Finally, there is the equivalent circuit model

(EqC) that represents a physical model in terms of ideal electrical components with easy to describe behavior. The combination of electrical components can describe physical systems simply, although not rigorously. The electrical impedance of the components is related to the physical nature of the real system. Utilizing a complex, nonlinear least square fit to the data, it is possible to confirm how well the EqC model mimics the sample being measured. Once the best model is determined, physical properties describing the sample can be derived.

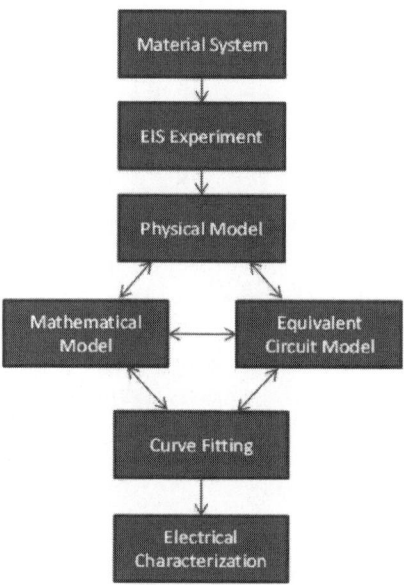

Figure 12.2 Flow chart of method for EIS.

12.1.5 Applications

EIS has been applied to a wide variety of electrochemical systems including conventional lead batteries,[11–14] newly developing fuel cell technologies,[15–17] simple electrochemical solutions to monitor reactions, and more recently it has been applied to biological systems for monitoring protein function and configuration and cell and tissue behavior.[18–22] EIS works with any biological system in which the changing movement of charge can be related to the impedance. For example, a single strand DNA molecule is capable of carrying a charge along its nucleotide sequence length. The binding of a complementary DNA strand will block the conduction along the chain showing an increase in impedance. Such a method has been used to detect the presence of Hepatitis B.[23] The rate of enzymatic activity can be determined, for example, by monitoring the electrical character of a collagen layer as it is exposed to collagenase, knowing the impedance of the layer is related to the thickness.[24] A more conventional non-biological application of EIS is in examining the impedance of a lead-acid battery.[11] The appearance of three distinct regions of the Nyquist plot demonstrates the multiple modes of conduction within the battery. We can already see the variety of applications for which EIS can be utilized. This chapter the focus will be mainly on biological applications.

12.2 Materials and Methods

To measure the impedance, a phase sensitive lock-in amplifier is used which compares the input AC signal to an output waveform. The difference between the two waves is characterized by the phase shift, represented as an angular measurement. Because the input potential is constant, the in-phase and out of phase components can be calculated through multiplying by the sine or cosine of the phase shift. The sum of the squares of these components equates to the square of the overall impedance. In this manner the distinct impedance measurements can be made over a range of frequencies. Each measurement can be taken within several milliseconds of one another at set intervals, which is important as the time scale of the measurement cannot exceed that of the reaction of interest.

In order to perform an EIS experiment, a voltage must be able to be applied the material system of interest between two electrodes which can be organized with various different geometries. Figure 12.3 shows three simple configurations for different applications: (a) liquids solutions, (b) solid–state solutions, (c) biological systems. For liquid systems electrodes can be placed within the solution and the electric field lines, along which charges move, are directed straight across between the electrodes. Similarly, a solid sample will be sandwiched between two circular electrodes of known diameter. For this configuration the sample must have a minimum aspect ratio between the diameter and thickness of the sample to assure that the electric field between the electrodes are all parallel to one another.[25] The last configuration uses interdigitated electrodes to measure a sample that has been deposited on the surface. It is important to emphasize the point regarding the formation of field lines through the material and how they must run through the material to be measured. With interdigitated electrodes the field lines form tall arcs that bridge between electrodes. If the prepared sample prepared is thin enough, the field lines can be approximated as travelling perpendicular through the sample as with the other two configurations. The interdigitated configuration has been widely used over the past 20 years for the study of biological phenomena.[18,20,22,26-35] This section will review the methods for running an EIS experiment using epithelial cells growing over interdigitated electrodes as an example.

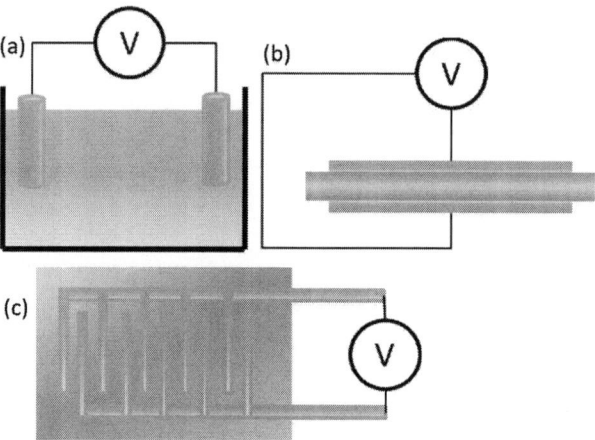

Figure 12.3 (a) Aqueous solution. (b) Solid material. (c) Interdigitated electrodes for biological samples.

12.2.1 Plotting the Data

Cells were grown on electrodes to confluence over several days and many impedance measurements were made during this time either manually or with some systems impedance measurements have been automated as to acquire data in a near continuous manner. As mentioned in the Section 12.1, the EIS measures the impedance (Z) of the system over a range of frequencies (f). The use of an AC signal produces an additional degree of freedom that becomes crucial for EIS. Any time delay between the voltage and current signal is referred to as the phase angle (θ). Instead of two degrees of freedom, there are three: impedance (Z), phase angle (theta), and frequency (f) called the polar measurements. The impedance is a vector and can be visualized on a complex plane (Fig. 12.4) with the impedance as the magnitude and phase angle as the degree of rotation. Just as the time delay in molecule rotation changes with frequency so will the measured impedance and phase angle. The impedance vector can be decomposed into its real and imaginary components to express in a more understandable, Cartesian form. The three variables: real impedance (Z_R), imaginary impedance (Z_I), and frequency (f), which are then plotted to look for patterns.

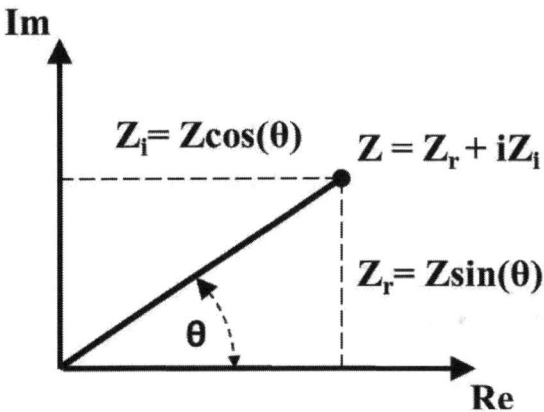

Figure 12.4   Complex impedance plane.

There are two main ways to plot the data received from an EIS experiment: the Nyquist plot and the Bode plot. The Nyquist plot is most commonly used because it allows those with even a little experience with EIS to glean intuitive information about the material system, thereby allowing for a more accurate first estimate at an equivalent circuit model. As shown in Fig. 12.5a, the Nyquist plot displays $-Z_I$ on the vertical and Z_R on the horizontal axis. Each point plotted corresponds to a different frequency with frequency increasing as the points approach the origin. Changes in curvature in the Nyquist plot suggest different time constants (inverse of resonant frequency) within the material system. This can be more easily seen in Bode plots, which show the magnitude of the impedance against the frequency. Figure 12.5b shows the Bode plot for a one time constant material. There is a distinct linear portion with a non-zero slope. This implies that there is one distinct mechanism for charge to move through the material. In Fig. 12.5d, the Bode plot shows two non-zero linear portions giving the impression that two mechanisms come into play, providing a starting point for developing the equivalent circuit model. Looking at the Nyquist plot in Fig. 12.5c, there are two peaks. The frequency that is at the peak is the inverse of the time constant that was described earlier (the time necessary for a molecule to react to a field). By a simple graphical analysis, the time constants of the mechanisms in our system can be determined. This is easy to figure out for simple, ideal models; however, real systems representing contemporary problems are rarely this simplistic. The ease of finding time constants disappears as the complexity of the model increases and as more complicated math is utilized.

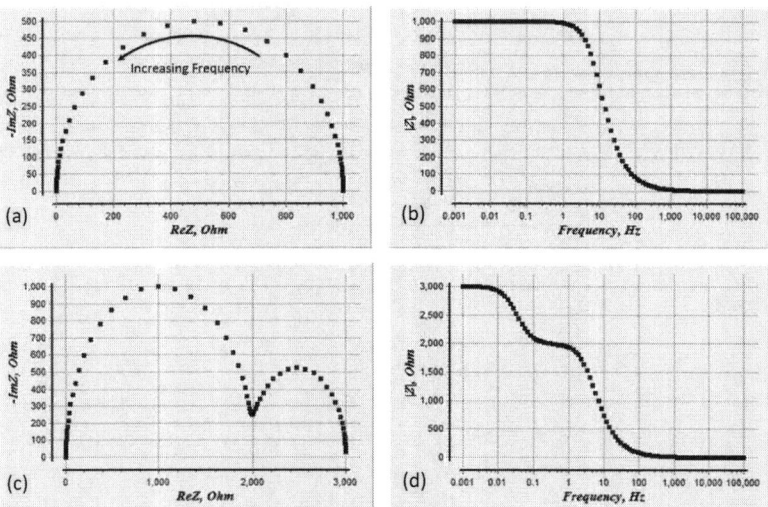

Figure 12.5   (a) Nyquist plot for one time constant system. (b) Bode plot for (a). (c) Nyquist plot for two time constant system. (d) Bode plot for (c).

12.2.2   Building an Equivalent Circuit Model

Building physical and equivalent circuit models that are good approximations for our real world systems is the next step. It is worth noting that the use of a model is only an approximation and will not fully describe the entire system. Any model must make use of a combination of ideal circuit components, common to electrical circuit design, and non-ideal components that were created in order to better explain impedance spectroscopy measurements. The three basic passive electrical components are a resistor, capacitor, and inductor, as seen in Table 12.1. A resistor acts to inhibit the flow of current through a material such as a slab of wood or a copper wire. Ideally, a resistor will have no frequency dependence and produce no lag between the current and voltage resulting in no phase angle and only a real component of the impedance. These characteristics of a resistor lead to a Nyquist plot that is a single point at the resistance of the resistor. An ideal capacitor, on the other hand, acts to store and release charge when exposed to an AC signal. Because energy is not released instantaneously, a lag between the current and voltage arises 90° out of phase creating a purely imaginary component of impedance. An inductor acts inversely to a capacitor where the voltage lags the current. This creates a negative phase angle leading to negative imaginary impedance.

Table 12.1 Table of equivalent circuit components

Component	Symbol	Diagram picture	Equation ($Z = ...$)	Possible physical meaning
Resistor	R	—/\/\—	R	Blocking flow
Capacitor	C	—||—	$\dfrac{1}{i\omega C}$	Double layer capacitance
Constant phase element	CPE	—[CPE]—	$\dfrac{1}{(i\omega)^n C}$	Diffusion, range of time constants
Wargburg element	Z_W	—[Z_W]—	$\dfrac{1}{(i\omega)^{1/2} C}$	Diffusion to surgace
Inductor	L	_/\/\/_	ωL	Adsorption

With a base for building an equivalent circuit model, an example of a biological system can be put together. For simplicity, an epithelial cell monolayer, such as Madine Darby canine kidney (MDCK) cells grown on interdigitated gold electrodes shown schematically in Fig. 12.6 is considered. When the monolayer has grown to confluence the complete coverage of the electrodes with distinct intercellular junctions is seen. The monolayer does not lie flat on the substrate but has several focal adhesion points of contact to keep it in place and can be modeled as "hovering" a certain distance above the electrodes. The field lines through our material system characterize the different components as they arise. For example, starting at A, there is an interaction at the electrode interface. This is an important interaction to remember to include, because it occurs in every EIS measurement. The interaction between the electrode and the material is systematic "noise" that needs to be taken out. As the movement of a charge across an interface is commonly modeled as a capacitor, the interface at A will be treated the same. At B the properties of the material of interest are present. There are two components in our cell monolayer. First are the cells themselves, which, due to their polar plasma membranes, can be treated as a capacitor. Electrons are unable to move instantaneously through a given cell and produce a lag between current and voltage. The second component is the junction between cells that inhibits but does not deny the passage of electrons. For this reason, the junctions can be treated as a simple resistor. The final component influencing the system is our cell media. Consisting of free ions and proteins, and mostly water, a polar molecule that responds temporally with the applied signal, will be treated as a capacitor. So now we have an equivalent circuit diagram that resembles our physical model.

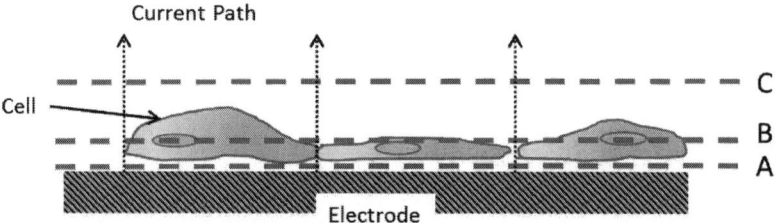

Figure 12.6 Schematic of epithelial cells grown on gold electrode.

Assembling a mathematical model can be much more difficult in that we are starting from first principles. Using the MDCK confluent monolayer grown on the electrode as an example and follow Giaver and Keese's derivation.[19,33] The derivation combines basic definitions of current as well as the various potential drops across the membrane in order to define the electronic flow as a function of electrical properties. Such inputs include the membrane capacitance, a constriction resistance between the cells and electrode, and a resistance due to the formation of tight junctions called the barrier function. The model assumes cells are discs, hovering a certain distance above the electrode. The process of looking at different cross sections of our physical model is the same; however, instead of using ideal elements Ohm's law is applied in cylindrical coordinates. Through some manipulation the resulting equation expressing the impedance of the material system as a function of the previously mentioned parameters is given. It is important to note is that these parameters are directly dependent on the physical properties of the cells and that by fitting Eq. (12.1) to our data the parameters and, therefore, the physical properties including the average cell radius and spacing from the substrate can be found.

$$\frac{1}{z_c} = \left(\frac{1}{z_n} \frac{z_n}{z_n + z_m} + \frac{\dfrac{z_m}{z_n + z_m}}{\dfrac{i\gamma r_c}{2} \dfrac{I_0(\gamma r_c)}{I_1(\gamma r_c)} + 2R_b \dfrac{1}{z_n + 1/z_m}} \right)$$

12.2.3 Validating and Improving the Model

The aforementioned model is useless for describing our material system if it does not match the experimental results. Fitting the model is easy using available software such as EIS analyzer,[36] which allows the user to input experimental data, our equivalent circuit and automatically find the fit parameters of interest. Figure 12.7 shows this process with data collected for a confluent monolayer of MDCK cells after 40 h of culture. The total error for this model exceeds 25% for each parameter involved for a total of over 150% error suggesting that the equivalent components might be changed, e.g., the plasma membrane. The membrane is not solely a dielectric material, but also contains surface proteins to allow transport of ions in and out of the cell. To some extent, this transport adds a resistive aspect to the membrane. Dielectrics that allow some conduction across them are considered "lossy." Instead of a capacitor, a constant phase element (CPE) can be used. The formula for impedance caused by a capacitor compared to a CPE (Table 12.1) contains an extra exponent n that provides a wider range of phase angles. The value of n reflects how "ideal" the component is. When

$n = 1$, we have an ideal capacitor. When $n = 0$ the component behaves as an ideal resistor. The CPE is substituted instead of the membrane capacitance. Examining the other capacitor, It is found that the using a capacitor to describe electrode interactions is common practice.[37-39] Finally, there is the contribution of the media. There are many things that cause a capacitor to be non-ideal including the presence of a distribution of time constants (multiple species) and diffusion of active species.[37-40] With the new model it is found that there is a significantly better fit (<10% error). Note that with any "fitting method" there is a risk of choosing a model that has the best fit despite its lack of relevance. Many different EqCs result in the same response; however, some are better at representing the physical system than others. It is important to keep in mind the significance of the model that is chosen to apply to the data.

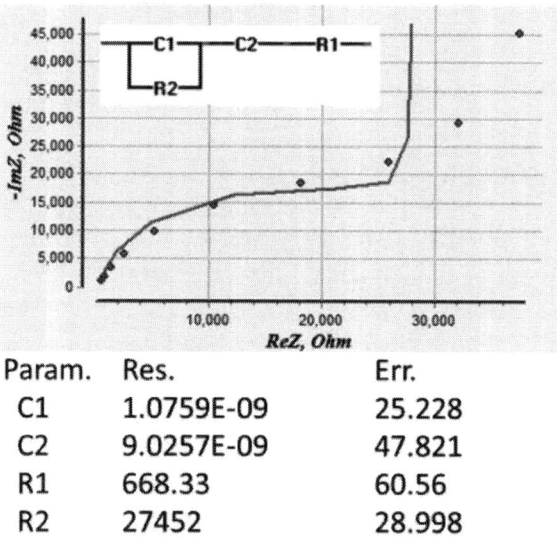

Param.	Res.	Err.
C1	1.0759E-09	25.228
C2	9.0257E-09	47.821
R1	668.33	60.56
R2	27452	28.998

Figure 12.7 First model results. Inset: Equivalent circuit model.

12.2.4 Physical Meaning of Our Parameters

Being able to quantify electronic biological behavior is important; however, without a connection to the physical meaning, the information will not have much significance. The capacitance at the electrode is not the material of interest and should be ignored or subtracted out. EIS analyzer is capable of subtracting electric circuit components with inputted parameters values from the data set. This is incredibly advantageous so that the components of interest are isolated. Physically, the capacitance describes electrons moving across the interface from gold to media and vice versa. This does not give any information about the cells and is viewed as an artifact of the measurement. Next, there is the CPE parameter of the media itself that describes how proteins and ions within the media diffuse toward the cell layer, or electrode surface. Although the composition of the media may be interesting, as it could measure the rate components are being metabolized by the cell, the changes of CPE parameters change very slightly over time and are not a strong signal. This will also be subtracted from the original data. This reduced model is now comprised of a CPE and R in parallel. This circuit should represent the confluent MDCK cell monolayer with the R describing the intercellular junctions and CPE describing the integrity of the membrane as a whole. The resistance of the intercellular junctions is directly proportional to the integrity of these cell–cell bonds, allowing us to monitor monolayer cohesiveness as we expose cells to different experimental environments. Examining the plasma membrane, as charge is allowed to flow through more easily, it is expected that n would reduce in value and the CPE to behave more like a resistor. This situation would arise if "holes" formed in the membrane or, in general, when the cell was allowing easier transport of ions across the membrane. This model does not hold for all time-points, however. One assumption of conventional impedance spectroscopy is that behavior does not change over time allowing us to form and maintain a single model. During cell growth, however, there are many changes occurring within the testing well, most notably the increased coverage of the electrodes. Until the cell culture reaches confluence, our model is invalid. A model for our confluent MDCK monolayer was just described; however, using ideal components is inexact. Although such a model helps to understand what mechanisms come into play, a rigorous examination of the charge transfer requires the use of a mathematical model.

12.3 Applications

It has been shown that EIS analysis can be applied to a simple monolayer biological sample. The ability to exploit the electronic characteristics of our sample for practical applications has yet to be explored. The remainder of the chapter will cover the different uses of EIS as applied to biologically derived materials for the creation of biosensors. There are three main fields for which biosensors can be applied: understanding fundamental behavior of biological samples, clinical diagnosis, and applications in the healthcare industry, and environmental monitoring for pollutants or biological warfare agents. Table 12.2 summarizes the four common biological materials that act as the sensor in our device. There are advantages and drawbacks for using each material as a sensing agent. For example, antibody based biosensors are highly sensitive, being able to detect target molecules down to the nanogram level; however, their highly specificity limits their use in detecting unknown unpredicted factors.

Table 12.2   Types of biosensors

Type of sensor	Sensing component	Applications
Enzyme based	Enzyme reactive layer	Enzyme kinetic studies[24] Cancer drug screening[74]
Immunosensors	Antibodies/ antigens bound to the surface	Antibody/antigen binding kinetics Pathogen detection Body fluid analysis[75]
Nucleic acid Based	ssDNA, RNA, dsDNA	Viral detection[23] Genetic testing Hybridization kinetics
Whole cell sensing	Live cell culture	Cancer[18] Pathogen detection Cell mobility[29,33]

12.3.1   Fundamental Biology

EIS is useful for quantifying common cellular behaviors in real-time, including cell–substrate adhesion and spreading, cell proliferation, and cell–cell contact. Most commonly, the experimental setup consists of a suspension of cells plated in culture wells containing an electrode on the bottom, coated with defined substrate molecules, such as purified extracellular matrix proteins; the reference electrode is placed in a side wall of the chamber to establish the circuit. Cell adhesion results in a rapid (seconds to minutes), but small rise in impedance, and spreading induces a more sustained (minutes to hours) rise in impedance as more of the electrode surface is covered by cell membranes.[28,41,42] This approach is useful for discriminating between different cell morphologies as cells remodel their cytoskeleton and/or express new cell surface receptors.

The impedance technique is suitable for a wide range of eukaryotic cell biosensing applications because of its sensitivity to detect subtle changes to cell membranes. This includes aspects of cell motility such as chemotaxis, migration in wound healing or trans-endothelial migration in metastasis, cytotoxicity assays for chemicals that disrupt cell adhesion or cell–cell junctions, similar genetic and epigenetic impact on cell membranes via internal or extracellular-based receptor signaling events, and responsiveness of cells to different extracellular matrix substrates. For the latter it has been shown, for example, that MDCK cells spread most rapidly when fibronectin is used to coat electrodes versus vitronectin or laminin. Environmental factors that affect the micromotion of cells will also influence the impedance signal, including temperature, pH, and media composition, and therefore must be evaluated in experiment and device design. EIS provides critical insights into how cellular information pathways direct dynamic changes to cell behavior-processes that are at the core of embryonic to adult development and homeostasis.

12.3.1.1 Cell adhesion

One clever approach uses anti-integrin receptor antibodies as the adhesive substrate, and cells transfected to express this integrin receptor adhere as a function of gene expression and protein transport to the cell surface.[43] By varying the adhesive substrate and/or receptors expressed, it is possible to directly compare the functional differences between specific substrate/receptor combinations. Cells that form highly adhesive interactions with one substrate/receptor will typically form adhesion complexes with smaller gaps between membrane and substrate, compared to weakly adhesive substrate/receptor combinations. This provides a quantitative measurement of membrane-substrate proximity, which is traditionally observed with light microscopy techniques such as differential interference contrast.[44]

12.3.1.2 Cell mobility

If cells are initially plated at subconfluent densities, cell proliferation is reflected by a further rise in impedance over hours to days.[45] Epithelial/endothelial cells capable of forming tight junctions induce an additional increase in impedance as they reach confluence and establish these junctions. Alternatively, cell migration is easily quantified by allowing cells to grow to confluence, then selectively killing the cells growing on top of the electrode with a fatal pulse of current. Migration is reflected by an increase in impedance as the surviving cells migrate to cover the resulting gap in the monolayer.[46–48] An additional application of polarized cell monolayers is measuring the loss of impedance as cancer or immune cells migrate through the monolayer. For example, we used this approach to explore the role of specific cell–cell adhesion proteins in transmigration of breast cancer cells through an endothelial monolayer; the rate and magnitude of impedance drop caused by the invading cancer cells is sensitive to function-blocking antibodies that target specific cell surface proteins.[49]

12.3.1.3 Cell signaling

Lastly, EIS techniques provide an opportunity to explore the functional roles of specific cellular signaling pathways in control of these cellular behaviors.[50–53] Signaling molecules, drugs, blocking antibodies/peptides, etc., can be added to these assays at any time, to compare treated and untreated conditions in the same sample. Indeed, virtually all of the modern techniques for exploring the function of specific molecules in cells are applicable to EIS analysis. For example, when coupled with genetically engineered cells and conventional fluorescence microscopy techniques, one can couple the expression and localization of specific proteins with the quantitative real-time data EIS provides.

12.3.2 Healthcare Applications

In EIS applications, the versatility of electrical cell impedance sensing to be able to detect multiple aspects of cell behavior makes it an integral affordable tool for basic research as well as more advanced health care therapeutic applications. The membranes of cells exhibit capacitive properties allowing for the use of changes in electrode impedance to monitor key aspects of cell behavior with relevance to normal development or disease states. Current flow through the cell is frequency dependent, which also allows fluid volume inside and outside of cells to be monitored. In addition to applications in body composition, other measurable parameters include cycling cell behavior-such as morphological transitions of growth and mitotic cell division, the cell attachment terrain, cell spreading in monolayers, and changes to the robustness of cell–cell barrier contacts at tight junctions. Diseases, whether acute or chronic, that affect tissue systems of the body generally impact some aspect of these processes. The utility of EIS is that it offers an alternative in adherent assay applications to traditional tissue culturing in vitro or animal testing in vivo. As well, the concept of cell impedance sensing is a familiar component of the rapidly emerging field of bio-nano-electromechanical and bio-micro-electromechanical systems (bioNEMS and bioMEMS) that underlie many of the lab-on-a-chip, organ-on-a-chip, or sensor-on-a-chip medical devices aimed at addressing more personalized, real-time, affordable medical care and point-of-care diagnostics.[54–56]

Herein, we highlight two applications here that utilize EIS in DNA or cell based biosensing applications along with selected reviews that provide both detail and overview of these fields of research. There is a wealth of applications and publications in health and environment sensing from a large field of contributors in this rapidly moving important technology.

12.3.2.1 Hepatitis-B DNA sensors

Accidental exposure in handling of blood and tissue samples, as well as through sexual contact with an infected individual, can transmit the hepatitis B virus (HBV), which if persistent will disable liver function via hepatitis, cirrhosis and cancer. Although a variety of approaches have been used to detect and monitor infection,[23] genetic diagnosis of serum HBV DNA levels, applied over the last decade, continues to offer an advantage of a simple, sensitive, low-cost and quantitative approach when coupled with amplification methods of standard polymerase chain reaction (PCR) or newer technologies, such as the TaqMan fluorogenic detection method[57] as well as electrochemical sensing. Taqman offers the advantage of avoiding cumbersome gel electrophoresis while enabling high-throughput screening. In electrochemical sensing, combining PCR amplification with detection on gold electrodes has been shown to allow detection of femtogram material once amplified to 10^4 copies for detection.[23] In this sensor approach, single stranded DNA (ssDNA) molecules are immobilized onto a thioglycolic acid (TGA) monolayer on a gold electrode surface. The binding of complementary HBV DNA molecules creates a double stranded DNA (dsDNA) molecule that is detected by the electroactive indicator ferrocenium. The use of PCR and DNA hybridization in electrochemical sensors is being rapidly explored by many groups that incorporate a variety of immobilization strategies for the ssDNA that include physical absorption, electrostatic binding, self-assembly, covalent bonding and common biotin–streptavidin interactions. Important parameters include the accessibility of the ssDNA strand in terms of its positioning and its freedom of motion for promoting hybridization reactions to form dsDNA. To enhance these parameters, a recent study attaches ssDNA covalently via a multi-atomic arm cross-linker, in this case glutaraldehyde, to a highly conductive composite film containing multi-walled carbon nanotubes (MWNTs).[58] In this device, the electron transfer kinetics of the biosensor was dramatically improved. The biosensor via EIS revealed a detection limit of 10^{-14} M and a linear range of 10^{-13} M to 10^{-10} M, and revealed both faster hybridization rate and frequency than other reported DNA biosensors.

12.3.2.2 Cancer metastasis

Cancer metastasis represents one of the most devastating features of cancer. EIS technology can be applied toward enabling early detection of cancer as well as understanding the underlying genetic, epigenetic and environmental influences on metastatic potential.[59] A defined signaling environment normally regulates differentiated cells, preventing over-proliferation of neighboring cells or migration from this site. The underlying changes facilitating acquired migration, intravasion into blood or lymph systems and extravasion into new "foreign" and unrestricted physiological environments are under investigation. This includes understanding how cells in up to 25% of cancers eventually metastasis beyond the blood brain barrier to reach the brain through the central nervous system vasculature. EIS is an important technology resource for detecting and measuring key parameters of cell–cell and cell–substrate interactions as well as cell–cell communication networks. Single cell resolution information can be obtained over days on a scale of minutes and when implemented with fluorescent measurements and miniaturization provides an invaluable tool underlying high-content studies that enable drug discovery. Normal cells form a monolayer that impedes the electrical output. Invading metastatic cells are able to disrupt cell–cell interactions in a mesothelial cell layer to penetrate this monolayer. This results in a significant and measurable drop in resistance. EIS is also invaluable as an in vitro means to monitor real-time the cytotoxicity effect as a function of compound and dose-dependent changes on various cell types for rapid screening of anti-cancer therapeutics. This information, coupled with knowledge of the cell biology, cell cycle, and cell signaling aspects, makes for rapid reliable and cost effective advancements.

12.3.3 In the Field: Environmental Monitoring

Rogers discusses the need for us to have a constant monitor on our environment for pollutants.[60,61] The selectivity and sensitivity that biosensors offer make them attractive for detecting low concentrations of pollutant including herbicides,[62] heavy metals,[63,64] and aromatic hydrocarbons.[65] With the growth in use of biological and chemical weapons, detection of such agents has also become a significant concern. There are many technical and practical challenges of creating such a device, including questioning the ability for a biosensor to detect the incredibly broad spectrum of pollutants; however, these are being addressed utilizing BioMEMS technology.

12.3.3.1 Detection of environmental pollutants

The impact of waste that is produced during everyday life is of major concern. In a product's life cycle, from manufacturing processes to treating crops with herbicides, any production step can lead to waste that if left unchecked has the potential to seep into our water supply and create a health risk. Wong et al. discussed the real-time sensing of the herbicide imazethapyr using a fiber optic immunosensor capable of detecting the herbicide in the nM scale range within 30 sec after exposure.[62] Although this work demonstrates the sensitivity of such immunoassays, the sample needed to be prepared by fluorescently tagging the analyte. This prevents it from being practically used as a mobile monitoring device. Panasyuk-Delaney et al. employed an impedance based chemosensor operating under the same principles of specific binding and noticed distinct changes in capacitance after exposure to varying amounts of the herbicide desmetryn.[66] The core-sensing concept of the device, however, can be applied to impedance spectroscopy. The sensor is an antibody–antigen reaction, relying on the binding of imazethapyr to immobilize the antigen and cause fluorescence. The intensity of the fluorescence is directly correlated to the concentration of herbicide present. If the antibodies are bound to an electrode setup, instead of a quartz disc, the overall impedance can be measured over time. When the herbicide selectively binds to the antibody, the resulting configuration change will act to restrict current flow and an overall increase in impedance would be seen. This eliminates the necessity for fluorescent tags, and therefore sample preparation, while maintaining desired sensitivity and real-time results.

12.3.3.2 Detection of foreign pathogens

Biological warfare agents have been used throughout history; however, recent advances in technology have increased the potential threat over the past seven decades. The means of toxin introduction into the body can vary by ingestion, absorption through the skin, or inhalation into air passages. The latter that involves aerosolized particles is the most probable and most dangerous. The development of mobile monitoring systems for detection of unknown toxins will increase defense against such potential threats. A number of assays have been proposed to accomplish such a task. A similar setup to Ye's work with ssDNA based biosensors could be used in order to seek out specific virus based threats.[23] One could imagine a microfluidic device with multiple steps to lyse a virus, amplify the DNA by polymerase chain reaction, then pass the processed solution over electrodes coated in complementary DNA to determine the presence of specific toxins.[67,68] This could be taken a step further and with recent micropatterning processes, create an array of various antibodies on specific electrode sites, allowing for a wider range of pathogens to be detected. Although the sensitivity is such a described system is high, the problem still remains that unpredicted pathogens in our environment will go undetected in this setup. For this reason, cell-based biosensors have been more thoroughly investigated leading to the development of several portable electronic sensors.[61,69-71] Briefly, a cell-based biosensor is created by culturing cells onto an adherent electrode. No special preparation of cells is necessary in contrast to DNA and immune based sensors. These cell sensors act as a canary in the coal mine, reacting to any adverse conditions that it may be exposed to. Looking at the example of epithelial MDCK cells from section two, when exposed to sufficient levels of toxin we would observe that the cells undergo apoptosis (programmed cell death) that results in the tight junctions at cell–cell interfaces as well as the focal adhesions to the substrate being degraded. As the junctions are removed, current more easily passes through these interfaces and the resistance drops. DeBusschere et al. developed a portable cell-based biosensor so that it may be carried into hazard zones to detect potential cell-toxic threats.[71] The device uses a cartridge loaded with a cell culture sample of cardiomyocytes that have been grown over an integrated circuit board. The beat rate of the sample was monitored over time. Changes in beat rate were found to be indicative of adverse conditions. More recent advances on integrating cell cultures into circuitry have been investigated[69,70] although there are still significant challenges including maintenance of cell viability over long periods of time and the introduction of sufficient toxin into the sensor itself. These are engineering problems that are expected to be addressed by integration of additional technologies.

12.4 Conclusion

Electrical impedance spectroscopy is based on measuring the behavior of charge carriers while being exposed to a time varying voltage signal over a range of frequencies. By properly modeling our material system, we can develop mathematical and electrical relationships that correspond to physical and chemical properties. This has been used for electrolyte characterization and corrosion detection and recent years have shown impedance spectroscopy being applied in successful and exciting applications to biological samples ranging from fundamental biological characterization to clinical diagnostic point of care techniques.

There is still much work to be done to improve the use of impedance spectroscopy with biological systems. The amount of information currently generated by the system to relate to biological functions is limited in parameters; however, with the use of imbedded conductive particles placed into the cellular membrane a readout of reactions happening within the cell can be obtained.[72] Such strategies will lead to a greater degree of characterization of cellular behavior, while continuing to allow monitoring cells non-destructively in real-time and avoid the use of fluorescent tags to provide information. Clinical applications are moving toward point-of-care treatment methods that allow early detection by real time monitoring as well as "upon need" dispensing of medicines to limit the need for frequent hospital visits. Microfluidic devices such as lab-on-a-chip[73] have already moved in this direction for specific monitoring and diagnostics. By combining microfluidic engineering with impedance based sensing, fast and reliable point-of-care devices can improve clinical care through personalized medicine. Similar techniques will be applied to field-able biosensors for pollutant and warfare agent detection. There have already been several prototypes of portable sensors built. The limitations they currently face, including viability of cells and toxin introduction methods, are engineering problems that can be addressed by better integration of existing technologies including microfluidics, resistive heaters, and air sampling methods. Solving these engineering problems will allow the maturity of biological impedance spectroscopy to be fully realized as a standard diagnostic tool in health care, medicine, and biosensing.

References

1. Macdonald, D. (2006). Reflections on the history of electrochemical impedance spectroscopy. *Electrochim. Acta,* **51**, 1376–1388.

2. Warburg, E. (1899). Uber das Verhalten sogenannter unpolarisirbarer Elektroden gegen Wecheslstrom. *Ann. Phys. Chem*, **67**, 493–499.

3. Randles, J. E. B. (1947). Kinetics of rapid electrode reactions. *Discuss. Faraday Soc.,* **11**, 11–19.

4. Dolin, P. I., and Ershler, B. V. (1940). The kinetics of discharge and ionization of hydrogen adsorbed at Pt-electrode. *Acta Physicochim URSS URSS*, **13**, 747.

5. Smith, D. (1971). Recent developments in alternating current polarography. *CRC Crit. Rev. Anal. Chem.,* **2**, 247–343.

6. Gathright, W., Jensen, M., and Lewis, D. (2011). Phase field model of chemical reactions with an example of a solid electrolyte gas sensor. *Electrochem. Commun.,* **13**, 520–523.

7. Gathright, W., Jensen, M., and Lewis, D. (2012). A phase field model of electrochemical impedance spectroscopy. *J. Mater. Sci.,* **47**, 1677–1683 LA.

8. Applied_Biophysics, http://www.biophysics.com/. at <http://www.biophysics.com/>.

9. Metrohm,http://www.metrohmusa.com/Products/Electrochemistry/? gclid=COnCzrL6364CFYbe4AodXGtDXA.

10. Zurich_Instruments, http://www.zhinst.com/products/hf2is?gclid=C IbWn6H6364CFcbc4Aodcgtcag.

11. Mauracher, P. (1997). Dynamic modelling of lead/acid batteries using impedance spectroscopy for parameter identification. *J. Power Sources,* **67**, 69–84.

12. Blanke, H., et al. (2005). Impedance measurements on lead–acid batteries for state-of-charge, state-of-health and cranking capability prognosis in electric and hybrid electric vehicles. *J. Power Sources,* **144**, 418–425.

13. Rodrigues, S., and Munichandraiah, N. (2000). A review of state-of-charge indication of batteries by means of ac impedance measurements. *J. Power Sources,* **87**, 12–20.

14. Salkind, A. J., et al. (2003). Impedance modeling of intermediate size lead–acid batteries. *J. Power Sources,* **116**, 174–184.

15. Springer, E., Zawodzinski, A., Wilson, M. S., and Golfesfeld, S. (1996). Characterization of polymer electrolyte fuel cells using AC impedance spectroscopy. *J. Electrochem. Soc.,* **143**, 587–599.

16. Lefebvre, M. and Martin, R. (1999). Characterization of ionic conductivity profiles within proton exchange membrane fuel cell gas diffusion electrodes by impedance spectroscopy. *Electrochem. Solid-State,* **2**, 259–261.

17. He, Z., Wagner, N., Minteer, S. D., and Angenent, L. T. (2006). An upflow microbial fuel cell with an interior cathode: assessment of the internal resistance by impedance spectroscopy. *Environ. Sci. Technol.,* **40**, 5212–5217.

18. Keese, C. R., Bhawe, K., Wegener, J., and Giaever, I. (2002). Real-time impedance assay to follow the invasive activities of metastatic cells in culture. *Bio.Techniques,* **33**, 842–844, 846, 848–850.

19. Lo, C. M., Keese, C. R., and Giaever, I. (1995). Impedance analysis of MDCK cells measured by electric cell–substrate impedance sensing. *Biophys. J.,* **69**, 2800–2807.

20. Aberg, P., Nicancer, I., and Ollmar, S. (2003). Minimally invasive electrical impedance spectroscopy of skin exemplified by skin cancer assessments. *Engineering in Medicine and Biology Society, 2003. Proceedings of the 25th Annual International Conference of the IEEE* **4**, 3211–3214.

21. Pirovano, F., Piazza, I., and Brambilla, F. (1995). Impedimetric method for selective enumeration of specific yoghurt bacteria with milk-based culture media. *Le Lait,* **75**, 285–293.

22. Naumann, R., et al. (2002). Proton transport through a peptide-tethered bilayer lipid membrane by the H(+)-ATP synthase from chloroplasts measured by impedance spectroscopy. *Biosens. Bioelectron.,* **17**, 25–34.

23. Ye, Y. K., Zhao, J. H., Yan, F., Zhu, Y. L., and Ju, H. X. (2003). Electrochemical behavior and detection of hepatitis B virus DNA PCR production at gold electrode. *Biosens. Bioelectron.,* **18**, 1501–1508.

24. Saum, A. G. E. (1998). Use of substrate coated electrodes and AC impedance spectroscopy for the detection of enzyme activity. *Biosens. Bioelectron.,* **13**, 511–518.

25. Hsieh, G., Mason, T., Garboczi, E., and Pederson, L. (1997). Experimental limitations in impedance spectroscopy: part III. Effect of reference electrode geometry/position. *Solid State Ionics,* **96**, 153–172.

26. Pancrazio, J. J., Whelan, J. P., Borkholder, D. A., Ma, W., and Stenger, D. A. (1999). Development and application of cell-based biosensors. *Ann. Biomed. Eng.,* **27**, 697–711.

27. Lo, C. M., Keese, C. R., and Giaever, I. (1993). Monitoring motion of confluent cells in tissue culture. *Exp. Cell Res.,* **204**, 102–109.

28. Ren, J., et al. (2006). Lysophosphatidic acid is constitutively produced by human peritoneal mesothelial cells and enhances adhesion, migration, and invasion of ovarian cancer cells. *Cancer Res.,* **66**, 3006–3014.

29. Giaever, I., and Keese, C. R. (1984). Monitoring fibroblast behavior in tissue culture with an applied electric field. *Proc. Natl. Acad. Sci. U. S. A.,* **81**, 3761–3764.

30. Mcauley, E., Plopper, G. E., Corr, D. T., and Chrisey, D. B. (2011). Evaluation of electric cell–substrate impedance sensing for the detection of nanomaterial toxicity Bhavana Mohanraj and Theresa Phamduy. *Arthritis Res.,* **2**, 136–151.

31. Wegener, J., Keese, C. R., and Giaever, I. (2000). Electric cell–substrate impedance sensing (ECIS) as a noninvasive means to monitor the kinetics of cell spreading to artificial surfaces. *Exp. Cell Res.,* **259**, 158–166.

32. Keese, C. R., and Giaever, I. (1986). Electric to monitor the dynamical behavior in tissue culture. *Trans. Biomed. Eng.,* **BME-3**, 242–247.

33. Giaever, I., and Keese, C. R. (1991). Micromotion of mammalian cells measured electrically. *Proc. National Acad. Sci. U. S. A.,* **88**, 7896–900.

34. Sargent, A., and Sadik, O. A. (1999). Monitoring antibody–antigen reactions at conducting polymer-based immunosensors using impedance spectroscopy. *Electrochim. Acta,* **44**, 4667–4675.

35. Savolainen, V., et al. (2011). Impedance spectroscopy in monitoring the maturation of stem cell-derived retinal pigment epithelium. *Ann. Biomed. Eng.,* **39**, 3055–3069.

36. Bondarenko, A. S., and Ragoisha, G. A. EIS Analyzer. at <http://www.abc.chemistry.bsu.by/vi/analyser/>.

37. Raistrick, I. D. (1986). Application of Impedance spectroscopy to material science. *Ann. Rev. Mater. Sci.,* **16**, 343–370.

38. Barsoukov, Evgenij Macdonald, J. R. (2005). *Impedance Spectroscopy: Theory, Experiment, and Apllication*, 2nd Ed. 615 p. (Wiley: Hoboken, NJ, USA).

39. Amirudin, A., and Thierry, D. (1995). Application of electrochemical impedance spectroscopy to study the degradation of polymer-coated metals. *Prog. Org. Coatings,* **26**, 1–28.

40. Pan, J., Thierry, D., and Leygraf, C. (1996). Electrochemical impedance spectroscopy study of the passive oxide film on titanium for implant application. *Electrochim. Acta,* **41**, 1143–1153.

41. Heijink, I. H., et al. (2010). Characterization of cell adhesion in irway epithelial cell types using electric cell–substrate impedance sensing. *Eur Respir J,* **35**, 894–903.

42. Bouafsoun, A., et al. (2007). Electrical probing of endothelial cell behavior on a fibronectin/polystyrene/thiol/gold electrode by Faradaic electrochemical impedance spectroscopy (EIS). *Biochemistry,* **70**, 401–407.

43. Lin, C.-Y., et al. (2011). Real-time detection of beta1 integrin expression on MG-63 cells using electrochemical impedance spectroscopy. *Biosens. Bioelectron.,* **28**, 221–226.

44. De Blasio, B. F., Laane, M., Walmann, T., and Giaever, I. (2004). Combining optical and electrical impednce techniques for quantitative

mesurement of confluence in MDCK-I cell cultures. *Bio. Techniques,* **36**, 650–654.

45. Newbold, C., et al. (2010). Changes in biphasic electrode impedance with protein adsorption and cell growth. *J. Neural Eng.,* **7**, 056011.

46. Chan, C. M., et al. (2010). Effects of (-)-epigallocatechin gallate on RPE cell migration and adhesion. *Moll Vis.,* **16**, 586.

47. Wu, N. L., et al. (2010). Zeaxanthin inhibits PDGF-BB-induced migration in human dermal fibroblasts. *Exp Dermatol.,* **19**(8), e173–e181.

48. Hsu, C.-C., Tsai, W. C., Chen, C. P.-C., Lu, Y.-M., and Wang, J.-S. (2010). Effects of negative pressures on epithelial tight junctions and migration in wound healing. *Am. J. Physiol. Cell Physiol.,* **299**(2), C528–C534.

49. Earley, S., and Plopper, G. E. (2006). Disruption of focal adhesion kinase slows transendothelial migration of AU-565 breast cancer cells. *Biochem. Biophys. Res. Commun.,* **350**, 405–412.

50. Sun, C., et al. (2010). ADAM15 regulates endothelial permeability and neutrophil migration via Src/ERK1/2 signalling. *Cardiovasc. Res.,* **87**, 348–355.

51. Yang, J., Duh, E. J., Caldwell, R. B., and Behzadian, M. A. (2010). Antipermeability function of PEDF involves blockade of the map kinase/GSK/β-Catenin signaling pthway and uPAR expresion. *Invest Ophthalmol. Vis Sci.,* **51**, 3273–3280.

52. Han, J., Liu, G., Profirovic, J., Niu, J., and Voyno-Yasenetskaya, T. (2009). Zyxin is involved in thrombin signaling via interaction with PAR-1 receptor. *FASEB,* **23**, 4193–4206.

53. Grab, D. J., et al. (2009). Protease activated receptor signaling is required for African trypanosome traversal of human brain microvascular endothelial cells. *PLOS,* **3**, e749.

54. Bashier, R. (2004). BioMEMS: state-of-the-art in detection, opportunities and prospects. *Adv. Drug Deliv. Rev.,* **54**, 1565–1586.

55. Lam, P. (2008). Electric cell–substrate impedance sensing. *Encyclopedia Biomater. Biomed. Eng.,* 908–914. doi:10.3109/E-EBBE-120015020.

56. Aruda, D. L. (2009). Microelectrical sensors as emerging platforms for protein biomarker detection in point-of-care diagnostics. *Expert Rev. Mol. Diagn.,* **9**, 749–755.

57. Hawrami, K., and Breuer, J. (1999). Development of a fluorgenic polymerase chain reaction assay (TaqMan). for the detection and quantification of varicella zoster virus. *J. Virol. Meth.,* **79**, 33–40.

58. Wang, Q., Zhang, B., Lin, X., and Weng, W. (2011). Hybridization biosensor based on the covalent immobilization of probe DNA on chitosan-multiwalled carbon nanotubles nanocomposite by using glutaraldehyde as an arm linker. *Sens. Actuators,* **156**, 599–605.

59. Hong, J., Kandasamy, K., Marimuthu, M., Choi, C. S., and Kim, S. (2011). Electrical cell–substrate impedance sensing as a non-invasive tool for cancer study. *Royal Soc. Chem. Anal.,* **136**, 237–245.

60. Rogers, K. (1995). Biosensors for environmental applications. *Biosens. Bioelectron.,* **10**, 533–541.

61. Rogers, K. R. (2006). Recent advances in biosensor techniques for environmental monitoring. *Anal. Chim. Acta,* **568**, 222–231.

62. Wong, R., and Anis, N. (1993). Reusable fiber-optic-based immunosensor for rapid detection of imazethapyr herbicide. *Anal. Chim. Acta,* **279**, 141–147.

63. Thompson, R. B., and Jones, E. R. (1993). Enzyme-based fiber optic zinc biosensor. *Anal. Chem.,* **65**, 730–734.

64. Tescione, L., and Belfort, G. (1993). Construction and evaluation of a metal ion biosensor. *Biotechnol. Bioeng.,* **42**, 945–952.

65. Vo-Dinh, T., et al. (1987). Antibody-based fiber optic biosensor for the carcinogen benzo(a)pyrene. *Appl. Spectrosc.,* **41**, 735–738.

66. Panasyuk-Delaney, T., Mirsky, V., and Ulbricht, M. (2001). Impedometric herbicide chemosensors based on molecularly imprinted polymers. *Anal. Chim.,* **435**, 157–162.

67. Easley, C. J., et al. (2006). A fully integrated microfluidic genetic analysis system with sample-in-answer-out capability. *Proc. Natl. Acad. Sci. U. S. A.,* **103**, 19272–19277.

68. Pal, R., et al. (2005). An integrated microfluidic device for influenza and other genetic analyses. *Lab on a chip,* **5**, 1024–1032.

69. Graham, A H. D., et al. (2012). Modification of standard CMOS technology for cell-based biosensors. *Biosens. Bioelectron.,* **31**, 458–462.

70. Di Capua, R., et al. (2012). Towards the realization of label-free biosensors through impedance spectroscopy integrated with IDES technology. *Eur. Biophys. J. EBJ,* **41**, 249–256.

71. DeBusschere, B. D., and Kovacs, G. T. (2001). Portable cell-based biosensor system using integrated CMOS cell-cartridges. *Biosens. Bioelectron.,* **16**, 543–556.

72. Ye, J.-S., Ottova, A., Tien, H. T., and Sheu, F.-S. (2003). Nanostructured platinum-lipid bilayer composite as biosensor. *Bioelectrochemistry,* **59**, 65–72.

73. Toner, M., and Irimia, D. (2005). Blood-on-a-chip. *Ann. Rev. Biomed. Eng.,* **7**, 77–103.

74. Thielecke, H., Mack, A., and Robitzki, A. (2001). A multicellular spheroid-based sensor for anti-cancer therapeutics. *Biosens. Bioelectron.,* **16**, 261–269.

75. Moissl, U. M., et al. (2006). Body fluid volume determination via body composition spectroscopy in health and disease. *Physiol. Meas.,* **27**, 921–933.

Index